AF760953

Photostasis and Related Phenomena

Photostasis and Related Phenomena

Edited by

Theodore P. Williams and
Anne B. Thistle

Florida State University
Tallahassee, Florida

Plenum Press • New York and London

Library of Congress Cataloging-in-Publication Data

Photostasis and related phenomena / edited by Theodore P. Williams and
Anne B. Thistle.
p. cm.
"Proceedings of the Neuroscience Program Symposium on Photostasis
and Related Topics, held February 21-23, 1997, in Tallahassee,
Florida"--T.p. verso.
Includes bibliographical references and index.
ISBN 0-306-45806-3
1. Photoreceptors--Congresses. 2. Rhodopsin--Congresses.
I. Williams, Theodore P. II. Thistle, Anne B. III. Neuroscience
Program Symposium on Photostasis and Related Topics (1997 :
Tallahassee, Fla.)
QP481.P53 1998
573.8'836--dc21 97-52808
CIP

Proceedings of the Neuroscience Program Symposium on Photostasis and Related Topics, held February 21–23, 1997, in Tallahassee, Florida

ISBN 0-306-45806-3

© 1998 Plenum Press, New York
A Division of Plenum Publishing Corporation
233 Spring Street, New York, N.Y. 10013

http://www.plenum.com

10 9 8 7 6 5 4 3 2 1

All rights reserved

No part of this book may be reproduced, stored in a retrieval system, or transmitted in any form or by any means, electronic, mechanical, photocopying, microfilming, recording, or otherwise, without written permission from the Publisher

Printed in the United States of America

To the Memory of Werner K. Noell

(1913-1992)

Scientist, Colleague, and Friend

FOREWORD

Over the weekend of 21–23 February 1997, a small group gathered in Tallahassee, Florida, at the invitation of Ted Williams, to discuss "photostasis and related topics." The majority of participants were former students and colleagues of Ted's, but an occasional outsider such as myself was generously included. The papers presented there are collected in this volume.

The theory of photostasis was first outlined in a landmark paper by John Penn and Ted, published in 1986 in *Experimental Eye Research.* They provided compelling data showing that, in the albino rat eye, levels of rhodopsin, outer-segment length, rhodopsin regeneration rate, and even, perhaps, rhodopsin packing density all depend on the levels of cyclic illumination (12 hours light, 12 hours dark) in which the animal was reared. So, for example, there is fourfold less rhodopsin in a retinal extract derived from an animal reared at 400 lux than in an extract from the retina of an animal reared at 3 lux. Animals reared at intermediate levels of light show intermediate amounts of rhodopsin that are correlated with illumination level.

What these data immediately suggested is that the photoreceptor cell can adjust its photon-catching ability in response to the levels of light in which an animal is reared, and they also provided a compelling rationale for outer-segment turnover, a phenomenon discovered 20 years earlier by Richard Young but whose function has remained obscure. Perhaps most surprising was the finding that the number of photons caught per eye over the 12-hour light period appeared quite similar for all animals regardless of the level of light available. In other words, the photoreceptor cell appears to regulate its photon-catching ability so that the number of photons caught each day by an eye (about 10^{16}) is constant. This is the theory of photostasis—that the visual system controls the photon catch of the eye.

Since the 1986 paper, numerous studies have confirmed the general findings of Penn and Williams, and they have now been extended considerably—by studies varying light and day-length conditions widely and testing other animals for evidence of the phenomenon. Of particular interest is the evidence that even some invertebrates exhibit photostasis. For two and one-quarter days, the group in Tallahassee listened to formal presentations and then discussed the issues. It is probably fair to say that there was less discussion of the theory of photostasis itself than of the critical issue of the effects of light on photoreceptors and other parts of the retina and of the factors that underlie, modify, or modulate these effects. There was clear agreement among all participants that tremendous changes do take place in rod structure and function (at least in rodents) in response to the lighting conditions under which the animals are reared or live.

A fundamental issue brought up again and again (and not resolved) is where photostasis ends and light damage begins. It has been known since the mid-1960s that continuous exposure of albino rats to even low levels of light causes dramatic damage to photoreceptors. The action spectrum for these damaging effects corresponds to the absorption spectrum of rhodopsin itself, indicating that these effects are not due to thermal effects (i.e. heating of the tissue). Light-induced retinal damage has now been shown to occur in a variety of animals, from invertebrates to primates.

One reason that photostasis and light damage are confounded is that animals reared under dim (5-lux) lighting conditions or in darkness are much more susceptible to light damage than are animals reared at moderate (500-lux) or bright (800-lux) light intensities, and indeed this pattern makes sense if low-level light damage is mediated by rhodopsin—i.e. photon catch—but the story when finally unraveled may be much more complicated and interesting and even extend beyond the photoreceptors! Ted Williams provided data at the conference showing that inner-plexiform-layer thickness varies with light exposure (day length in this case). What might this mean? The inner nuclear layer did not appear to change in thickness in these experiments, suggesting that ambient light levels affect process size and/or number in the inner plexiform layer. Is this damage or a regulatory phenomenon? Is it reversible? How is it mediated? All are fascinating questions that need study.

A key and fun feature of the conference was its broadness. Papers were presented that discussed a variety of relevant topics, including the effects of genetic manipulations on rod function, circadian-rhythm regulation of photosensitivity, the effects of light on seasonal depression in humans and on oxygen-induced retinopathy in rats, and finally, the regulation of photon catch (photostasis) in plants. On Saturday afternoon, a fascinating tour of nearby Wakulla Springs State Park was followed by a marvelous dinner for the participants at the Williams' retreat in the tiny Florida town of Sopchoppy. Ruth-Ann's famous chili was the mainstay of the menu and greatly enjoyed by all. It was a wonderful weekend, hosted warmly by Ted and his colleagues at Florida State University, and we all left having learned much and having made or renewed warm friendships.

John E. Dowling

PREFACE

This volume is a compilation of papers presented as a symposium entitled "Photostasis" and held at The Florida State University on February 21–23, 1997. The symposium was funded by the Program in Neuroscience, Rushton Lecture Series, which has sponsored other such events in the past.

The purpose of the symposium was to bring together those who would review the status of this subject with others who would discuss related topics. Its format was designed to stimulate discussion of the subjects at hand, and we are thankful that much of that discussion has spilled over into this book. We encouraged authors to be provocative and speculative, and without departing from solid science, they have been.

To some extent, the papers are organized in order of increasing system complexity. After an introduction to photostasis, for example, the next few papers deal with the manifestation of photostasis at the molecular level, and later ones move on to subjects at the cellular and organismal levels. In some cases, however, papers have been clustered by animal species studied rather than by level of biological complexity. A unique contribution to this volume is the chapter on photostasis in plants, which is presented last, as the only one that does not deal with the visual system in some way.

We thank all those who have participated as speakers and authors. Without their willingness to attend and to meet deadlines, this volume would not have been possible. Neither would it have been possible without the help, support, and encouragement of several others: Professor Michael Meredith, who suggested the title of this volume and, with artist Charles Badland, produced the artwork for the cover; Professor Michael Rashotte, codirector of the Program in Neuroscience, who helped with financial matters; Ms. Cheryl Lee of the Biological Science fiscal office, who made several rough places plain; local participants in the Neuroscience Program, faculty and students alike, who came to the symposium and who asked good questions. We thank all of them.

We proffer this book in the memory of Professor Werner K. Noell, whose discovery of and salient work on retinal light damage inspired many of us. Had he been alive at the time of this symposium, Werner surely would have been a valuable participant and contributor to this book. His appreciation that light history was the major factor protecting against light damage, or increasing vulnerability to it, was inherent in the thinking that led to the discovery of photostasis. In this spirit we hope that this volume will be useful as an introduction to photostasis and as a reference source for established researchers as well as novices in the field.

Theodore P. Williams
Anne B. Thistle

CONTENTS

1

EARLY STUDIES OF THE PHOTOSTASIS PHENOMENON

Retinal Adaptation to the Light Environment

John S. Penn

Arkansas Center for Eye Research
Jones Eye Institute
University of Arkansas for Medical Sciences
4301 W. Markham, Slot 523
Little Rock, Arkansas 72205

1. INTRODUCTION

In 1967, Young described the turnover of outer segment membranes in the retinal rods of rats. By administering radioactive amino acids as precursors for protein production, and sacrificing the animals after various intervals, Young illustrated a series of events from synthesis and incorporation of new membranous discs at the base of the rod outer segment (ROS) to the shedding, phagocytosis, and degradation of ROS tips. In the rat, this process was shown to take approximately 10 days. Several years later, Young (1976) had come to the conclusion that most of the energy that a rod expends goes not to the initiation of the visual message but to the repeated replacement of its own molecules. Since its discovery, although several researchers have addressed the question (Kaplan *et al.,* 1978; Baylor *et al.,* 1979; Andrews and Cohen, 1983; Williams, 1984), no one has found the reason for this energy-expensive turnover.

In 1977, Organisciak and Noell measured the relative opsin and phospholipid levels in ROS's of albino rats maintained for at least two weeks in darkness or cyclic light. The phospholipid-to-opsin ratio was 20–25% higher in cyclic-light animals. Also, total ROS phospholipid mass was low in these animals, suggesting to the authors a control of outer-segment length. The pigment concentration was presumably altered by changes in membrane synthesis, controlled by the relative rates of protein synthesis and incorporation, lipid synthesis and incorporation, and disc assembly. The outer-segment length was controlled by membrane synthesis, by membrane phagocytosis and degradation, or by a combination of the two processes. Battelle and LaVail (1978) studied ROS length and rhodopsin content in the albino rat after dark rearing or cyclic-light maintenance. They found both more rhodopsin per eye and greater outer-segment length in the dark-reared animals, and they attributed part of the increase in pigment to a change in rhodopsin concentration of newly synthesized membranes. Combined, these studies implied that rats were altering photoreceptor structure and membrane composition in response to environmental lighting. In both sets of experiments, the ROS adjustments were accomplished by changes in two seemingly dynamic parameters: outer segment length and (rhod)opsin concentration within the outer segment membranes. Either or both of these parameters could define the amount of rhodopsin in the dark-adapted retina. The continuous turnover of photoreceptor outer segment membranes provided a mechanism for this dynamic quality.

Penn and Williams (1986) suggested that one function of ROS renewal is to maintain a certain plasticity within the cell that allows for adaptation by photoreceptors to the environment across a wide spectrum of light conditions. Specifically, the hypothesis we formed was that albino rats reared under "dim" light conditions, where ambient photons were scarce, would have retinas with greater photon-catching ability than would those reared under "bright" light conditions, where ambient photons were abundant. ROS renewal would be responsible for these adjustments, at least in part. Synthesis of membrane components in the inner segment and their subsequent incorporation into the outer segment would control both the number of discs produced per day and the rhodopsin concentration per disc, whereas shedding at the tip, combined with synthesis, would determine the ROS length.

Implicit in our hypothesis was that the rat "needed" to absorb an optimal number of photons. Otherwise, all rats, regardless of their ambient light environment, would have ROS's of equal length and rhodopsin concentration. Other studies had already shown that this was not the case. Finally, we proposed that rats reared at various intensities of light would make retinal adjustments in such a way as to promote the capture of this hypothetical optimal number of photons per day. After estimating photon catch in rats reared at a number of different light levels, we went on to characterize further the ROS adjustments involved, to measure their effect on retinal function by ERG analysis, to determine whether the adjustments were reversible, and finally to investigate the role of ROS renewal in them (Penn and Anderson, 1987, 1992; Penn *et al.,* 1989).

2. METHODS

2.1. Photostasis

Several groups of albino rats were reared in environments in which the lights were on for 12 hours and off for 12 hours each day (12L:12D). Each group experienced a different light level during the light-on part of the cycle. The difference between the dimmest (3 lux, 12L:12D) and brightest (400 lux, 12L:12D) environments was more than two orders of magnitude. Animals were born and reared through 15 weeks of age in a given environment before sacrifice. Because each rat was born under the experimental lighting conditions and remained there throughout its life, its entire "light history" was known and controlled. At the time of sacrifice, the animals' retinas were subjected to the following analyses: (1) dark-adapted and steady-state (daytime) whole-retina rhodopsin levels, (2) *in vivo* regeneration rate of rhodopsin, (3) ROS rhodopsin "concentration" by microspectrophotometry, and (4) ROS length and photoreceptor cell density. The last two parameters were measured in several locations along the retinal vertical and horizontal meridians.

ROS rhodopsin "concentration" was measured in fixed frozen sections of retinal tissue prepared and sectioned under dim red light. A microspectrophotometer was used to measure rhodopsin absorbance in the ROS layer of the sections, which were of known thickness. In this way absorbance per micrometer of section thickness could be calculated, and from ROS length measured from plastic-embedded retinas of identically treated rats, "axial absorbance" of rhodopsin could be calculated. This measure stood as our first estimate of rhodopsin concentration, or "packing density," within outer segment membranes.

In order to calculate photon catch for a given light regime, we measured the rhodopsin regeneration rate *in vivo*. At the end of the light period, rats were allowed to remain in the dark for specific lengths of time before sacrifice. Rhodopsin levels were then measured, and absorbances were plotted against time in the dark. The slope of the natural log of the fraction of total rhodopsin found in the bleached state plotted against time in the dark was considered the rate constant for regeneration, assuming first-order kinetics.

A simplified expression for the loss of rhodopsin from the retina with time is:

$$-\mathrm{d}R/\mathrm{d}t = [(k_\mathrm{f} \bullet I) \bullet R] - [k_\mathrm{r} \bullet B] \tag{1}$$

I is the intensity of the bleaching agent, R is the concentration of unbleached rhodopsin, B is the concentration of bleached rhodopsin, and k_f and k_r are the rate constants for bleaching and regeneration, respectively. At steady state (during a typical day in a given light habitat) the rate of rhodopsin loss equals the rate of gain from regeneration. Therefore,

$$\mathrm{Num} = (k_\mathrm{f} \bullet I) \bullet R_\mathrm{ss} = k_\mathrm{r} \bullet B_\mathrm{ss} \tag{2}$$

where R_ss equals unbleached rhodopsin at steady state, and B_ss equals rhodopsin bleached at steady state (or total dark-adapted rhodopsin minus R_ss). The number of photons absorbed during a light cycle is simply the product of k_r and B_ss converted to appropriate time units.

2.2. ROS Membrane Composition and Phospholipid Synthesis

Encouraged by the photon-catch calculations (see 3. RESULTS), we conducted additional experiments designed to characterize more completely photoreceptor cell density and function (by ERG analysis) and ROS composition from rats reared in different light environments. The lighting regimens for these new treatments are shown in Figure 1. Light histories of 5, 300, and 800 lux 12L:12D cyclic light were chosen. Some albino rats born and reared under each of these regimens were sacrificed at 12 weeks of age. At that time several specific characteristics of ROS membrane composition were measured, including (1) rhodopsin packing density within the membrane (this time by molar ratio analysis), (2) phospholipid class composition, (3) fatty-acid profile of membrane phospholipids, and (4) the membrane cholesterol levels. Having noticed specific reductions in docosahexaenoic-acid–containing molecular species of ROS phospholipid in bright-light rats, we investigated the synthesis and turnover of these membrane constituents in additional rats reared from birth in cyclic light of 5 or 800 lux. At 12 weeks of age, they were injected intravitreally with [2–^{3}H]glycerol and sacrificed 1, 2, 3, 4, 6, or 10 days later. ROS's were prepared and phosphatidylcholine (PC), phosphatidylserine (PS), and phosphatidylethanolamine (PE) were isolated, derivatized, separated into molecular species by HPLC, quantified, and counted and their specific radioactivities determined.

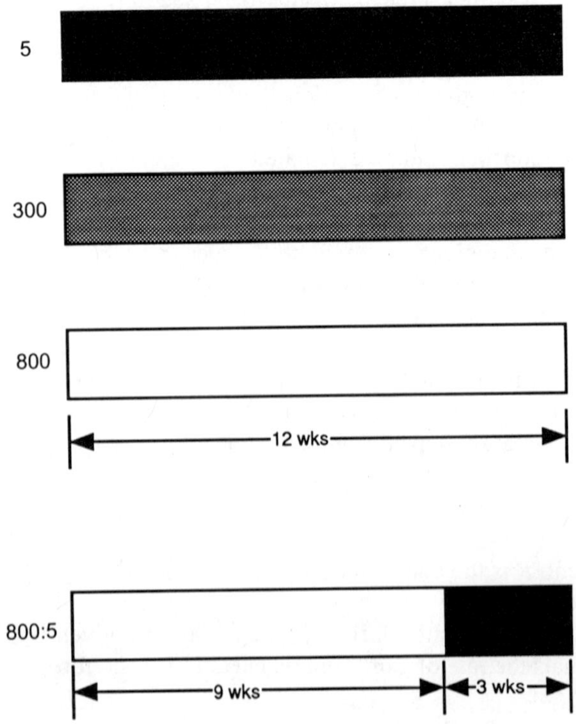

Figure 1. A schematic representation of the light treatments used to investigate the effect of light history on several retinal parameters. All treatments consisted of 12L:12D exposures to illuminance levels designated on the left side of the figure. The effect of a mid-life change in light history was examined in rats moved from 800 lux 12L:12D to 5 lux 12L:12D at 9 weeks of age.

2.3. Effect of Mid-Life Change in Light Environment

The nature of the ROS membrane alterations measured in rats from different light histories continued to support the idea of an "adaptive" capability. Even though the retinas of rats reared under the various illuminance levels were different in nearly every respect, the daily photon catch was the same in all groups. Still, especially because one result of bright-light experience was the irreversible loss of postmitotic neurons, the hypothesis that these contrasting retinal features were adaptive required additional experiments. One true test of adaptation is the plasticity of the adaptive change, and it remained to be determined whether retinal characteristics would be altered by a mid-life change in lighting environment. Further, if retinal characteristics were altered, would the new profile reflect an adaptation to the new regime? In an attempt to address this question, we reared another group of rats (hereafter called the "800:5" group) in 800-lux cyclic light for 9 weeks and then moved them to 5-lux cyclic light for an additional 3 weeks. At the end of this 12-week period, the animals were sacrificed, and their retinas were studied for the effects of light history by the same means employed for previous treatment groups. A few rats from each group were sacrificed at the end of 9 weeks so that the true effect of the subsequent 3-week period could be determined.

2.4. Time Course of Retinal Adjustments

Finally, in order to determine the contribution of precocious or retarded development to the retinal alterations, we reared two additional groups of rats, one at 800 lux

Table 1. Retinal parameters contributing to photostasis

	Molecules per eye × 10^{-15}			Rhodopsin (nmol per eye)		Rate × 10^4	Photons absorbed per eye	
I (lux)	R_o	R_{ss}	B_{ss}	Dark-adapted	Steady-state	(s^{-1}) k_t	per 12 hours × 10^{-16}	
3	1.34	1.10	0.24	2.23 ± 0.13	1.83 ± 0.30	8.67	0.90	
80	0.86	0.22	0.64	1.43 ± 0.12	0.37 ± 0.15	3.90	1.07	Mean
200	0.77	0.17	0.60	1.28 ± 0.13	0.38 ± 0.09	3.72	0.97	1.02 ± 0.108
400	0.36	0.01	0.35	0.60 ± 0.17	0.03 ± 0.07	7.45	1.12	
80′	1.01	0.42	0.59	—	—	4.38	1.11	
200′	0.61	0.23	0.38	—	—	5.81	0.96	Mean
400′	0.37	0.02	0.35	—	—	9.99	1.51	1.19 ± 0.28

I = intensity (primed values refer to data from a duplicate experiment); R_o = total dark-adapted rhodopsin; R_{ss} = rhodopsin present at steady state; B_{ss} = rhodopsin bleached at steady state; k_t = regeneration rate constant. For easier comparison, rhodopsin values are also listed in nmol per eye for the dark-adapted and steady-state conditions.

12L:12D and the other at 5 lux 12L:12D. Animals were sacrificed at 1, 2, 3, 6, 9, 12, 16, and 28 weeks of age, and the outer nuclear layer (ONL) area and ROS length were measured. Additionally, a few rats that had remained from birth through 21 weeks at 800 lux 12L:12D were removed to 5 lux 12L:12D for 3 weeks and sacrificed.

3. RESULTS

3.1. Photostasis

Table 1 shows the photon-catch values for two separate experiments conducted on rats reared in four separate light environments. Levels of rhodopsin in dark-adapted and steady-state bleached retinas both decreased with increasing habitat illuminance. The decline in rhodopsin with increasing illuminance was found empirically to be logarithmic over the range from 3 to 400 lux for both the dark-adapted and steady-state conditions. Furthermore, although rats experienced illuminance levels differing by more than two orders of magnitude, they all caught the same number of photons during the 12-hour light period.

The decrease in dark-adapted rhodopsin levels with increasing environmental illumination was a reflection of several other retinal parameters acting in concert: (1) the number of photoreceptors was reduced as a function of increasing illuminance, (2) the length of the outer segments of remaining photoreceptors decreased with increasing illuminance, and (3) the rhodopsin concentration of outer-segment membrane (as determined by microspectrophotometry or by phospholipid-to-rhodopsin molar ratio) decreased with increasing illuminance. Given the latter two findings, it appeared the ROS renewal process was operating differently in different light environments.

By the nature of our method, we took into account all three of these parameters when measuring axial absorbance. Figure 2 shows the retinal topography of axial absorbance for four different habitat light levels. When integrated over the vertical meridian, this measure was highly correlated with habitat illuminance ($r^2 = -0.998$, $p < 0.002$). This integrated measure was also correlated with dark-adapted rhodopsin levels ($r^2 = 0.961$, $p < 0.05$). It should be noted that the axial absorbance profiles illustrate an area within the central superior retina of dim-light–reared (3-lux) rats that contains significantly more rhodopsin than any other. This finding is important in the context of constant-light–induced retinal damage, like that first characterized by Noell *et al.* (1966).

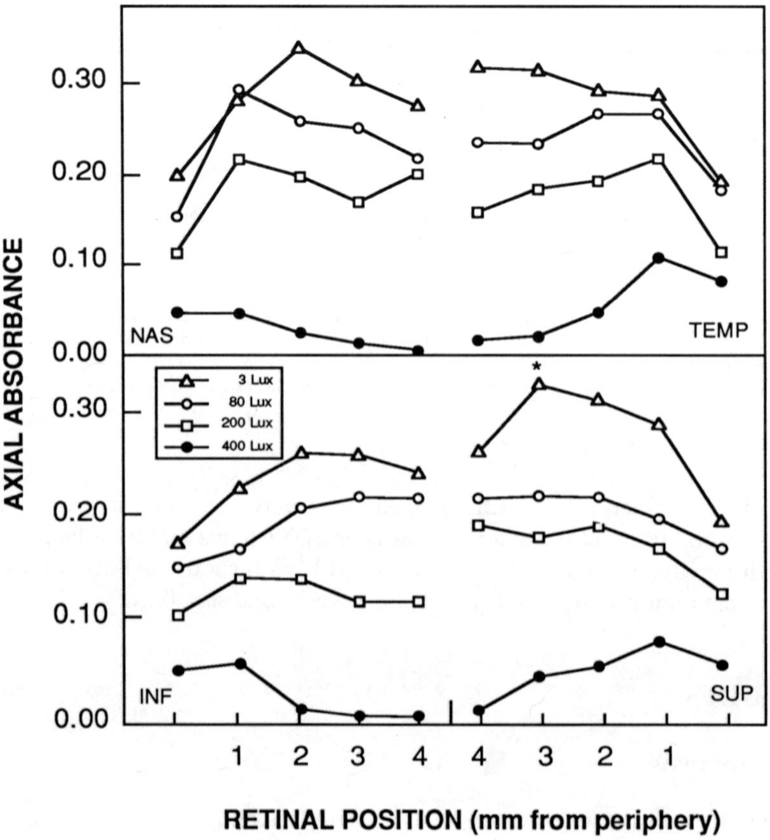

Figure 2. Axial absorbance plotted against retinal position for 10 locations each on the horizontal and vertical meridians. Axial absorbance is the absorbance at 500 nm wavelength per micrometer of section thickness through the ROS, multiplied by the average length of ROS's in that area. An area of the retina between 1 and 2 mm superior to the optic nerve (*) has an inordinately high rhodopsin concentration in rats reared in dim light. This is the location of greatest susceptibility to photoreceptor loss from constant light exposure. All rats were reared in 12L:12D regimes. From Penn and Williams (1986) with permission of Academic Press Limited.

3.2. ROS Morphology

The effect of the three light histories on ROS's and overall retinal morphology is shown in Figure 3. Rats from the three light levels (5, 300, and 800 lux) were dark adapted overnight before sacrifice and enucleation. It is clear from the light micrographs that ONL thickness decreased with increasing habitat illuminance, reflecting a decrease in photoreceptors. Also of note are the swollen and vesiculated ROSs in rats reared in bright cyclic light (800 lux, 12L:12D). Our first reaction was to consider these cells pathologically altered and dying. Later results would show that this ROS appearance represented a steady state from which the organelles could adjust to a more typical outer segment structure.

The morphometrical measures of ONL area and ROS length are shown in Table 2. The area of the ONL decreased with increasing habitat illuminance, and the decrease reflected a retinal pattern inconsistent with the constant-light-induced retinal damage described by Noell *et al.* (1986). Loss of cells was slightly greater in the *inferior* half of the

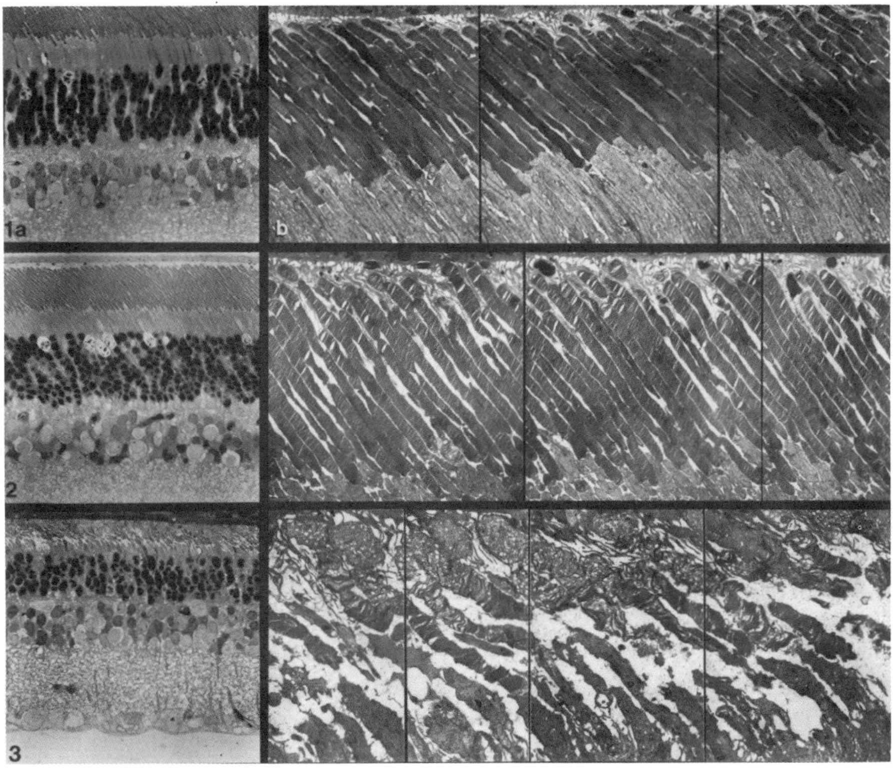

Figure 3. The left-hand side of the plate consists of photomicrographs of a central portion of the inferior retina of rats reared at 5 lux (top), 300 lux (middle), and 800 lux (bottom) after an overnight dark adaptation (225×). The right-hand side shows electron-micrograph montages illustrating the ultrastructure of photoreceptor outer segments in the same area of the retina for each group (1 and 2, 1210×; 3, 1840×). From Penn and Anderson (1987) with permission of Academic Press Limited.

Table 2. Effect of light history on retinal morphology

Cyclic rearing illuminance (lux)	Outer nuclear layer area (mm^2)		Rod outer segment length (μm)			
			Superior retina		Inferior retina	
	Superior	Inferior	Central	Mid-peri	Central	Mid-peri
5 lux, 12 weeks	0.133 ± 0.010	0.14 ± 0.008	31.3 ± 1.6	26.5 ± 1.2	26.8 ± 2.3	22.6 ± 2.4
300 lux, 12 weeks	0.124 ± 0.006	0.127 ± 0.002[a]	27.9 ± 2.0	27.2 ± 1.2	23.3 ± 1.2	20.5 ± 2.6
800 lux, 12 weeks	0.090 ± 0.007[b]	0.077 ± 0.00[b]	0.09 ± 1.0[b]	9.5 ± 2.0[b]	0.4 ± 0.8[b]	10.6 ± 2.7b
5 lux, 9 weeks	0.124 ± 0.003	0.129 ± 0.007	29.7 ± 2.2	24.5 ± 0.2	23.5 ± 1.6	21.5 ± 0.4
800 lux, 9 weeks	0.099 ± 0.002[c]	0.094 ± 0.007[c]	9.7 ± 1.6[c]	13.2 ± 0.4[c]	8.1 ± 0.4[c]	11.5 ± 0.5[c]
800 lux, 9 weeks 5 lux, 3 weeks	0.092 ± 0.009[c]	0.091 ± 0.013[c]	32.0 ± 2.6	29.3 ± 4.1[d]	28.8 ± 2.1[d]	25.3 ± 3.6[d]

Values are listed as the mean ± standard deviation.Measurements of ROS's were taken 1 mm from the optic nerve (central) or 3 mm from the optic nerve (mid-peri).
[a]Significantly different ($p < 0.05$) from 5-lux values. [b]Significantly different ($p < 0.001$) from 300-lux values. [c]Significantly different ($p < 0.001$) from 5-lux values. [d]Significantly different ($p < 0.05$) from 5-lux values.

vertical meridian. In rat retinas damaged by constant light exposures, on the other hand, more cells are lost in the *superior* hemisphere (Rapp and Williams, 1980). ROS length also decreased as a function of increasing habitat illuminance; a drastic decrease was evident in 800-lux–reared animals.

3.3. Effect of Mid-Life Change in Light Environment

As expected, the lower ONL area that was observed after 9 weeks in 800-lux cyclic light in 800:5 rats was unchanged by the subsequent 3-week stay in 5-lux cyclic light, but the 800:5 rats showed remarkable augmentation of ROS length during the dim-light period. The outer segments of these animals were longer than those of rats reared from birth in 5-lux cyclic light in all retinal locations. Measurements taken from electron microphotographs (3600× magnification) revealed a uniformly greater ROS diameter in the inferior central retina of 800:5 rats (1.49 ± 0.06 μm) than in 5-lux rats (1.34 ± 0.04 μm; $p < 0.01$), as well.

3.4. Retinal Function

ERG data are displayed in Table 3. After 9 weeks in the 800-lux regime, rats had b-wave maxima of about 800 μV and a-wave maxima of about 110 μV. Three additional weeks at 800 lux resulted in still lower values. When the 3 additional weeks were spent in 5-lux cyclic light, the average a-wave maximum increased by a factor of almost three, and the b-wave increased to a level statistically indistinguishable from the b-wave of rats reared from birth at 5 lux. Part of the decrease in ERG wave amplitudes between 9 and 12 weeks in bright-light–reared rats was certainly due to age, because a similar, but smaller, decrease was seen between these two ages in dim-light–reared animals. The decrease of a- and b-wave thresholds was as dramatic as the increase in maximum amplitudes. The b-wave threshold in the 800:5 group was lower than that of rats reared in dim cyclic light for 12 weeks. Some restoration of the electroretinogram was expected, given the increase in ROS length during the 3 weeks of 5 lux. An additional contribution certainly came from the elevated rhodopsin concentration depicted in Table 4.

3.5. ROS Membrane Composition and Phospholipid Synthesis

Rod outer segment rhodopsin and lipid phosphorus levels reflected the expected decrease in bright-light rats caused by lower numbers of photoreceptors and shorter outer segments. Membrane packing density was also lower in rats reared in bright light. The extraction method used in these experiments corroborated the previous estimation of rhodopsin concentrations by microspectrophotometry (Penn and Williams, 1986). At 9 weeks, the whole-retinal rhodopsin level of rats reared at 5 lux was 2.5 times that of rats reared at 800 lux (Table 4). After 3 weeks in dim light, the 800:5 group had whole-retina rhodopsin levels statistically indistinguishable from those of rats reared from birth in the dim environment. As shown in Table 4, a significant increase was also seen in rhodopsin *concentration* (phospholipid-to-rhodopsin molar ratio) during the 3 weeks.

The retina is unique in that a large portion of the molecular species of PE, PS, and PC contains two molecules of docosahexaenoic acid (22:6 n-3) (Wiegand and Anderson, 1983). The fatty-acid profile of ROS membrane phospholipids reflected a pattern of preferential decrease of this polyunsaturated fatty acid in bright regimes; rats from 5 lux and 800 lux habitats differed by more than threefold (Table 5). The apparent increase of satu-

Table 3. Effect of light history on retinal function

Cyclic rearing illuminance (lux)	a-wave		b-wave	
	Threshold	Max amp (μv)	Threshold	Max amp (μv)
5 lux, 12 weeks	5.3	535 ± 58	8.5	1355 ± 163
800 lux, 12 weeks	—	25 ± 10[a]	5.7	220 ± 49[a]
5 lux, 9 weeks	5.6	610 ± 90	9.0	1530 ± 169
800 lux, 9 weeks	4.1	110 ± 80[a]	7.7	815 ± 256[b]
800 lux, 9 weeks; 5 lux, 3 weeks	4.2	283 ± 35[a]	8.7	1300 ± 82

Values are listed as mean ± standard deviation. Maximum a-wave response of 800-lux (12-week) rats was below our arbitrary threshold level.
[a]Significantly different ($p < 0.001$) from 5-lux (12-week) values.
[b]Significantly different ($p < 0.002$) from 5-lux (12-week) values.

rated fatty acids in 800-lux rats reflects the inherent nature of mole % as a measure; the absolute mass of all fatty acids was lower in these animals. The phospholipid class showed a small, but significant, decrease of PS and PE in rats reared in bright light. These are the two classes that account for most of the docosahexaenoic acid in dim-light–reared rats. It was reasoned that the lower levels of di-22:6 n-3 phospholipids in the bright-cyclic-light–reared animals could be due to an increase in turnover of these molecules in the bright-reared group or a decrease in their synthesis.

One means of assessing specific loss of polyunsaturates is to calculate the molar ratio of 22:6 n-3 to 16:0. The latter, palmitic acid, is relatively stable and resistant to such insults as lipid peroxidation, and it serves as an accurate denominator because values for the two fatty acids can be obtained from the same chromatogram. These ratios serve to emphasize both the dramatic effect of bright-light rearing and the extent of retinal modification upon transfer to dim light. Values for the 5-lux–reared rats were 5.0 at 9 weeks and 6.5 at 12 weeks, whereas 800-lux ratios were 2.1 and 0.6 for the two ages and studies, respectively. The 800:5 rats had a ratio of 4.7, a greater than twofold increase during the 3 weeks in dim light.

The logarithms of the specific activities of di-22:6-PC, -PE, and -PS were plotted against time, and the *turnover* rate of each was calculated by linear regression analysis.

Table 4. Effect of light history on rhodopsin concentration of ROS membrane

Cyclic rearing illuminance (lux)	ROS lipid Pi per retina (nmol)	ROS rhodopsin per retina (nmol)	Whole retina rhodopsin (nmol)	ROS phospholipid: ROS rhodopsin (nmol:nmol)
5 lux, 12 weeks	98.8 ± 9.3	1.60 ± 0.18	2.33 ± 0.18	61.8
300 lux, 12 weeks	52.2 ± 12.4[a]	0.52 ± 0.07[a]	1.75 ± 0.11[a]	100.4
800 lux, 12 weeks	32.2 ± 7.3[b]	0.23 ± 0.03[b]	0.74 ± 0.07[c]	140.0
5 lux, 9 weeks	—	—	2.27 ± 0.26	—
800 lux, 9 weeks	—	—	0.92 ± 0.07[d]	—
800 lux, 9 weeks; 5 lux, 3 weeks	77.5 ± 6.2[e]	1.47 ± 0.23	2.11 ± 0.18	52.7

Values are listed as mean ± standard deviation. Differences in ROS and whole-retina rhodopsin values reflect the yield of ROS membrane-purification procedures.
[a]Significantly different ($p < 0.001$) from 5-lux values. [b]Significantly different ($p < 0.05$) from 300-lux values. [c]Significantly different ($p < 0.001$) from 300-lux values. [d]Significantly different ($p < 0.001$) from 5-lux value at matched age. [e]Significantly different ($p < 0.01$) from 5-lux value at matched age.

Table 5. Effect of light history on lipid composition of ROS membrane

Cyclic rearing illuminance (lux)	Fatty acid mole % 16:0	18:0	18:1	20:4	22:6	Cholesterol molar ratio	22:6–16:0 (nmol-nmol)
5 lux, 12 weeks	8.6 ± 0.7	28.0 ± 1.0	2.0 ± 0.2	2.4 ± 0.2	56.0 ± 1.0	6.6 ± 0.86	6.5
300 lux, 12 weeks	10.7 ± 1.6	32.0 ± 3.1	2.5 ± 0.7	3.0 ± 0.4	44.0 ± 6.6[a]	16.6 ± 1.98[b]	4.1
800 lux, 12 weeks	25.8 ± 3.2	38.1 ± 4.3[c]	5.6 ± 1.5[a]	6.9 ± 1.7a	16.0 ± 6.4[d]	19.7 ± 3.69[a]	0.6
5 lux, 9 weeks	10.1 ± 0.8	29.0 ± 1.9	5.0 ± 0.4	3.1 ± 0.1	50.0 ± 5.5	—	5.0
800 lux, 9 weeks	15.7 ± 2.4[e]	36.4 ± 5.2[c]	7.7 ± 2.0	3.6 ± 0.4	32.7 ± 4.3[f]	—	2.1
800 lux, 9 weeks; 5 lux, 3 weeks	9.9 ± 0.9	25.5 ± 0.5	4.6 ± 0.6	3.5 ± 0.7	46.8 ± 4.8	—	4.7

Values are listed as mean ± standard deviation.
[a]Significantly different ($p < 0.05$) from 300-lux values. Statistical differences in molar ratio of all fatty acids but 22:6 are inherent in the calculation. No absolute increase occurred in any fatty acid with increasing habitat illuminance. [b]Significantly different ($p < 0.001$) from 5-lux values. [c]Significantly different ($p < 0.05$) from 5-lux values. [d]Significantly different ($p < 0.001$) from 300-lux values. [e]Significantly different ($p < 0.05$) from 5-lux values. [f]Significantly different ($p < 0.001$) from 5-lux values.

The turnover rates of these molecular species in PC and PS did not differ, although di-22:6-PE turned over faster in the bright-light–reared group. Thus, in PC and PS, the altered 22:6 n-3 composition in bright-light–adapted animals was not due to an effect on turnover of di-22:6 molecular species. However, increased turnover of PE could play a role in light-history–induced membrane composition differences.

The effect of light history on the *synthesis* of 22:6-containing molecular species was determined in the following way. It was assumed that the radioactivity in ROS's measured one day after the injection of the radiolabeled tracer reflected newly synthesized phospholipids. The relative amount of each phospholipid molecular species that was synthesized could then be determined by calculation of the relative distribution of the label in the various molecular species of ROS's at day 1. As seen in Table 6, the relative amount of label in di-22:6 molecular species of PE, PS, and PC was lower in the ROS's of animals reared in bright cyclic light. It therefore appears that the lower 22:6 n-3 levels in these ROS phospholipids in the bright-cyclic-light–reared animals is primarily a result of decreased synthesis of phospholipid molecular species that contain two molecules of 22:6 n-3.

3.6. Time Course of Retinal Adjustments

Results of ROS length and ONL area measurements from rats sacrificed at intervals in bright- and dim-light environments are shown in Figure 4. Differences in ROS structure between the two groups began as soon as outer segment development commenced; length differences became significant at 6 weeks of age and persisted throughout the length of

Table 6. Relative mass and radioactivity distribution of di-22:6-containing lipids one day after [2–^{3}H]glycerol injection

Lipid class	Mass (molar ratio) Dim	Mass (molar ratio) Bright	p value	Radioactivity (DPM%) Dim	Radioactivity (DPM%) Bright	p value
PC	5.5 ± 0.9	2.2 ± 0.66	<0.0001	14.6 ± 0.7	7.0 ± 1.5	<0.0001
PE	17.8 ± 1.7	8.9 ± 0.6	<0.0001	22.7 ± 1.4	14.7 ± 2.5	<0.0003
PS	22.6 ± 2.9	11.6 ± 1.6	<0.0001	19.5 ± 4.2	14.5 ± 2.0	>0.08

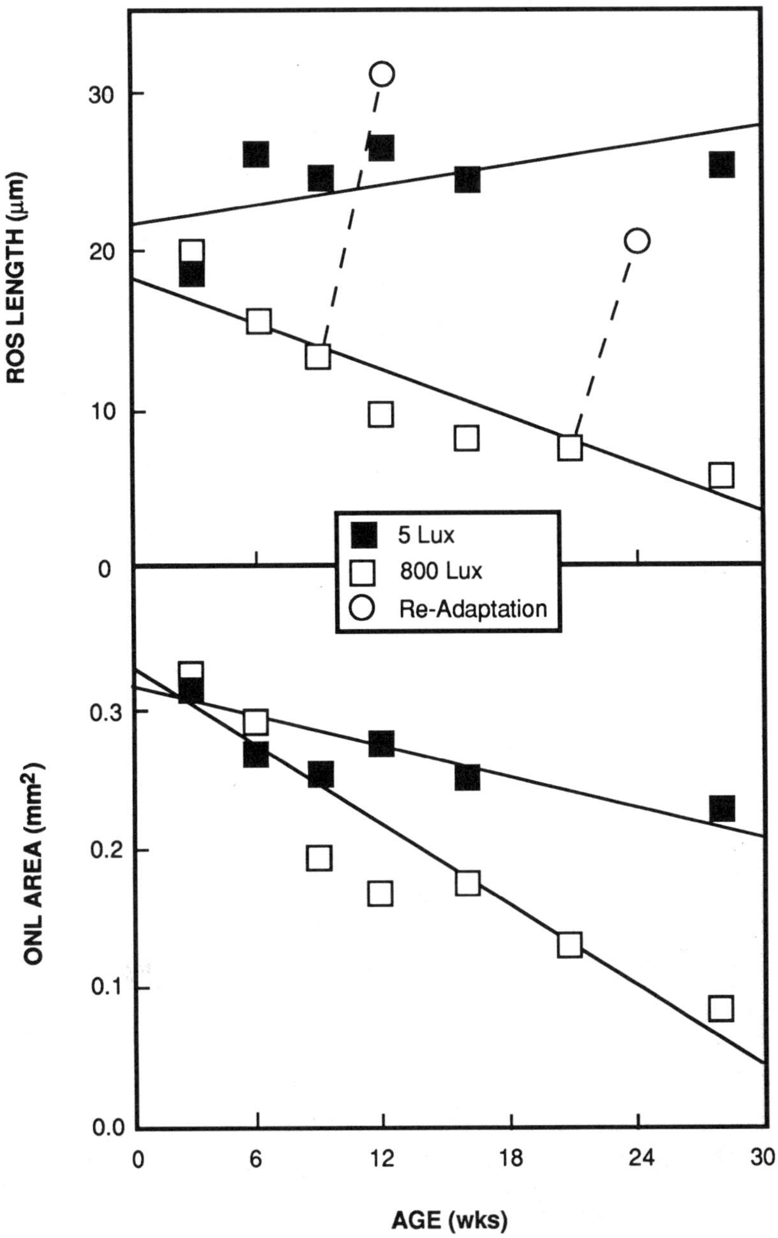

Figure 4. ROS length (top panel) is depicted over the time course of the study. Measurements were taken in the superior central portion of the retina. Circles represent ROS length in animals that were reared in 800-lux cyclic light for 9 or 21 weeks and then transferred to 5 lux for 3 weeks. ONL area (bottom panel) represents the number of photoreceptors in a section through the vertical meridian and is an estimation of photoreceptor loss with age.

the study (28 weeks). When the data were fitted to linear regressions, the loss of ROS length was 0.5 μm/week in 800-lux rats. Rats reared at 5 lux maintained a constant ROS length from 6 weeks of age to the end of the experimental period. The fact that 800-lux rats never displayed fully developed ROS lengths implies that developmental processes are altered in these animals.

By 9 weeks, differences in ONL area had become apparent. The loss of photoreceptors in bright-cyclic-light–reared rats was a relatively slow process that occurred over the course of weeks, rather than hours, as is the case for classical light damage from acute exposures (Noell *et al.,* 1966). Dim-light–reared animals showed a loss of ONL area with time that presumably reflected a normal aging process (3.5 μm^2/week). If this loss is subtracted from that in 800-lux rats (10.0 μm^2/week), the true rate of environmentally induced cell loss in the latter group can be appreciated. It remains to be determined whether bright cyclic light simply accelerates the aging process, but this seems unlikely to be the case given the abnormal early development of ROS's evident in rats reared at 800 lux (Penn *et al.,* 1992).

At two ages, rats were moved from 800 lux to 5 lux for 3 weeks before sacrifice. In both cases the change resulted in a rapid restructuring of the outer segments of remaining photoreceptors. Animals transferred at 9 weeks of age attained ROS lengths that exceeded those of age-matched, dim-light–reared rats within only 3 weeks (Table 2, Fig. 5). Rats transferred at 21 weeks also increased ROS length dramatically. In the rats transferred at 9 weeks, this restructuring of ROS's resulted in total replacement of rhodopsin lost as a result of the reduced number of photoreceptors that is a natural consequence of bright-light rearing. In rats transferred at 21 weeks, rhodopsin was not measured, but increased ROS length and diameter in these retinas would suggest that the ROS volume was great enough to accommodate high rhodopsin levels. The lower photoreceptor density that resulted from bright light rearing led to more room in the ROS layer; the remaining cells apparently took

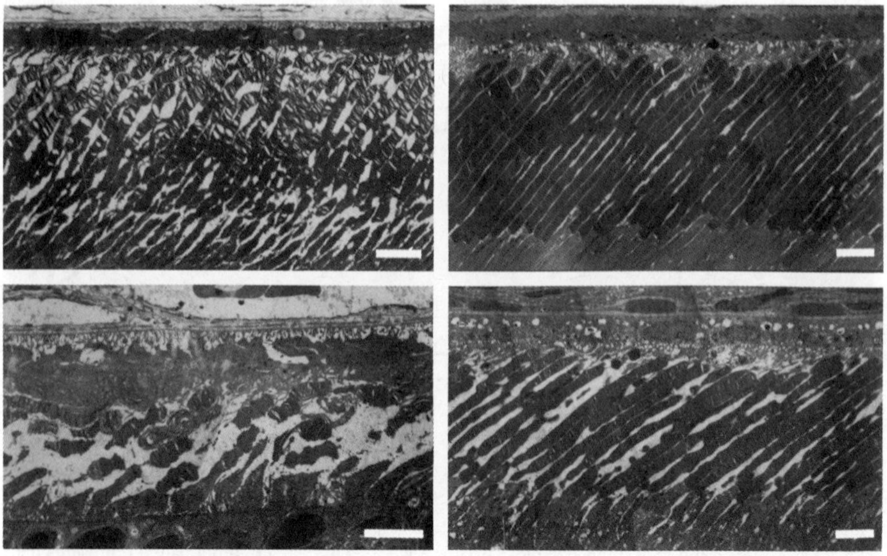

Figure 5. These electron micrographs illustrate the capacity of the rat retina to readapt to a new lighting condition. Rats were reared for 9 (top, left) or 21 (bottom, left) weeks in 800-lux cyclic light and then transferred to 5 lux for an additional 3 weeks (corresponding right-hand montages). Calibration bars are 5 μm in length. From Penn *et al.* (1992) with permission of Plenum Publishing Corp.

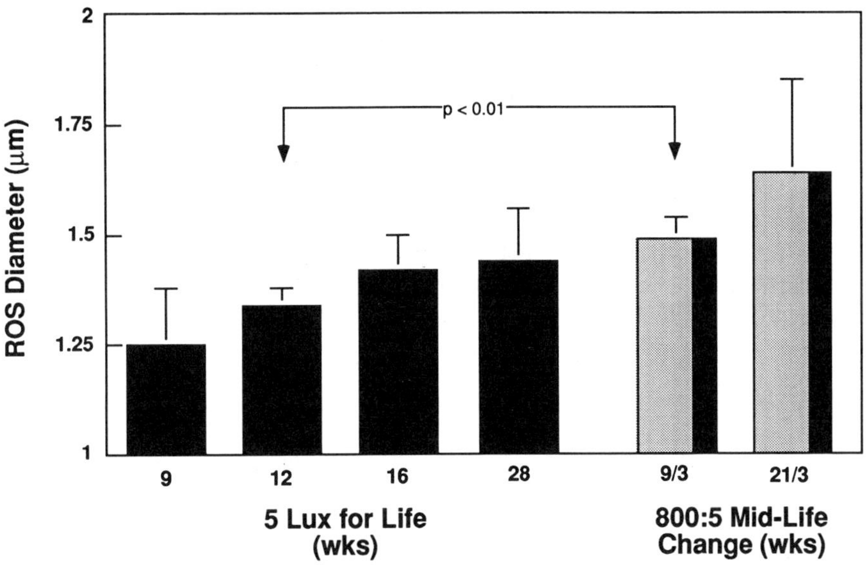

Figure 6. Average ROS diameter is depicted for several of the treatment groups. Values represent four measurements from each of 12 EM negatives from three rats in each treatment group. Correction was made for the inherent error of determining the diameter of a cylinder from longitudinal sections in different planes. The shading of the bars indicates 5-lux and 800-lux environments. Several groups differed significantly. Specifically, ROS diameter increased with age in dim-light–reared rats and increased drastically as rats adapted to dim light after bright-light rearing.

advantage of this space during readaptation to the new environment. Measures of ROS diameter for several treatments are shown in Figure 6. High diameters in older rats reared in dim light are presumably due to cell loss resulting from aging. More than any other results of our studies, the prodigious restructuring of the ROS's illustrated by Figures 5 and 6 verifies the remarkable plasticity and adaptability of the vertebrate photoreceptor.

4. DISCUSSION

Following Young's conclusions (1976) regarding the energy expense of continuously replacing ROS membrane, Williams designed a study (1984) to test the hypothesis that molecular senescence was the reason for the renewal process. Rhodopsin accounts for 80–90% of the total protein of ROS's (Hall *et al.,* 1969), so Williams concentrated on the rhodopsin molecule in his experiments. Because of the nature of ROS turnover, rhodopsin is always laid out in linear chronology along the length of the outer segment (Kaplan, 1982). Williams measured several important functional properties of the molecule as a function of its position (and therefore its age) with a microspectrophotometer. He found no difference between "old" (distal) and "new" (proximal) rhodopsin in the ROS's of *Bufo marinus.* Young's (1976) earlier suggestion that ROS renewal is a process of "preventative maintenance" and that "parts are replaced before they wear out" appeared well grounded at this point.

On the basis of our measurements of photon catch (Penn and Williams, 1986), we advanced the explanation that at least one function of ROS renewal was to facilitate adaptation to the light environment. This was not a new explanation for membrane turnover per se but a new application of an old idea. Finean *et al.* (1978) had written previously about cells in general: "a potential advantage may be that continuous turnover [of membrane] enables a cell to respond rapidly to any changing demand on its membrane functions by maintaining its membranes in a state of dynamic equilibrium." The photostasis phenomenon implied a principal role for ROS renewal in the process of adaptation to light environment—an implication that was partly legitimized by our phospholipid synthesis data (Penn and Anderson, 1992). If the difference in 22:6 n-3 between dim- and bright-light–reared rats were due solely to a different rate of turnover of the di-22:6 n-3-containing molecular species, it could be argued that events in the ROS (peroxidation, phospholipase activation, etc.) dictated the steady-state level of 22:6 n-3 in the two light regimes, but this was not the case. The differences were due in large part to a difference in the rate of *de novo* synthesis of di-22:6 n-3—containing molecular species. This finding supports the idea of adaptation in these retinas and points to the inner segment of the photoreceptor as a contributor to the adaptation process, because that is where the message for synthesis of these molecules and the machinery for their production and transport are located.

More recent experiments have addressed the mechanism by which the ROS renewal process adjusts to suit the requirements of the light environment. Particularly important is the work of Schremser and Williams (1995a, b), who presented strong evidence that post-translational alterations in ROS protein and alterations of ROS disc shedding and phagocytosis were both affected by changes from bright to dim or dim to bright light environments. These authors also reported the elegant observation that, when rats have recently been moved from one light environment to another, the rhodopsin concentration of distal ROS tips (constructed under the old illuminance) differs from that of the proximal base (constructed under the new).

The argument has often been raised that the ROS adjustments described by us and others in bright-light–reared rats are pathological rather than adaptive. The single most convincing piece of evidence for adaptation is the photostasis phenomenon itself. The notion that light-induced retinal pathology, typically poorly regulated in its progression and extent, results in control of photon catch is difficult to accept, particularly in light of an additional report by Williams (1990) that "the photon-catching ability of the retina [can be] adjusted up or down to compensate for [a] new intensity." In other words, Williams showed that even when rats are moved from one light regime to another, they maintain the optimum photon catch (ca. 1.0×10^{16} photons per eye per 12-hour light period).

Let us consider the case of rats moved at 9 weeks of age from 800 lux into 5 lux for 3 weeks before assessment. These rats have 76% of the photoreceptor number of a rat reared from birth in dim light (as estimated from ONL area, Table 2), yet they have 17% longer ROSs and 17% higher rhodopsin concentration in those ROS (Tables 2, 4). The remaining photoreceptors therefore have 37% more rhodopsin than do photoreceptors of a rat reared at 5 lux from birth ($1.17 \times 1.17 = 1.37$). If 76% of the photoreceptors remain, and each has 37% more rhodopsin than its 5-lux–reared counterpart, the total amount of rhodopsin should be approximately the same ($0.76 \times 1.37 = 1.04$). Because the ambient photons are also the same for these two groups, the potential for similar photon catch numbers is high. Of course, rhodopsin regeneration rate, which was not measured in the transferred rats, plays a critical role in the photostasis calculations.

Retinal rhodopsin level and the parameters that control it (ROS length and rhodopsin membrane concentration) probably are not capable, alone, of maintaining balance between the retina and its light environment. If they were, the polyunsaturated fatty acid content of ROS membranes, unless its change was associated with accommodating different packing densities of rhodopsin molecules, would have been the same for all groups. Because this was not the case, it is more likely that tailoring of the retina to suit the light environment is not solely a rhodopsin-mediated phenomenon but involves a complex combination of photoreceptor mechanisms acting in concert. Still, all but one (photoreceptor number) of the mechanisms discovered to date are tied to the process of ROS renewal. We therefore believe that one function of renewal is to permit animals to adapt to changes in ambient illumination. Of course, we should not overlook the primary means of light regulation in most mammals—the pigmented iris. Experiments employing chronically dilated and nondilated pigmented rats will address the true relevance of the photostasis phenomenon to adaptation in the wild. Pigmented rats may well catch the optimal number of photons, but drastic alterations of the renewal process probably are not required to facilitate photostasis. The use of albino rats allowed us to alter retinal irradiance to the extent that extreme measures were necessary to control photon catch, and especially because the environmentally induced differences in ROS renewal could be reversed in mid-life, the phenomenon provided us and others with an opportunity to explore the renewal process and its controlling factors. Recent studies continue to address ROS renewal at the molecular level, and there is ample cause for additional study. Such investigations are facilitated in the photoreceptor, because it is a cell whose function is well characterized and whose membranes can be isolated in large quantities and examined by established methods. Questions related to photoreceptor renewal are particularly compelling because malfunctions of the process are strongly implicated in a number of blinding conditions.

LITERATURE CITED

Andrews, L. D., and Cohen, A. I. (1983). Freeze-fracture studies of photoreceptor membranes: new observations bearing upon the distribution of cholesterol. *Journal of Cell Biology* 97:749–755.

Battelle, B. A., and LaVail, M. M. (1978). Rhodopsin content and rod outer segment length in albino rats: modification by dark adaptation. *Experimental Eye Research* 26:487–497.

Baylor, D. A., Lamb, T. A., and Yau, K. W. (1979). The membrane current of single rod outer segments. *Journal of Physiology (London)* 288:589–611.

Finean, J. B., Coleman, R., and Michell, R. H. (1978). *Membranes and Their Cellular Functions,* 2nd edition. Halstead Press, New York.

Hall, M. O., Bok, D., and Bacharach, A. D. E. (1969). Biosynthesis and assembly of the rod outer segment membrane system. Formation and fate of visual pigment in the frog retina. *Journal of Molecular Biology* 45:397–406.

Kaplan, M. W. (1982) Do birefringence bands in rod outer segments record daily disk synthesis? *Investigative Ophthalmology and Visual Science (Suppl.)* 22:204.

Kaplan, M. W., Deffebach, M. E., and Liebman, P. A. (1978). Birefringence measurements of structural inhomogeneities in *Rana pipiens* rod outer segments. *Biophysical Journal* 23:56–69.

Noell, W. K., Walker, V. S., Kang, B. S., and Berman, S. (1966). Retinal damage by light in rats. *Investigative Ophthalmology* 5:450–473.

Organisciak, D. T., and Noell, W. K. (1977). The rod outer segment phospholipid/opsin ratio of rats maintained in darkness or cyclic light. *Investigative Ophthalmology and Visual Science* 16:188–190.

Penn, J. S., and Anderson, R. E. (1987). Effect of light history on rod outer-segment membrane composition in the rat. *Experimental Eye Research* 44:767–778.

Penn, J. S., and Anderson, R. E. (1992). Effects of light history on the rat retina. In: Osborne, N. N., and Chader, G. J., eds. *Progress in Retinal Research, Vol. 11.* Pergamon Press, New York. pp. 75–98.

Penn, J. S., and Williams, T. P. (1986). Photostasis: regulation of daily photon-catch by rat retinas in response to varying cyclic illuminances. *Experimental Eye Research* 43:915–928.

Penn, J. S., Thum, L. A., and Naash, M. I. (1989). Photoreceptor physiology in the rat is governed by the light environment. *Experimental Eye Research* 49:205–215.

Penn, J. S., Tolman, B. L., Thum, L. A., and Koutz, C. A. (1992). Effect of light history on the rat retina: time course of morphological adaptation and readaptation. *Neurochemical Research* 17:91–99.

Rapp, L. M., and Williams, T. P. (1980). A parametric study of retinal light damage in albino and pigmented rats. In: Williams, T. P., and Baker, B. N., eds. *The Effects of Constant Light on Visual Processes.* Plenum Press, New York. pp. 135–159.

Schremser, J. L., and Williams, T. P. (1995a). Rod outer segment (ROS) renewal as a mechanism for adaptation to a new intensity environment. I. Rhodopsin levels and ROS length. *Experimental Eye Research* 61:17–24.

Schremser, J. L., and Williams, T. P. (1995b). Rod outer segment (ROS) renewal as a mechanism for adaptation to a new intensity environment. II. Rhodopsin synthesis and packing density. *Experimental Eye Research* 61:25–32.

Wiegand, R. D., and Anderson, R. E. (1983). Phospholipid molecular species of frog rod outer segment membranes. *Experimental Eye Research* 37:159–173.

Williams, T. P. (1984). Some properties of old and new rhodopsin in single *Bufo* rods. *Journal of General Physiology* 83:841–852.

Williams, T. P. (1990). Regulation of retinal light damage. Abstracts of the 9th International Congress of Eye Research. *Proceedings of the International Society for Eye Research* 6:145.

Young, R. W. (1967). The renewal of photoreceptor outer segments. *Journal of Cell Biology* 33:61–72.

Young, R. W. (1976). Visual cells and the concept of renewal. *Investigative Ophthalmology* 15:700–725.

2

LIGHT HISTORY AND PHOTOSTASIS

What Is a "Normal" Rat Retina?

Theodore P. Williams

Department of Biological Science
Florida State University
Tallahassee, Florida 32306-4340

1. INTRODUCTION

The origins of our work on photostasis can be traced back to 1976, when we began studying retinal light damage in the albino rat. Ten years earlier, Werner Noell's group (Noell *et al.,* 1966) had shown that photoreceptors die if they "absorb too many photons." After some delay, we began to study this phenomenon. My interest was twofold: First, why should these cells die from absorbing too much light? Noell was studying rats, and rods predominate in their retinas, so death of the rods was the main topic of interest. Rods function at night but were being killed by daytime light exposures, even though they were not signalling vision at the time. Was this a clue to the damage mechanism? Second, if the rods were dying because they were absorbing too many photons, how many is too many? Because the second question seemed the more tractable, and because I knew how to count photons, my group slowly edged its way into studies of light damage, hoping to arrive someday at an answer to the second question.

It became clear in those early experiments that, if we wanted to obtain reproducible results, we had to rear our own litters of rats from birth. The rats we obtained from various suppliers (and we tried several; see Williams, 1989) came to us in various states of what, at that time, we called light damage. That is, they showed great variability in the thickness of the outer nuclear layer (ONL), a hallmark of light damage. If our randomly selected "control" rats did not have uniform ONL thicknesses, how could we hope to learn anything from the experimental groups? Engrossed as we were in this newly discovered phenomenon of damage to the rods by excessive light, we did not take time to study the *cause* of the variability in the retinas of the commercial animals. We simply guessed that it stemmed from their light history. Our guess proved to be correct, and by rearing our own rats in special boxes from birth we were able to control their light history and eliminate the interanimal variability. By this time, the effect of previous light exposure was being examined by Noell and coworkers (Organisciak and Noell, 1977), and by stages, our own

Photostasis and Related Phenomena, edited by Williams and Thistle
Plenum Press, New York, 1998

Table 1. Rod properties and light history

	ROS	Rhodopsin per disk	Regeneration rate	Rod count	Photons per day	Photons rod^{-1} s^{-1}
Low-light habitat or short day	Long	High	High	High	$\sim10^{16}$	~14,000
High-light habitat or long day	Short	Low	Low	Low	$\sim10^{16}$	13,000–14,000

light-damage studies, along with those of Noell and others, led us to study light-history effects per se. These studies led, in turn, to the discovery of photostasis.

Table 1 summarizes the qualitative effects of light history. It serves as a reminder that rats make substantial adjustments in several retinal properties but keep fixed the number of photons absorbed per eye per day: $10^{16} \pm \sim20\%$.

My main purpose in this chapter is to review certain features of the light-induced retinal adjustments presented in Table 1 and to relate them to photostasis. I also present previously unpublished results that answer some of the questions frequently posed about photostasis and about our studies of it. For example, why do we use albino animals, and what happens to photon absorptions when the rat is asleep during the day? Finally, I hope to answer the question, "What is a *normal* rat retina?" To do so, I make an assumption: Because the "photostasis number" is constant regardless of the retinal adjustments summarized in Table 1, I assume that the *function* of the adjustments is the maintenance of that number.

2. DARK-ADAPTED RHODOPSIN LEVELS, R_0

R_0 is the entire amount of rhodopsin in a dark-adapted eye. It has units of nanomoles per eye but is sometimes given as molecules per eye. This is the amount of rhodopsin the animal has after a full night of dark adaptation—the amount of rhodopsin the animal has to "start its day." R_0 is set by ROS renewal processes in the eye, as described by Schremser-Berlin (Schremser and Williams, 1995a, b; Schremser-Berlin, this volume). These rhodopsin levels are measured by quantitative extraction, with detergent, of the rhodopsin from whole retinas. This extraction procedure gives the total amount present but, used alone, does not describe the concentration of rhodopsin in the ROS, nor does it give information about the regional distribution of that pigment in the retina.

R_0 values vary inversely with light intensity in the habitat and with day length. Penn (this volume) presents the effects of intensity, and below I give some results on day length from the work of Charles Parker in our lab (see also Penn and Williams, 1986). First, however, some description will be given of the cellular mechanisms that produce these R_0 values.

The total amount of rhodopsin in a dark-adapted rat eye, R_0, depends on the (1) the number of rods in that eye, (2) the length of the ROS, and (3) the concentration of rhodopsin in the ROS disks. That the number of rods in a rat eye depends on the lighting environment in which the rat lives frequently comes as a surprise to those outside the field; after all, retinal rod cells are central nervous system neurons, and their loss in mammals is irreversible. Penn and coworkers (e.g., Penn and Thum, 1987; Penn *et al.,* 1992) have addressed this question of the adaptive nature of rod cell count, and their conclusions are, in summary: (i) In adaptation to light-rearing situation, rods are lost mainly in the inferior retina and cause no disruption of retinal organization, whereas light damage removes rods mainly from the superior retina and causes extensive disorganization. (ii) During the first 2–3 weeks of life, groups of infant rats lose rods at the same rate regardless of light intensity in the habitat (Is this programmed cell death?), but beyond this time, rats living in dim cyclic light stop losing rods even though rats living at higher intensities continue to lose them. These features suggest that, when rats are reared under various lighting conditions, rods are lost as an *adaptation* to those conditions.

We measure the area of the ONL and take the result to be proportional to the number of rods in a histological section that has been stained with hematoxylin-eosin. Then we integrate over the entire retinal area, having measured the dimensions of the partial sphere subtended by the ONL in the histological section. (The rat retina extends to about 80° in all directions from the posterior pole.) Using the *area* of the ONL has the advantage that the cell count does not depend on whether these nuclei are compacted into a small retina or spread out over a larger one (as occurs during development). Alternatively, we measure the ONL thickness at 8–12 locations across the retinal section and then integrate, using the average thickness of the ONL as the average increase in eye radius, dr. We have found that the two methods give good agreement and that the number of cells in a 12L:12D, 3-lux rat is 1.7×10^7/eye. This estimate agrees well with those made by Lashley (1932) and by Cone (1963), who reported 1.5 and 2.0×10^7 rods per eye, respectively. Note that even if the number of cells, calculated in our way, is in error, the error cannot be too great because agreement with the work of others is good and none of the uses to which these cell counts are put depends on exact numbers—our conclusions on this subject would be valid even if we were in error by a factor of almost two, as will be seen below. The dependence of rod cell count on habitat lighting is given in Table 2.

Rod cell count is not the only factor that determines the dark-adapted rhodopsin level. R_0 values are also affected by the length of the ROSs and by the concentration of rhodopsin in them. ROS length is measured by means of a standard light-microscopic technique: Sections are prepared and stained with the lipophilic dye Sudan Black, and lengths are measured in well-oriented sections.

ROS length and rhodopsin concentration in the ROS disks are adjusted by renewal of the ROSs, a process discovered by Young (1967), who suggested that renewal was a way to maintain a constant ROS length while at the same time permitting the removal and replacement of "old" molecules. This hypothesis was modified slightly when Young suggested that the cell anticipates approaching senescence and gets rid of the "old" molecules before dysfunction occurs.

Table 2. Effects of habitat light intensity and day length on rod cell count

	Rod cell count $\times 10^{-7}$
Lux (12L:12D)	
3	1.7 ± 0.08
80	1.7 ± 0.09
200	1.5 ± 0.04
400	0.8 ± 0.08
Day length (hours)	
8	2.5 ± 0.10
12	1.6 ± 0.05
16	1.2 ± 0.07

Despite efforts to verify Young's senescence hypothesis (Baylor *et al.,* 1979; Baylor and Lamb, 1982; Schnapf, 1983; Williams, 1984), no compelling evidence has been found for the existence of "aging" either in the nature of rhodopsin itself or in the ability of rods to transduce light in the "old" distal tips. Kaplan (1984) has reported a slightly lower disk fluidity in the distal tips of ROS than in the proximal bases, but this difference was not observed by Williams (1984). Furthermore, Battelle and LaVail (1978) discovered that ROS lengths in rat retinas are not constant but vary with the lighting environment. I asked myself at one point whether very *long* ROS's contain very *old* rhodopsin in their tips, but I found no evidence that they do.

The lack of evidence for the senescence hypothesis, combined with the difficulties associated with trying to test a hypothesis whose basis lies in the possibility that molecules are discarded *before* they became aged, prompted me to suggest another hypothesis to explain the existence of the renewal process (Williams, 1984). I proposed that renewal was adaptive and that it permitted rats to make long-term adaptations to habitat light intensity (or to day length). According to my hypothesis, the adaptation was to manifest itself in the length of the ROS and in the concentration of rhodopsin in the ROS disks, so we set out to examine this possibility.

Ever since Penn's work in my laboratory, we had assumed that renewal was responsible for adaptation of ROS lengths and rhodopsin concentration per disk. Penn's (1984) dissertation, Bush's (unpublished) pilot studies, and Schremser's (1990) master's thesis all showed that these rod properties varied with light intensity in the habitat. Schremser showed that the retina was plastic, capable of reversing ROS lengths when intensities were reversed (Fig. 1). In her first studies, she allowed the animals to adapt to the "new" environment for two weeks.

The results of Bush's study of plasticity in young rats are shown in Figure 2. He examined the effects of age on the plasticity of R_0 and on maintenance of the "photostasis number," N. He showed that, when the rats were moved to a "new" lighting environment, R_0 values were up- or down-regulated with habitat light intensity as predicted, but the parameter that did not change was the number of photons absorbed per eye per day. Bush examined rats as young as 5 weeks of age and as old as 15 weeks. He found no apparent "critical period," over this age range, at which the rats failed to adapt to the new lighting. He exposed the rats to the new environment for 2 weeks before making the required measurements, and we now know from Schremser's dissertation work that exposing them for 3 weeks would have been slightly better.

What process confers this plasticity on the rods? It seemed obvious that only renewal of the outer segment could be responsible for the variation in these rod properties. What other mechanism could make a long or short rod, pack more or less rhodopsin on the disks? A definitive study making a connection between rod plasticity and renewal of the

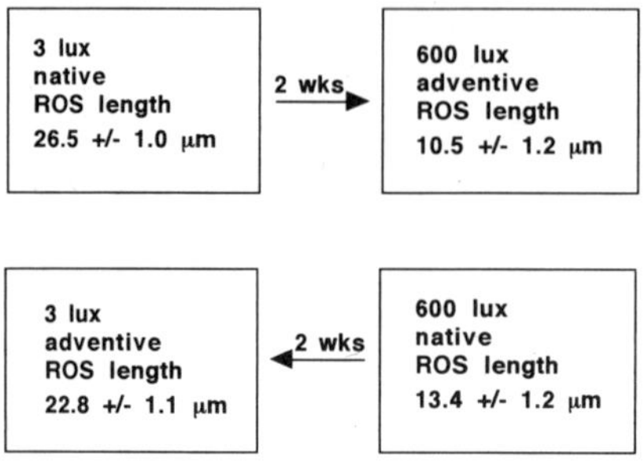

Figure 1. Plasticity of rod outer segment length. Albino rats were reared in either 3- or 600-lux cyclic light. At 12 weeks of age, some from each group were moved into the opposite intensity; others were allowed to stay where they were. After 2 weeks in the "new" environment, the ROS's were measured. ROS lengths increased and decreased inversely with habitat light intensity. Length is therefore a dynamic property of the rod.

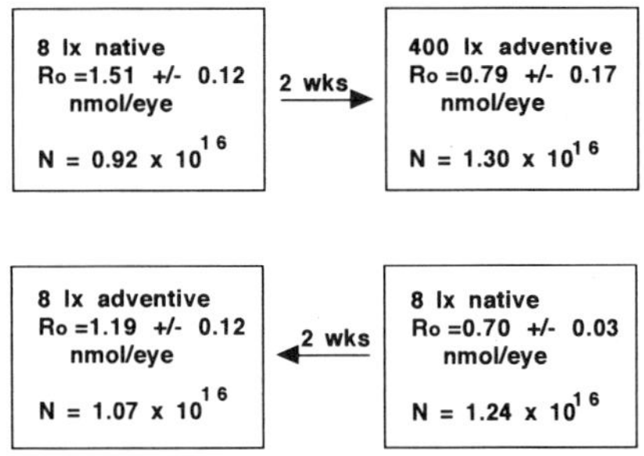

Figure 2. Plasticity of dark-adapted rhodopsin level (R_0) and constancy of the number of photons absorbed per eye per day (N). R_0 values increase and decrease inversely with intensity, but N remains constant to within 12%, despite a 50-fold change in light intensity. At 5 weeks of age, some rats were moved to "new" lighting; others stayed in their "native" lighting habitats. After 2 weeks, rhodopsin levels were assayed and N's were calculated.

ROS was needed. By what *means* did an ROS get longer or shorter? What is the *time course* of changing the concentration of rhodopsin in disks? Schremser-Berlin (this volume) has answered these questions. She has shown how renewal is adaptive and how it affects R_0. In a particularly elegant part of her work, Schremser-Berlin has caught some rods in the process of changing their rhodopsin concentration upon a switch from one light intensity to another. Her results are shown schematically in Figure 3.

The conclusion that visual-cell membrane turnover in invertebrates is adaptive enjoys support from work by White and Lord (1975) by Blest (1978, 1980) and recently by Chamberlain and coworkers (this volume and references cited there). They have studied a number of marine invertebrates and have presented evidence that certain of these animals, living in constant lighting environments (either complete darkness or dim, phosphorescence-produced, constant light), do not turn their light-absorbing membranes over, whereas those living in environments in which the intensity is variable renew these membranes daily. Although the turnover mechanisms involve different cell types, vertebrates and invertebrates apparently share the common need to match light-absorbing capabilities inversely with habitat light intensity and, if changes in these intensities occur, to modify their visual cells by renewal.

We propose that events in the outer segments of rods, namely the number of photons absorbed per eye per day, set the level of R_0 in the eye. Therefore, information about the occurrence of those outer segment events must somehow be conveyed to the inner segment, where the synthesis and maturation of rhodopsin occur. *A priori,* I see two possibilities: Either the membrane potential of the rod is, itself, a signal, or material from the outer segment, produced or modified by bleaching events, arrives at the inner segment in proportion to the number of events that has occurred in the outer segment. The latter is attrac-

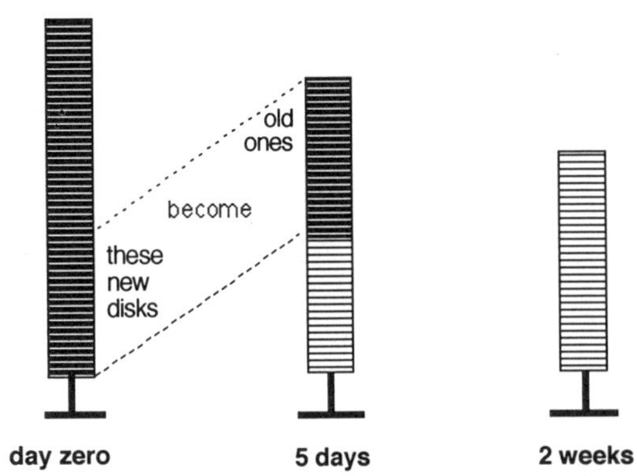

Figure 3. Plasticity of rhodopsin concentration in ROS disks shown schematically. Schremser-Berlin (this volume) caught some rods about halfway through the process of changing their concentration.

tive because molecular transport between the inner and outer segments is known to occur (Philp *et al.,* 1987; Whelan and McGinnis, 1988; Mangini *et al.,* 1994). The transported "material" could be retinoid, and its concentration would probably be in proportion to the number of bleaching events because vertebrate rhodopsin dissociates into retinal and opsin upon bleaching. Linking the *concentration* of this or any other "signal" molecule to the number of bleaching events is, of course, desirable if not mandatory for photostasis considerations. Furthermore, retinoids are known to be regulatory molecules in many gene expressions. Although attractive, and perhaps a prime suspect in this regulation, retinoid involvement in controlling, e.g., transcription has not been supported unequivocally by all investigators (see Katz *et al.,* 1991; Huber *et al.,* 1994), nor do we see evidence that the regulation in rats is at the transcriptional level (Fig. 4, below).

I believe membrane potential holds promise as a conveyor of ROS information. Not only are membrane potentials in nerves sometimes used as gene regulatory signals (Sheng *et al.,* 1990; Birren *et al.,* 1992), but one possible function of photostasis, discussed below (see section 6), is the maintenance of rod cells at saturation.

The concentration of rhodopsin in the disc membranes varies inversely with intensity. Making a start toward determining where this regulation occurs, we have measured the expression of opsin mRNA as a function of habitat light intensity. A northern blot of the message levels, depicted in Figure 4, shows that rats raised at 3 lux and at 200 lux do not differ in levels of opsin message. Nevertheless, the rhodopsin levels in these rats differ by 60%; the low-light-intensity animals have more pigment, as expected. This result was found independently in my laboratory by both Huber and Schremser-Berlin. Furthermore, Huber *et al.* (1994) have shown that regulation of rhodopsin levels in fly eyes is posttranscriptional, apparently taking place at the point where the chromophore is added. Perhaps such a mechanism obtains in the rat.

In summary, the loss of rod cells in rats reared at high light intensities is apparently not a "damage" mechanism but an adaptive one. Rigorous proof is still lacking but could come from studies that determine whether or not the cell loss is apoptotic. Renewal of the ROS's, on the other hand, is definitely an adaptive process and regulates, inversely with intensity, ROS length as well as ROS rhodopsin concentration. The function of renewal is to allow rods to adapt to slow, daily, or seasonal changes in ambient lighting. Whether re-

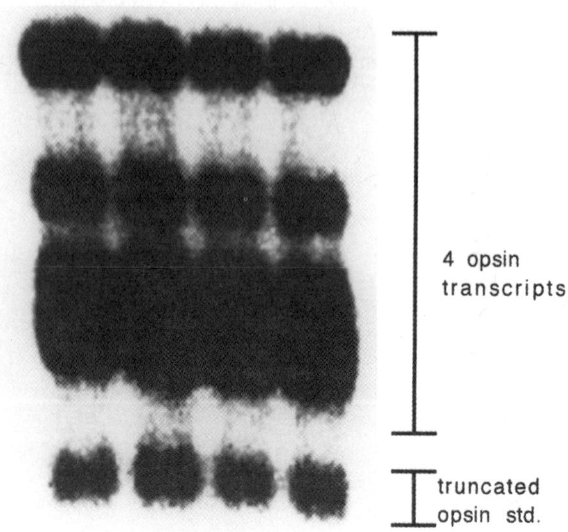

Figure 4. Northern blot of opsin mRNA from rats reared either in 3- or in 200-lux cyclic light. From left to right, first two lanes, 3 lux; second two lanes, 200 lux. Rats have 4 opsin transcripts, and Betagen analysis of counts in these lanes shows no differences in total transcript levels in the two intensity groups. The bottom-most blots are truncated opsin sense-standards that had been added to the retinal homogenates. Use of such external standards avoids the problem of constitutively expressed "standards," which may themselves change with lighting conditions.

newal also serves to dispose of molecules that would otherwise become dysfunctional with age has yet to be shown. Finally, regulation of R_0 in rats appears to be posttranscriptional, but the details of the mechanism remain to elucidated.

3. CHOICE OF ANIMALS: PIGMENTED OR ALBINO?

My laboratory uses albino rats because their irises are virtually transparent. Even if constricted, these tissues occlude almost no light. We demonstrated this point by dilating the pupils of living albino rats and comparing the levels of bleach in their eyes to those in albino eyes with no pupil dilation (Rapp and Williams, 1980). Animals of both groups were allowed to move around freely in a well-regulated, uniformly lighted box, and their bleach levels were measured after sacrifice. The levels of bleach were indistinguishable, permitting the experimenter, without the daily use of mydriatics, to expose living, freely moving rats to a desired dose of light and be sure that the light gets to the retina.

This is a straightforward, even powerful, reason for using albinos, and it finds general acceptance but nevertheless prompts the further question: Are the retinal changes we observe peculiar to albino rats? Do pigmented rats show them? Yes, they do. Long-Evans pigmented rats do show the retinal adaptations, but either mydriatics or much higher light intensities in the habitat must be employed to induce them. Some of the results from pigmented rats are given by Fulton (this volume). Developmentally, albino and Long-Evans pigmented rats, both without mydriasis, accumulate rhodopsin in their eyes with almost identical maturational time courses, and importantly, the time courses of intensity effects are indistinguishable, but the reader should carefully note the intensities used in these comparative studies: 3500 lux are needed to achieve approximately the same effect on R_0 in pigmented rats that 200 lux has on albinos. That is, intensities approaching outdoor levels are need to achieve these R_0 values. In north Florida, even on a winter day, 10,000–50,000 lux is not uncommon at midday outdoors, so 3500 lux is not extraordinary.

Dodge and Baker, in our laboratory, have reared groups of pigmented rats at various intensities of cyclic light, and their results are given in Table 3. They found that, *without* pupil dilation, the R_0 values were all about the same, regardless of light intensity in the habitat (up to 1200 lux), but if pupils of these mature Long-Evans rats, living in 200-lux cyclic light, were dilated daily for 10 days, their R_0 values were 36% lower (last row of Table 3) than those of animals whose pupils were not dilated. Combining these results with those given by Fulton (this volume) makes clear that rhodopsin levels in these pigmented eyes are inverse functions of the habitat light intensity. That pigmented rats up- or down-regulate R_0 inversely with habitat intensity is an important clue that they perform

Table 3. R_0 in pigmented rats

Lux	R_0
3	2.73 ± 0.14
80	2.49 ± 0.19
200	2.55 ± 0.10
400	2.47 ± 0.09
1200	2.51 ± 0.14
200 (with dilation)[a]	1.64 ± 0.16

[a]The value for 200 lux with dilation is given for comparison.

photostasis, too, but they apparently do so by regulating the intensity of light falling on their retinas by means of their heavily melanized, light-occluding irises.

Even with mydriatics, the R_0 values in pigmented eyes are not precisely those found in albino rats reared at the corresponding light intensities (Table 3), because, given equal ambient intensities, the effective intensity inside the pigmented eye is lower than that in the albino eye. This effective-intensity reduction was proposed by Rapp and Williams (1980) to be a factor of about two, because about twice as much light was needed in dilated, pigmented rat eyes to get the same level of damage found in albino eyes. We suggested that light scatters inside albino eyes, effectively doubling the intensity. Later work here showed that this conclusion is correct (see Fig. 5). These results were obtained with the ocular transmission photometer, the function of which will be described below. The important feature of these results is that, at equal input intensity, the bleaching rate across the retina is about twice as great in albino eyes as in pigmented ones.

These results strongly imply that, like albinos, pigmented rats perform photostasis. A direct demonstration that they do is difficult, because there is no measurable steady state bleach, B_{ss}, in these pigmented animals at any of the intensities we normally use (1–400 lux), and it is therefore impossible to determine the regeneration rate constant, k_r, needed for the calculation of the photostasis number.

Relevant here are our attempts to determine frequency and duration of eye closure as well as the effectiveness of the albino eyelids in occluding light. Eye closure could have influenced our results if albino rats slept a great deal during the day and if not much light penetrated to the retina during those times. Accordingly, Bush monitored with closed circuit TV the behavior of rats during daylight hours and found that they sleep only fitfully during the day, displaying many hours of wakefulness and movement (Bush and Williams, 1991). Furthermore, they do not always close their eyes even if they appear to be asleep.

To determine how much bleaching occurs in an albino rat eye that is truly closed, Reiser (unpublished data) killed dark-adapted rats with CO_2 in dim red light and sutured the eyelids of one eye together with clear suture to give the appearance of a normally closed eye. The retina of the other eye was removed, set aside, and used as a control. The prepared rat was then positioned in a standard way in front of a uniform bank of lights, which had been pre-adjusted to deliver 200 lux at the level of the rat eye, and the lights were turned on for a preset number of minutes, during which the rhodopsin in the sutured eye underwent bleaching. The extent of bleaching (always kept below 40%) was determined by comparison with the control, fellow eye and with unsutured eyes from other rats

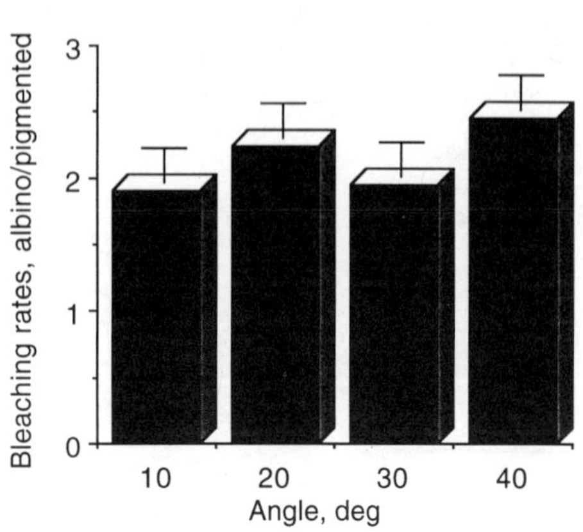

Figure 5. Comparison of bleaching rates in albino and pigmented rats. Measurements across the retina were made with OTP. With identical light-intensity inputs, albino rods bleach about twice as fast as do those in pigmented eyes.

Table 4. Beaching through rat eyelids

	n	Minutes	Percent bleach	Percent per minute
Sutured	4	5.0	35 ± 5	7.0 ± 1.0
Sutured	6	4.5	22 ± 2	4.1 ± 0.4
Sutured	3	4.0	21 ± 9	5.3 ± 2.3
Unsutured	6	2.0	25 ± 3	12.5 ± 1.3

placed in front of the same bank of lights. The results of this study (Table 4) show that the albino eyelids reduce the bleaching rate by only about half, even when fully closed. Our results seem sufficiently reproducible from one rat to the next to justify neglecting the effects of eyelid closure.

We continue to use albino rats in our studies because, although the same changes can be induced in pigmented rats, they can be induced in albinos at intensities normally encountered indoors (e.g., 1–400 lux) and without the use of mydriatics.

4. RATES OF RHODOPSIN REGENERATION

The maximum amount of rhodopsin in a dark-adapted eye, R_0, is set by regulation of the renewal process, but the daytime, moment-to-moment availability of the pigment for light absorption is regulated by the rate at which rhodopsin is regenerated from the bleached state. Perlman (1978) demonstrated that the regeneration rate in albino rats was affected by the extent of bleaching: the greater the extent of bleach, the slower the regeneration, and vice versa. Therefore, the rods, by regulating their regeneration rates, regulate their photon absorption rates. Indeed, at low to moderate intensities, as a general rule, the rate of regeneration is inversely related to intensity (Penn and Williams, 1986). However, at high intensity, where rod cell number is reduced significantly, regeneration rate has been observed to deviate from this simple inverse relationship (see Penn and Williams, 1986; 400-lux rats).

The reader is reminded that the only regeneration rate relevant to photostasis calculations is the one that results from bleaching due to the animal's living in its own habitat. Because regeneration rates depend upon the extent of bleaching, any given rate can be induced by exposure of the animal to arbitrarily chosen beaching conditions.

The method for measuring rhodopsin regeneration rates in rats requires a slight modification of the detergent-extraction procedures described above. At the end of their normal day, we kill animals at various intervals after the lights go off in their habitats and assay their retinas for rhodopsin. By tracking the accumulation of rhodopsin with time in the dark, by constructing first-order, integrated rate plots, and by finding the slopes, we determine k_r.

It appeared from our initial work that k_r "fine tuned" photostasis. The large cellular adaptations that affect R_0 provided the "coarse tuning," but the moment-to-moment availability of rhodopsin during daylight hours was controlled by regeneration rate. What, then, regulates k_r? The extent of bleaching is involved in some way in the regulation of the regeneration rate (Perlman, 1978), and it is known that the lifetime of the metarhodopsin II–arrestin complex is also involved (Broekhuyse *et al.,* 1985; Hofmann *et al.,* 1992), but otherwise little is known. Nevertheless, it is clear that regeneration rates vary with habitat intensity in ways that permit photostasis.

We measured the dark accumulation of rhodopsin in our rats after the lights were shut off at the end of the normal daylight hours. The rate constant for regeneration measured this way is the same as that at steady state, when the lights are on: Alpern and Krantz (1981) showed that regeneration rates in human eyes are the same whether the adapting lights are on or off. Thus, in the rat, regeneration rates in the light and in the dark are likely to be the same.

5. STEADY-STATE LEVELS OF RHODOPSIN, R_{ss}

Steady-state levels of rhodopsin are also determined by detergent extraction of whole retina and are given in units of nanomoles per eye or, sometimes, molecules per eye. The amount of rhodopsin present in the living rat eye is lower during daylight hours than in the dark-adapted eye. Bleaching occurs in light, of course, and the dark-adapted rhodopsin level is reduced accordingly, but regeneration of the pigment occurs concurrently and works to restore the rhodopsin to some extent. When bleaching and regeneration occur *in vivo,* they can give rise to a steady-state level of rhodopsin in the eye. In our laboratory, Rapp (Rapp and Williams, 1980) showed that, in albino rats, such a steady-state exists: At intervals throughout the day he euthanized, with barbiturate, rats in their home cages and as soon as they had succumbed to the drug, he quickly removed them from the light and extracted the rhodopsin from their retinas. He found that, after about 1 hour of light exposure in the various animals' habitats, the rhodopsin in the eyes arrived at a steady-state level, which then persisted for the rest of the day. The level depended upon the light intensity of the habitat relative to the regeneration rate. This observation, made over a range of intensities (3–400 lux), permits our calculation of the photostasis number (see below).

Recently, have found excellent agreement between Rapp's method and a simpler one. The new method uses an extrapolation of linear, integrated first-order rate plots to time zero. Because time zero is the moment the lights were shut off, the rhodopsin existing at that instant is the steady-state level. This method eliminates the necessity for barbiturates and the uncertainty in time of death. The Y-intercept gives R_{ss} values that agree to within 10% with those obtained by Rapp's method. Thus, in the process of measuring k_r we can obtain R_{ss}, the other parameter needed for photostasis calculation. Figure 6 gives an example of this extrapolation technique.

6. PHOTOSTASIS CALCULATIONS

The rate equation that describes bleaching and regeneration *in vivo* is:

$$-dR/dt = \Phi I \bullet R - k_r \bullet B \quad (1)$$

where Φ is the photosensitivity of rhodopsin, a molecular property and assumed here to be constant, and B is the concentration of bleached molecules. At steady state, the net change in rhodopsin with time is zero, and therefore

$$\Phi I \bullet R_{ss} = k_r \bullet B_{ss} \quad (2)$$

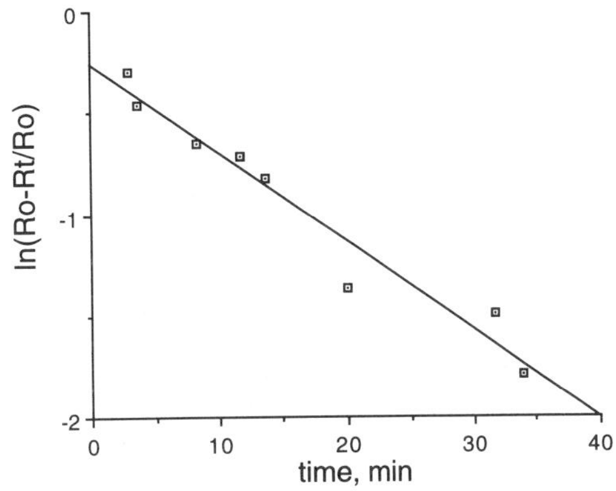

Figure 6. Integrated, first-order rate plot of rhodopsin regeneration. Data are for 8-hour day length, 200-lux cyclic environment. The straight line extrapolated to time zero gives the rhodopsin level just when the lights went out, i.e. R_{ss}. This method of determining R_{ss} gives the same values (to within 10%) obtained by the more difficult earlier method of Rapp and Williams (1980). In this case the intercept, −0.258, is set equal to $\ln(R_o - R_t/R_o)$, and $R_t = R_{ss}$ is calculated. The intercept is obtained from least-squares fit of the data.

Before we (Penn and Williams, 1986) knew how to describe the intensity distribution inside an albino rat eye, we solved Eq. 2 by measuring the two parameters on the right-hand side. The logic of this approach is based on Rapp's demonstration at various intensities of habitat lighting in our laboratory that steady states in fact exist. As a result, the number of regeneration events (right-hand side) occurring per unit time at steady state is equal to the number of bleaching events (left-hand side) per unit time. By counting the number of regeneration events per unit time, we are, *de facto,* counting the number of bleaching events per unit time. The first results of such calculations are given by Penn (this volume). The regulation of this number of absorptions per eye per day is remarkable: 10^{16} ± about 20%, even though the intensity of habitat varied by more than 100-fold.

For legitimate use of the steady-state equation, Eq. 2, the habitat lighting has to bleach an appreciable fraction of R_o and at the same time leave enough R_{ss} to be measured accurately. Because, if the bleach is too little or too great, one cannot measure the necessary variables, does photostasis exist at extremes of intensity, either very low or very high? At present we have no definitive answer, but we have been able to calculate the intensity minimum below which it is not possible for rats to absorb 10^{16} photons in (for example) 12 hours of light. It lies between 1 and 3 lux. It is interesting that the lowest intensity at which Penn and Williams could measure a steady-state bleach is exactly in this range.

7. OCULAR TRANSMISSION PHOTOMETRY

Since discovering that rats regulate the total number of photons absorbed each day, we have asked ourselves what the biological purpose of the process might be. In searching for an answer to this question, I reexamined old data and in the process calculated the rate of absorption per rod. This calculation is very simple:

$$\text{photons/rod/second} = 10^{16}/(\#\ \text{rods/retina}) \times (\text{day length in seconds})$$

The number of rods per retina (12L:12D cycle, 1–3 lux) is, in our hands, 1.7×10^7. Thus for 12 hours of daylight, the *average* rate of absorptions is 13,100 $\text{rod}^{-1}\ \text{s}^{-1}$. At first, this rate seemed too high, but a literature search disclosed that it lies near the middle of the range of rates that are known to saturate human rods (Aguilar and Stiles, 1954;

Pirenne, 1962). Thus, it seems possible that photostasis holds the rate of photon absorption at just the right level to keep rods saturated during daylight hours, but only if photon absorptions are spread evenly across the retina. If a small fraction of the rods were absorbing at very high rates but the remainder were absorbing at near-zero rates, the rod *system* would not be saturated. We therefore set out to determine the distribution of photon absorption rates across the rat retina.

The rate at which photons are absorbed by rod cells is given by

$$-dR/dt = \Phi I \bullet R \tag{3}$$

If the product IR is measured at several locations across the retina, the distribution of absorption rates can be plotted. We have designed and built an ocular transmission photometer (OTP) to do just that (Williams and Webbers, 1995). It exposes excised, dark-adapted eyes to a uniformly lighted *Ganzfeld* and monitors, during exposure, rhodopsin bleaching at several stations across the rat retina. The OTP is shown schematically in Figure 7. Light from the laser-illuminated *Ganzfeld* enters the eye and begins to bleach rhodopsin everywhere in the eye. Some of the light exits through the sclera, where it is picked up by fine-tipped optical fibers and its intensity is monitored with a photon-counting system. The optical fibers are positioned on the back of the eye at specific locations by means of a goniometer, calibrated in degrees. The stump of the optic nerve is taken as zero degrees, and recordings are made at various angles relative to this reference point. The *extent* of intensity change due to local bleaching gives the local rhodopsin concentration and the *rate* of that bleaching gives (inversely) the intensity falling on the rods at that location. Thus, using this device, we determine both the local I and the local R. The product of these, IR, obtained in this way is proportional to the rate of photon absorption at each of the various locations. The OTP resolves about 2% of the rods at any one location in the rat eye, so this method provides information that whole retina extractions do not: it describes the distribution of photon absorptions across the rat retina.

An example of these results is given in Figure 8, and the reader is referred to Williams and Webbers (1995) for details. Note that the rhodopsin concentration and intensity are fairly uniformly distributed across the retina, and their product, IR, is even more uni-

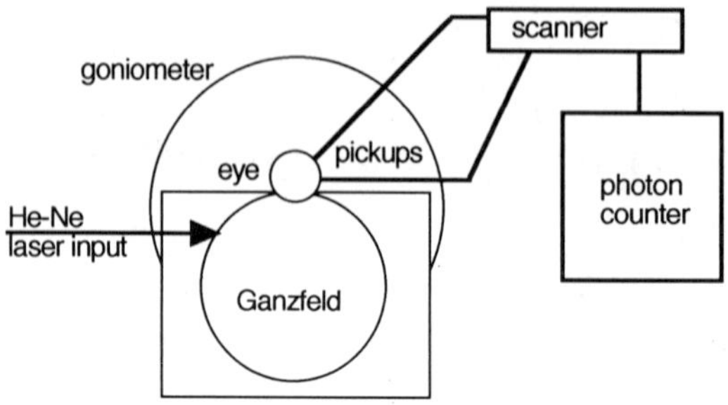

Figure 7. Schematic drawing of the ocular transmission photometer, OTP. An excised eye is positioned so as to view the *Ganzfeld* through the normal optics of that eye. Laser light is turned on, and rhodopsin inside the eye begins to bleach, causing the intensity of light in the eye to rise. Some of the light is transmitted through the sclera and picked up by fine optical fibers, and the intensity conveyed by those fibers is monitored by the photon-counting system. The scanner is a stepper-motor-driven device that allows the monitoring in succession of up to four independently positioned pick-ups. The pick-ups are positioned on the back of the eye by means of a goniometer, calibrated in degrees. The optic nerve stump is used as the zero-degree reference point.

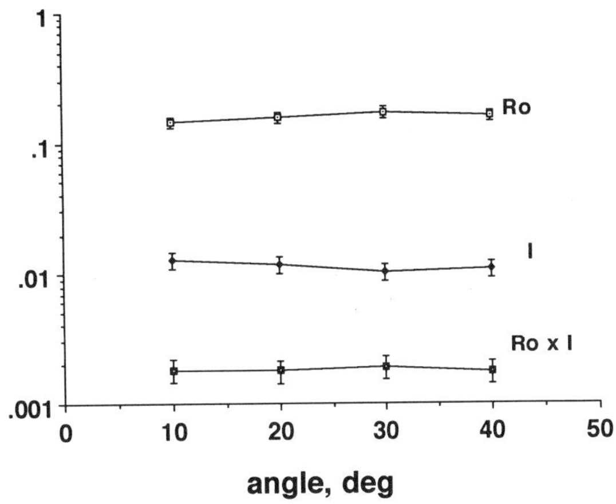

Figure 8. Rhodopsin concentration, intensity, and their product plotted against position across the albino rat retina. All three functions are nearly flat, showing that the distribution of photon absorptions is almost uniform across this region of the superior retina. Additional results show this is also the case for pigmented rats.

formly distributed. We therefore conclude that (1) the average rate of photon absorption, ~13,000/rod/second, calculated above, is in fact representative of absorption by most, if not all, of the rods in the rat retina; (2) because this average rate lies within the range of those known to saturate the rods, one function of photostasis is the maintenance of rod saturation during photopic conditions; and (3) the rate of photon absorption per rod is apparently regulated, because these rates were calculated on the basis of 10^{16} photons absorbed per eye per "day," a number that is itself regulated regardless of intensity and day length.

8. EFFECTS OF DAY LENGTH

Thus far, then, the results indicate that two parameters are being held constant in the rat retina: (i) total number of photons absorbed per eye per day and (ii) number of photons absorbed per rod per second. Could one be simply the consequence of the other? What happens if, for example, one holds intensity constant but varies day length? We know that rats living in a 12L:12D, 200-lux habitat have about 1.4 nanomoles of rhodopsin per eye. If we rear rats in a short-day, 200-lux habitat, what will happen to their R_0? Will it increase so that the photostasis number of photons can be absorbed in the short day, or will it remain at 1.4 nanomoles so as to keep the rate of absorption at about 13,000 per rod per second?

In our laboratory, Charles Parker has found that R_0 is up- or down-regulated inversely with day length. These results are given in Table 5, but note that this table also shows that rod cell count varies inversely with day length such that *the rate of absorption*

Table 5. Effects of day length on characteristics of the rat eye

Day length (hours)	R_0 (nmol/eye)	Rods/eye	Photons/eye/day	Photons/rod/sec
8	2.27	2.5×10^7	0.99×10^{16}	13,700
12	1.41	1.6×10^7	1.07×10^{16}	15,400
16	0.51	1.2×10^7	0.93×10^{16}	13,500

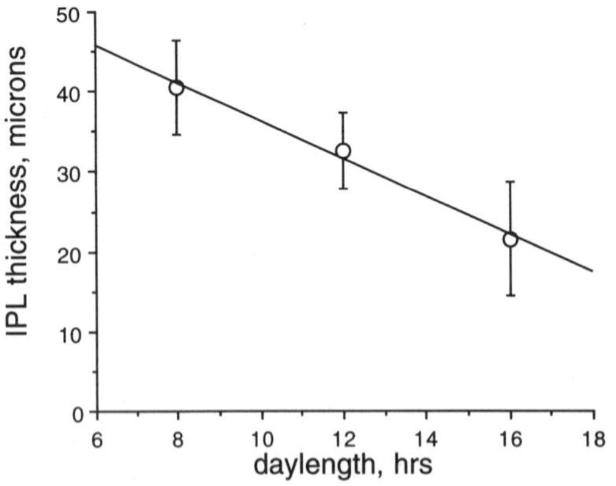

Figure 9. Inner plexiform layer thickness in the albino rat eye is inversely proportional to day length. The slope of this line indicates that the thickness decreases by about 2.5 μm per hour of added day length over this range of day lengths.

per rod is also constant regardless of day length. We therefore conclude that rats regulate *both* rate and total number of absorptions and that the number of rods in a retina plays a direct role in maintaining the stasis.

Another surprising result from Parker's work is that raising rats at different day lengths causes different thicknesses of inner plexiform layer (IPL). This result is shown in Figure 9. The thickness of the IPL is directly proportional to the number of rod cells in these eyes. The consequence of variable IPL thicknesses for photostasis or for vision is not obvious.

9. WHAT IS A NORMAL RAT RETINA?

Modifications of the rat retina in response to intensity of light in the habitat or day length are adaptive. Regardless of light intensity or of day length, all rats absorb the same total number of photons every day. Furthermore, because the number of rods in the retina also depends on these lighting conditions, the rate of photon absorption per rod is almost constant. This rate lies within the range of rates reported to saturate the rods, and this observation lends support to the notion that one of the functions of photostasis is to hold rods at saturation during the day. Perhaps photostasis evolved simultaneously with the duplex retina and is the mechanism whereby rods are kept from responding under photopic conditions.

How does renewal of the ROS enter this picture? If future work proves that photostasis keeps rods saturated during daylight hours, then the prospect might be good that rod membrane potential is a signal that is "read" by the inner segment. For example, if, on any given day, the rods fail to be saturated by the ambient intensity (i.e., if their potentials are less than maximal), the inner segments might get a signal to up-regulate the rhodopsin in their newly synthesized disks. This change would slightly increase the rate of photon absorptions the next day. Such daily "titrations" by light of the rhodopsin in a rod against its membrane potential would be one way to make small modifications in the light-absorbing properties of that rod. Daily synthesis and shedding of rhodopsin-containing disks would therefore be an ideal way to adapt rod absorption properties to the slow intensity changes found daily and seasonally. Abrupt changes in lighting, such as we have imposed in the laboratory setting, produce the abrupt (2- to 3-week) changes we have reported. Experiments using programmed lighting that simulates daily and seasonal changes would help to reveal how rods adapt to normal environments.

How adaptive can the loss of rod cells be? This recurrent question finds no definitive answer to date. When we were studying intensity effects, we found almost the same number of rods per retina at 3 lux as at 200 lux, so we would have said that an intensity of 200 lux does not "stress" or down-regulate the cell count. As soon as we studied day length, however, we found that rats experiencing short days, 8 hours at 200 lux, have twice the rod count of those experiencing 16-hour days at the same intensity, whereas those living in a "normal" habitat with 12-hour days have intermediate numbers of rods. These results show, unequivocally, that the number of rods in a rat retina depends heavily on the lighting and hints that day length has a stronger influence on cell count than does intensity. What therefore is the "normal" number of retinal rods for a rat?

The same logic can be applied to ROS length, to R_0, to regeneration rates, to concentration of rhodopsin on the disk membranes. What are normal values for these rod properties?

My tentative answer is that a normal rat retina is one that has adapted to its lighting such that it is absorbing photons at a rate that saturates the rods and that, when the day is over, has absorbed 10^{16} photons per eye.

REFERENCES

Aguilar, M., and Stiles, W. S. (1954). Saturation of the rod mechanism of the retina at high levels of stimulation. *Optica Acta* 1:59–65.

Alpern, M., and Krantz, D. H. (1981). Visual pigment kinetics in abnormalities of the uvea-retinal epithelium interface in man. *Investigative Ophthalmology and Visual Science* 20:183–203.

Battelle, B. A., and LaVail, M. M. (1978). Rhodopsin content and rod outer segment length in albino rats: modification by dark adaptation. *Experimental Eye Research* 26:487–497.

Baylor, D. A., and Lamb, T. D. (1982). Local effects of bleaching in retinal rods of the toad. *Journal of Physiology (London)* 328:49–71.

Baylor, D. A., Lamb, T. D., and Yau, K.-W. (1979). The membrane current of single rod outer segments. *Journal of Physiology (London)* 288:589–611.

Birren, S. J., Verdi, J. M., and Anderson, D. J. (1992). Membrane depolarization induces p140trk and NGF responsiveness, but not p75LNGFR in MAH cells. *Science* 257:395–397.

Blest, A. D. (1980). Photoreceptor membrane turnover in arthropods: comparative studies of breakdown processes and their implications. In: Williams, T. P., and Baker, B. N., eds. *Effects of Constant Light on Visual Processes.* Plenum Press, New York. pp. 217–246.

Broekhuyse, R. M., Tolhuizen, E. F. J., Janssen, A. P. M., and Winkes, H. J. (1985). Light induced shift and binding of S-antigen in retinal rods. *Current Eye Research* 4:613–618.

Bush, R. A., and Williams, T. P. (1991). The effect of unilateral optic nerve section on retinal light damage in rats. *Experimental Eye Research* 52:139–153.

Cone, R. A. (1963). Quantum relations of the rat electroretinogram. *Journal of General Physiology* 46:1267–1286.

Hofmann, K. P., Pulvermüller, A., Buczyłko, J., Van Hooser, P., and Palczewski, K. (1992). The role of arrestin and retinoids in the regeneration pathway of rhodopsin. *Journal of Biological Chemistry* 267:15701–15706.

Huber, A., Wolfrum, U., and Paulsen, R. (1994). Opsin maturation and targeting to rhabdomeral photoreceptor membranes requires the retinal chromophore. *European Journal of Cell Biology* 63:219–229.

Kaplan, M. (1984). Rhodopsin lateral diffusion as a function of rod outer segment disk membrane axial position. *Biophysical Journal* 45:851–853.

Katz, M. L., Kutryb, M. J., Norberg, M., Gao, C.-L., White, R. H., and Stark, W. S. (1991). Maintenance of opsin density in photoreceptor outer segments of retinoid-deprived rats. *Investigative Ophthalmology and Visual Science* 32:1968–1980.

Lashley, K. S. (1932). The mechanism of vision. V. The structure and image-forming power of the rat's eye. *Journal of Comparative Psychology* 13:173–200.

Mangini, N. J., Garner, G. L., Okajima, T. L., Donoso, L. A., and Pepperberg, D. R. (1994). Effect of hydroxylamine on the subcellular distribution of arrestin (S-antigen) in rod photoreceptors. *Visual Neuroscience* 11:561–568.

Noell, W. K., Walker, V. S., Kang, B. S., and Berman, S. (1966). Retinal damage by light in rats. *Investigative Ophthalmology* 5:450–473.

Organisciak, D. T., and Noell, W. K. (1977). The rod outer segment phospholipid/opsin ratio of rats maintained in darkness or cyclic light. *Investigative Ophthalmology and Visual Science* 16:188–190.

Penn, J. S. (1984). *Adaptation in the Albino Rat Retina to Varying Cyclic Light Illuminances.* Ph.D. dissertation, Florida State University, Tallahassee.

Penn, J. S., and Thum, L. A. (1987). A comparison of retinal effects of light-damage and high illuminance light-history. In: Hollyfield, J. G., Anderson, R. E., and LaVail, M. M., eds. *Degenerative Retinal Disorders: Clinical and Laboratory Investigations.* Liss, New York. pp. 425–438.

Penn, J. S., and Williams, T. P. (1986). Photostasis: regulation of daily photon-catch by rat retinas in response to various cyclic illuminances. *Experimental Eye Research* 43:915–928.

Penn, J. S., Tolman, B. L., Thum, L. A., and Koutz, C. A. (1992). Effect of light history on the rat retina: time course of morphological adaptation and readaptation. *Neurochemical Research* 17:91–99.

Perlman, I. (1978). Kinetics of bleaching and regeneration in abnormal (RCS) and normal albino rats *in vivo. Journal of Physiology (London)* 278:141–159.

Philp, N. J., Chang, W., and Long, K. (1987). Light-stimulated movement in rod photoreceptor cells of the rat retina. *FEBS Letters* 225:127–131.

Pirenne, M. H. (1962). Liminal brightness increments. In: Davson, H., ed. *The Eye.* Academic Press, New York.

Rapp, L. M., and Williams, T. P. (1980). A parametric study of retinal light damage in albino and pigmented rats. In: Williams, T. P., and Baker, B. N., eds. *The Effects of Constant Light on Visual Processes.* Plenum Press, New York. pp. 135–159.

Schnapf, J. L. (1983). Dependence of the single photon response on longitudinal position of absorption in toad rod outer segments. *Journal of Physiology (London)* 343:147–159.

Schremser, J.-L. (1990). *Photoreceptor Plasticity in the Albino Rat Retina Following Unilateral Optic Nerve Section.* M.S. thesis, Florida State University, Tallahassee.

Schremser, J.-L., and Williams, T. P. (1995a). Rod outer segment (ROS) renewal as a mechanism for adaptation to a new intensity environment. I. Rhodopsin levels and ROS length. *Experimental Eye Research* 61:17–24.

Schremser, J.-L., and Williams, T. P. (1995b). Rod outer segment (ROS) renewal as a mechanism for adaptation to a new intensity environment. II. Rhodopsin synthesis and packing density. *Experimental Eye Research* 61:25–32.

Sheng, M., McFadden, G., and Greenberg, M. E. (1990). Membrane depolarization and calcium influx induce c-*fos* transcription via phosphorylation of transcription factor CREB. *Neuron* 4:571–582.

Whelan, J. P., and McGinnis, J. P. (1988). Light-dependent subcellular movement of photoreceptor proteins. *Journal of Neuroscience Research* 20:263–270.

White, R. H., and Lord, E. (1975). Diminution and enlargement of the mosquito rhabdom in light and darkness. *Journal of General Physiology* 65:583–598.

Williams, T. P. (1984). Some properties of old and new rhodopsin in single *Bufo* rods. *Journal of General Physiology* 83:841–852.

Williams, T. P. (1989). Ambient lighting and the integrity of the retina. In: Guttman, H. N., Mench, J. A., and Simmonds, R. C., eds. *Science and Animals: Addressing Contemporary Issues.* Scientists Center for Animal Welfare, Bethesda, Maryland.

Williams, T. P., and Webbers, J. P. P. (1995). Photometer for measuring intensity and rhodopsin distributions in intact eyes. *Applied Optics* 34:5720–5724.

Young, R. W. (1967). The renewal of photoreceptor outer segments. *Journal of Cell Biology* 33:61–72.

3

HINTS OF PHOTOSTASIS IN INVERTEBRATES

Steven C. Chamberlain and William W. Weiner

Institute for Sensory Research
Department of Bioengineering and Neuroscience
Syracuse University
Syracuse, New York 13244-5290

1. INTRODUCTION: EARLY EXPERIMENTS IN *LIMULUS*

Our interest in photostasis began promptly with the publication of the defining paper (Penn and Williams, 1986). Testing the photostasis hypothesis in adult horseshoe crabs, *Limulus polyphemus,* might indicate whether it was a widespread phenomenon. We already knew that the lateral eye of the horseshoe crab had several mechanisms for varying photon catch that were controlled and modulated by circadian efferent activity and environmental lighting: rhabdom shedding (Chamberlain and Barlow, 1979, 1984), aperture dimensions (Barlow *et al.,* 1980; Chamberlain and Barlow, 1987), rhabdom position (Barlow *et al.,* 1980; Chamberlain and Barlow, 1987), and screening pigment position (Kier and Chamberlain, 1990). A robust test of the hypothesis with the good possibility of being able to investigate underlying mechanisms became our objective.

Our first series of experiments (Hoenig and Chamberlain, 1989) used the same kind of lighting regimen that Penn and Williams (1986) had used. Six animals were maintained in an environment in which the lights were on for 12 hours and off for 12 hours each day (12L:12D cyclic lighting) for 5 days. On the sixth day, animals were transferred to constant darkness, and the intensity-response curves for the electroretinogram (ERG) measured from the lateral eye were determined at noon and midnight. The animals were then returned to 12L:12D cyclic lighting at 25%, 6.25%, or 1.5% of the original intensity and maintained for 5 more days—two animals for each condition. The animals were then placed in constant darkness for a second time, and the intensity-response curves were again measured at noon and midnight. All animals had a circadian rhythm during both periods of constant darkness. The changes in sensitivity from day to night were on average essentially the same after 5 days in full intensity cyclic lighting or 5 days in reduced-intensity cyclic lighting, but the intensity-response curves for each animal were shifted toward lower intensities; higher values of ERG were evident at the highest intensities after reduced-intensity cyclic lighting for measurements made both at noon and at midnight. These results are qualitatively consistent with photostasis, but the degree of shift in the ERG intensity-response curves did not correlate well with the degree of intensity reduc-

Photostasis and Related Phenomena, edited by Williams and Thistle
Plenum Press, New York, 1998

tion. That is, animals exposed to the dimmest cyclic lighting (1.5% of standard) did not consistently show the largest increases in sensitivity, but the degree of reduction in our reduced-intensity experiments may have been insufficient to separate treatment effects from animal-to-animal variability.

After acclimation to reduced-intensity cyclic lighting, the saturated ERG response measured at midnight was larger than the corresponding midnight response in animals maintained in cyclic lighting of normal intensity. We had already used our measurements of day-night changes in aperture dimensions and the three-dimensional positioning of the rhabdom (Chamberlain and Barlow, 1987) for quantitative modeling of the interaction of the optics of the cylinder lens (Fletcher *et al.,* 1954; Land, 1979) and retinal structure (Chamberlain, unpublished data). This modeling had clearly shown that, under our standard lighting conditions, the aperture dimensions and rhabdom dimensions are well matched, so that all light passing through the entrance aperture of the cuticular cone at night passes through the rhabdom. Further widening of the aperture or outward displacement of the rhabdom could not be the explanation for the apparent increased absolute sensitivity observed at midnight in animals acclimated to reduced-intensity cyclic lighting. This result led us to conclude that, as suggested by Penn and Williams, rhabdom turnover must be the homeostatic mechanism for maintaining photostasis, whether by increasing the amount of rhabdom or the packing density of rhodopsin in the rhabdom, at least in the horseshoe crab.

2. RHABDOM SHEDDING IN *LIMULUS*

The daily shedding of the light-sensitive membrane of the microvillar array (Chamberlain and Barlow, 1979, 1984) is a robust feature of circadian and diurnal changes in the structure and function of *Limulus* photoreceptors (see Barlow *et al.,* 1989; Battelle, 1991; Renninger and Chamberlain, 1993, for reviews). In the natural environment, rhabdom shedding occurs at dawn, during or just after the morning transition from the nighttime retinal structure, when the circadian retinal efferents are active, to the daytime retinal structure, when they are silent. It occurs synchronously across the retinal array even in the gradually increasing light intensities of dawn. The shedding event is initiated by light and requires a period of efferent activity during the preceding night. In the laboratory, shedding can be induced by first light exposure at any time during the subjective day. It requires at least 4 hours of efferent input since the previous day. Maintaining the animal in continuous darkness or removing efferent input by cutting the optic nerve prevents rhabdom shedding from occurring. The priming effects of efferent activity can be mimicked by shocking of the distal stump of the cut optic nerve for 4 or more hours.

Dawn triggers rhabdom shedding through the phospholipase C second messenger cascade, specifically by the formation of diacylglycerol and subsequent activation of a classical, calcium-dependent protein kinase C (PKC), perhaps βII (Jinks *et al.,* 1996). We assume that the target protein of the PKC might be a product of gene expression driven by efferent circadian activity, but we do not yet know how many steps may intervene between the activation of PKC and the depolymerization of the actin cores of the microvilli (Johnson and Chamberlain, 1989; Calman and Chamberlain, 1992) that probably initiates the formation of membrane whorls.

The anatomical changes that constitute the daily cycle of rhabdom shedding are summarized in Figure 1. At first light, whorls of microvillar membrane balloon into the subrhabdomeral palisades, where they rapidly break into a cloud of very small mul-

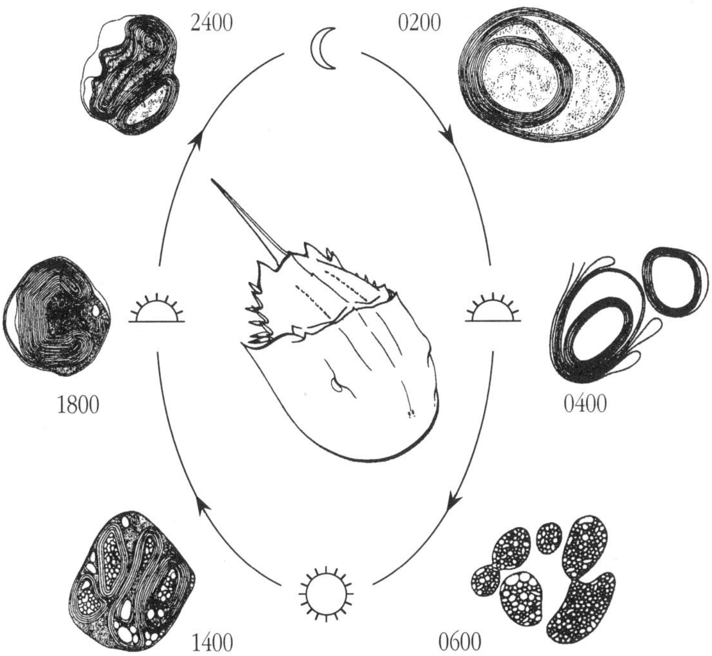

Figure 1. The daily cycle of rhabdom shedding in the lateral eye of *Limulus polyphemus*. Efferent activity primes rhabdom shedding each night when the circadian clock in the brain becomes active. The light of dawn initiates the formation of large whorls of photosensitive membrane from the microvillar array of the rhabdom. Over the next 24 hours, these whorls break up into vesicles, coalesce into multivesicular bodies, and evolve through the combination bodies and lamellar bodies typical of secondary lysosomes as shown. Some invertebrates shed rhabdomeral membrane as clouds of small vesicles, but the prominent multivesicular bodies, combination bodies, and lamellar bodies are persistent features in the cytoplasm of all microvillar photoreceptors that undergo cyclic turnover of photosensitive membrane (adapted from Chamberlain and Barlow, 1984, and Jinks *et al.*, 1996).

tivesicular bodies. Because the microvillar array of the rhabdom is reconstituted within 30 min of the initiation of shedding and we see no indications of rapid membrane synthesis, we assume that the multivesicular bodies represent some component of the rhabdom that has been stripped off while the rest of the membrane has been rapidly recycled back into the rhabdom. During the rest of the 24-hour cycle, the multivesicular bodies coalesce and migrate away from the rhabdom toward the arhabdomeral segment (A-segment) of each photoreceptor. The structural sequence is a classical secondary lysosomal pathway in which multivesicular bodies give way to combination bodies and then lamellar bodies as the contents are degraded and recycled. By the shedding event of the following day, a few largely empty lamellar bodies remain in the peripheral cytoplasm of the photoreceptor A-segment. Except for the smallest multivesicular bodies, all of these degradation structures are visible in 1-μm sections with light microscopy, although electron microscopy may be necessary to evaluate the onset of whorl formation in the first few minutes of light exposure and for an hour or so after the whorls have broken down into multivesicular bodies. The lateral eye of *Limulus* is thus a useful model system for studies of rhabdom shedding.

Until the advent of the photostasis hypothesis, we assumed that the massive daily shedding and rebuilding of the rhabdom was part of a general repair and maintenance procedure that allowed the light-sensitive membrane to alleviate the effects of light exposure and aging. We even considered the possibility that the turnover of invertebrate rhabdoms was necessary for regeneration of bleached rhodopsin, because experimental evidence indicated that, during rhabdom turnover, opsin is synthesized *de novo* and not just recycled (Stein *et al.,* 1979; Paulsen and Schwemer, 1983). Even as the photostasis hypothesis was gestating, data were accumulating that would call into question our ideas about the purpose of rhabdom turnover in arthropods. It had long been known that cephalopods supplied the 11-*cis* retinal necessary for insertion of rhodopsin into the membrane by the photoisomerization of all-*trans* retinal by the membrane-bound photoisomerase retinochrome (see Hara and Hara, 1972, for a review). The discovery and characterization of a cytosolic photoisomerase in honeybees (Pepe and Cugnoli, 1980; Pepe *et al.,* 1982; Schwemer *et al.,* 1984; Smith and Goldsmith, 1991a, b) was soon followed by the discovery of a similar photoisomerase in flies (Isono *et al.,* 1988; Schwemer, 1989). Then an orange-light cytosolic photoisomerase was discovered in *Limulus* (Smith *et al.,* 1992). Careful examination of the distribution of retinoids in the lateral eye over a 24-hour period revealed a system essentially in equilibrium, showing no statistically significant variation with light and darkness and no detectable change in the parcellation of retinoids at or after the time of the morning shedding event (Smith *et al.,* 1992). In fact, the *Limulus* lateral eye photoreceptors seem to have more all-*trans* retinal than 11-*cis* retinal throughout the daily cycle of light and darkness. Thus, it seems unlikely that repair or replenishment of the visual pigment is the purpose of this process.

Before we began to examine whether rhabdom turnover involved the repair or maintenance of other membrane constituents in *Limulus,* our study of the eye of the large deep-sea isopod *Bathynomus giganteus* (Chamberlain *et al.,* 1986) led our thinking in a different direction.

3. ABYSSAL INVERTEBRATES: IS RHABDOM SHEDDING UBIQUITOUS?

In 1984, we received several specimens of the largest known isopod, *Bathynomus giganteus,* from Jack and Ann Rudloe of Panacea, Florida. This impressive predator has large anterior eyes with a significant binocular field. Each eye has about 3500 ommatidia, whose packing is well ordered. *Bathynomus* lives at depths of 600 m in the zone where the faint cyclic lighting arriving from the surface is overwhelmed by the constant average light from bioluminescence. Indeed, *Bathynomus* is well equipped to find its bioluminescing prey by visual means. We were able to maintain animals in an apparently healthy state in cooled, darkened, marine-laboratory aquaria for several months. Despite repeated attempts, we were never able to record any ERG responses from the eyes, either immediately after the animals arrived or after several months of recovery in near darkness (R. B. Barlow, Jr., unpublished observations). We hypothesized that the exposure of the animals to daylight during collecting operations had irreversibly damaged their photoreceptors. Numerous other examples appear in the literature of similar permanent loss of visual sensitivity or structural damage to the rhabdom when deep-water animals are brought to the surface (e.g. Loew, 1976; Meyer-Rochow and Tiang, 1979; Meyer-Rochow, 1981; Nilsson, 1982; Nilsson and Lindström, 1983; Shelton *et al.,* 1985).

We examined the structure and ultrastructure of the compound eyes (Chamberlain *et al.,* 1986). Figure 2 summarizes the results. Despite the absence of measurable visual sensitivity, we found that the R-segments of the photoreceptors were large and filled with im-

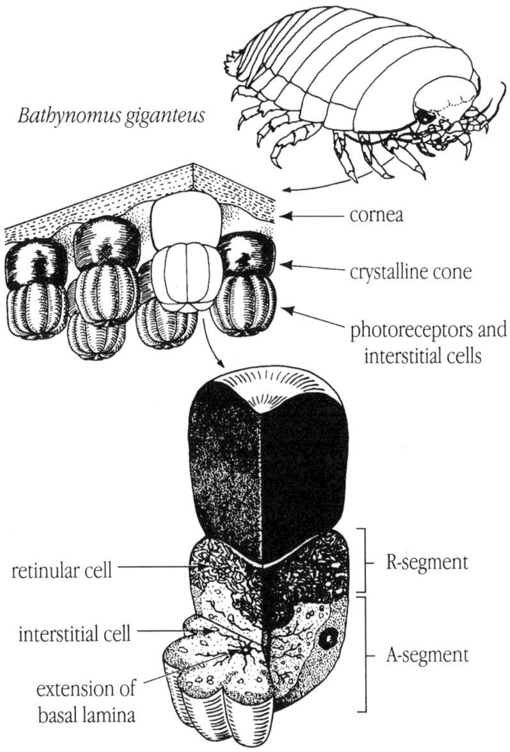

Figure 2. Retinal structure of the large isopod crustacean *Bathynomus giganteus.* This marine invertebrate lives at depths where light from bioluminescence swamps any cyclic daylight that penetrates from the surface. Its frontally directed eyes have large binocular regions in the visual field to enable this predator to locate and capture bioluminescing prey. Much of the space that might be occupied by the A-segment in the retina of a crustacean living at the surface is taken up in this retina by reflecting interstitial cells that serve to make the eye more sensitive. Significant daily shedding of rhabdomeral membrane is unlikely in this species because the R-segment is larger than the A-segment and is nearly filled with rhabdom. (Adapted from Chamberlain *et al.,* 1986).

pressive rhabdomeres, but the microvillar arrays were broken down so that, for example, we were unable to measure the volume density of the rhabdom. The photoreceptor A-segments contained clouds of multivesicular bodies, even after months of dark recovery from the daylight exposure. Furthermore, compared to the R-segment, the A-segment volume was diminished because it was displaced by reflecting pigment cells (interstitial cells) that form a large reflecting layer immediately behind the R-segments in the center of each ommatidium.

The continuing absence of visual sensitivity, the chronic disruption of the microvillar array, and the small size of the A-segment compared to the R-segment all suggested to us that rhabdom turnover, if it occurred at all, must proceed much more slowly than would be required to rebuild the rhabdom after exposure to daylight. This was our first hint that daily shedding of the photosensitive rhabdom might not be a universal characteristic of invertebrate photoreceptors.

We then searched the literature for other examples of photoreceptor structure in deep-sea animals to see whether the pattern of relatively large R-segments and small A-segments was characteristic of animals living in noncyclic lighting environments. The results shown in Figure 3 have been updated from our earlier work (Hornstein and Chamberlain, 1991; Chamberlain and Hornstein, 1993) to include new data from hydrothermal-vent shrimp. The plot shows the ratio of the volume of the R-segment to the volume of the A-segment on the ordinate plotted against total photoreceptor volume on the abscissa. The horizontal line simply divides those points below—a modest selection of animals that live in cyclic light environments—from those above—animals that live in noncyclic light environments. If many more species were tabulated, the apparent separa-

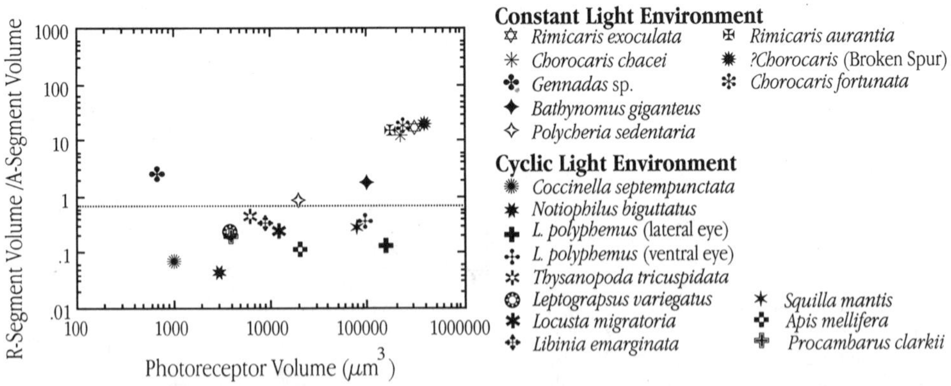

Figure 3. Plot of the ratio of the volume of the R-segment to the volume of the A-segment against total photoreceptor volume for a variety of invertebrates from cyclic and constant light environments. The horizontal line simply divides the two populations. In species that live in the cyclic lighting of the earth's surface, the arhabdomeral segment is always much larger than the rhabdomeral segment. In animals that live at great depths in the oceans, whether floating in the water column or colonizing hydrothermal vents, the relative sizes are reversed, and the A-segment is smaller than the light-sensitive R-segment. This trend is most extreme in the case of hydrothermal vent shrimp, which have very large photoreceptors, greatly hypertrophied R-segments, and atrophied A-segments. The data for these animals cluster at the upper right. (Updated from Chamberlain and Hornstein, 1993; data from Eguchi and Waterman, 1967; Home, 1975, 1976; Eguchi, 1976; Schönenberger, 1977; Meyer-Rochow and Walsh, 1977, 1978; Blest *et al.,* 1980; Calman and Chamberlain, 1982; Williams, 1982; Skalska-Rakowska and Baumgartner, 1985; O'Neill *et al.,* 1995; Nuckley *et al.,* 1996; Kuenzler *et al.,* in press; Lakin *et al.,* in press).

tion between these two groups might be reduced, but it seems likely that all animals living in cyclic lighting will have values of the ratio significantly less than 1 because they need large A-segments to provide the machinery for daily rhabdom turnover. Animals living in noncyclic lighting have values of the ratio greater than 1 and tend to have larger photoreceptors generally.

These observations suggested that the daily shedding and renewal of rhabdom membrane typical of photoreceptors in surface-dwelling invertebrates is reduced or absent altogether in species that live in noncyclic lighting. At this point, the evidence was indirect because the animals we had investigated (*Bathynomus giganteus,* Chamberlain *et al.,* 1986; *Rimicaris exoculata,* Van Dover *et al.,* 1989) had badly disrupted rhabdoms from prolonged exposure to daylight during collecting activities. We inferred that the small volumes of the A-segments in these species were insufficient to support a massive daily rhabdom turnover in the enlarged R-segments. We emphasized the significance of the correlation between cyclicity of the lighting environment and the inferred absence of rhabdom turnover rather than the alternative hypothesis that the absence of rhabdom shedding might be a consequence of dim-light environments. We discounted the dim-light hypothesis because species like *Bathynomus* are exposed to quite bright lights when they are close to their bioluminescing prey.

Thus, we reasoned that if rhabdom turnover were a homeostatic mechanism for modulating the amount of visual pigment to maintain photostasis, the absence of rhabdom turnover in photoreceptors of animals from constant-light environments was the predictable result. With this concept firmly in mind, we went to sea in 1993 to collect and study animals from a unique constant-light environment, oceanic hydrothermal vents.

4. HYDROTHERMAL VENT SHRIMP: VISION WITHOUT RHABDOM SHEDDING

In May 1993, Team Limulus (Barbara Battelle, Steve Chamberlain, Erik Herzog, Rob Jinks, Len Kass, and George Renninger) embarked from the Azores as part of R/V *Atlantis II* cruise 129–7 to collect and study bresiliid shrimp from black smoker chimneys at hydrothermal vents on the Mid-Atlantic Ridge (Fig. 4). Whereas, at most Pacific oceanic hydrothermal vents, the sessile vestiform tube worms form the largest metazoan population (and are eaten by mobile crustaceans), at vents on the Mid-Atlantic Ridge bresiliid shrimp are a predominant metazoan (and are eaten by sessile anemones) (see Van Dover, 1995, for a review of the ecology). Because the four vents we have studied were from 1700 to 3700 m below the ocean surface, no cyclic daylight penetrates to them.

Our original scheme for collecting and preserving photoreceptor ultrastructure in these shrimp was to use DSV *Alvin* to collect live animals and chop and fix them at depth before bringing them to the surface. This procedure, we hoped, would avoid any confounding effects of prolonged light exposure, temperature changes, and pressure changes. Although our apparatus for performing this operation worked flawlessly (O'Neill *et al.*, 1995), the rhabdom arrays in the photoreceptors of animals collected and fixed in this way were severely vesiculated, although the rest of the cytoplasm and organelles were very well preserved. We now believe this disruption in ultrastructure was caused by divalent cations in the seawater that unavoidably became part of the fixation solution. To our surprise, animals brought to the surface alive, exposed to the daylight of late afternoon during ascent, and fixed in standard double aldehyde fixation solution on deck had beautifully preserved microvillar arrays in their rhabdoms, although the cristae of the photoreceptor mitochondria were sometimes disrupted.

We collected and studied six species of bresiliid shrimp—*Alvinocaris markensis* (Williams, 1988; Wharton *et al.*, in press), *Chorocaris* (Broken Spur) (Kuenzler *et al.*, in press), *Chorocaris chacei* (Williams and Rona, 1986; Martin and Hessler, 1990; Lakin *et al.*, in press), *Chorocaris fortunata* (Martin and Christiansen, 1995; Kuenzler *et al.*, in press), *Rimicaris aurantia* (Renninger *et al.*, 1995; Nuckley *et al.*, 1996; Martin *et al.*, submitted), *Rimicaris exoculata* (Williams and Rona, 1986; Van Dover *et al.*, 1989;

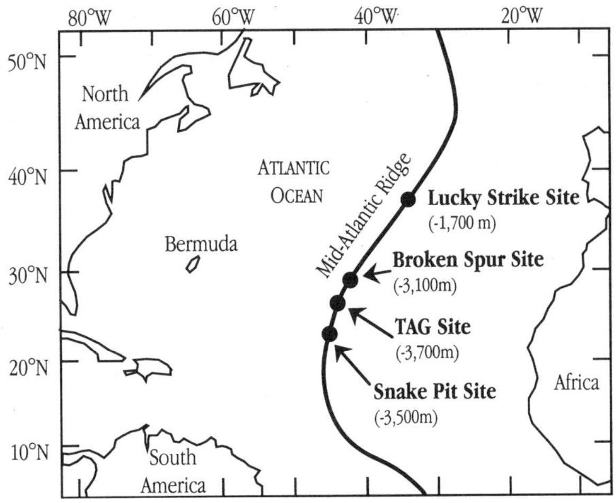

Figure 4. Map showing the locations and depth of four hydrothermal vent sites on the Mid-Atlantic Ridge where bresiliid shrimp with enlarged eyes have been collected and studied.

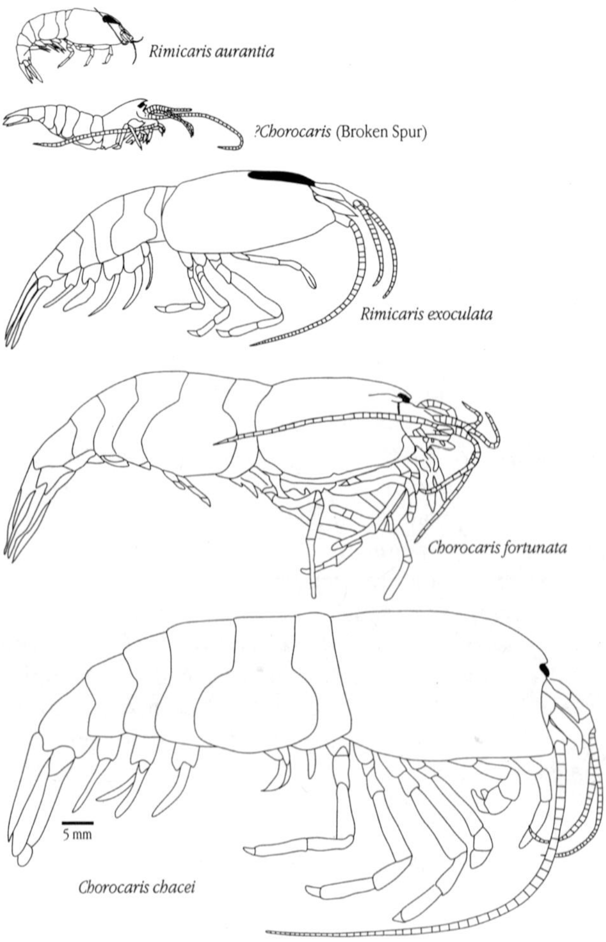

Figure 5. Five species of hydrothermal vent shrimp drawn to the same scale. The enlarged dorsal and anterior eyes are shown in black for emphasis. *Rimicaris aurantia* and *Rimicaris exoculata* live in massive swarms on the sides of black smoker chimneys and have dorsal eyes. The three *Chorocaris* species live in more solitary groups and have anterior eyes. (Composite made from O'Neill *et al.*, 1995; Nuckley *et al.*, 1996; Kuenzler *et al.*, in press; and Lakin *et al.*, in press).

O'Neill *et al.*, 1995)—five of which are shown in Figure 5. *Alvinocaris* lives at the bases of black smoker chimneys, the farthest from the light source at the chimney vent of any of the species. This large predator shrimp has gone blind, and the retina has no photoreceptors (Wharton *et al.*, in press). The two *Rimicaris* species live in large swarms along the sides of vent chimneys and have large dorsal eyes. The three *Chorocaris* species live in more solitary groups adjacent to the swarms and have large anterior eyes. All of these have probably evolved from a surface-dwelling caridean shrimp (Nuckley *et al.*, 1996) such as *Palaemonetes,* which has stalked spherical compound eyes (Itaya, 1976; Doughtie and Rao, 1984) adapted for pattern vision in a daylight environment.

We described the structure and ultrastructure of the eyes of these shrimp in a series of recent papers (*Chorocaris chacei,* Lakin *et al.*, in press; *Chorocaris fortunata* and ?*Chorocaris* (Broken Spur), Kuenzler *et al.*, in press; *Rimicaris aurantia,* Nuckley *et al.*, 1996; *Rimicaris exoculata,* O'Neill *et al.*, 1995). Figure 6 shows summary drawings of the retinal structure of the hydrothermal-vent shrimp and the surface-dwelling *Palaemonetes* for comparison. Although details vary, the eyes of all the vent shrimp have lost their individual lenses. The R-segments of the photoreceptors form a massive layer of rhabdom im-

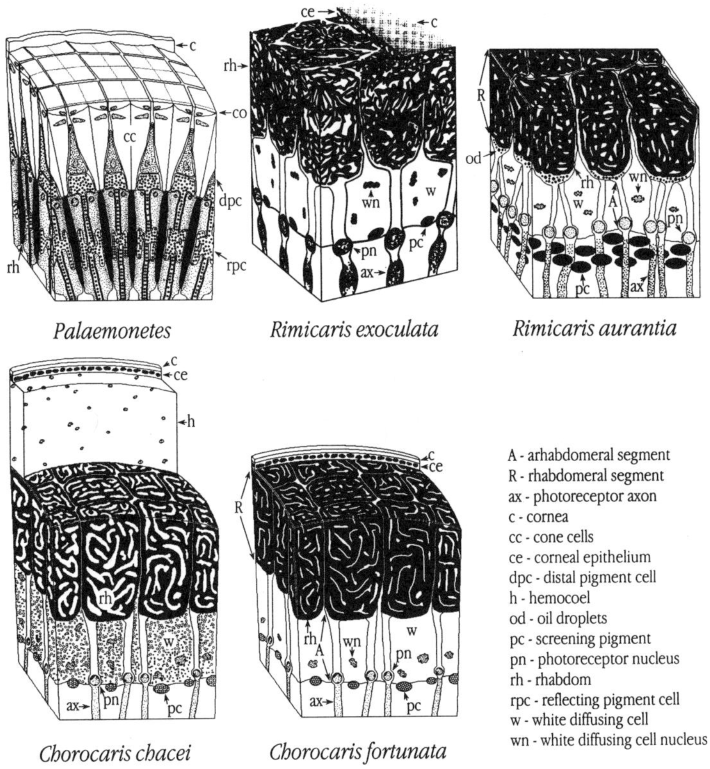

Figure 6. Semischematic drawings of retinal structure in four species of hydrothermal vent shrimp and one species of surface-dwelling shrimp from which the hydrothermal-vent shrimp may have evolved. The light-sensitive rhabdom is shown in black for emphasis. *Palaemonetes* has a prominent, square-packed dioptric array for high-resolution pattern vision. The rhabdom is small, and much of the volume in the retina is taken up by the A-segments of the photoreceptors and pigment cells that serve to isolate one ommatidium from another optically. In contrast, in all four species of hydrothermal-vent shrimp, the dioptric apparatus has been deleted and the cornea sits above a massive array of photosensitive rhabdom. The A-segments of the photoreceptors are mere strands of cytoplasm that penetrate a thick layer of white diffusing cells. The *Rimicaris* and *Chorocaris* retinas represent an extreme adaptation of a compound eye for detecting dim light at the expense of forming images. (Adapted from O'Neill *et al.*, 1995; Nuckley *et al.*, 1996; Kuenzler *et al.*, in press; and Lakin *et al.*, in press).

mediately adjacent (except in *C. chacei*) to a smooth cornea. Beneath the slab of R-segments is a matte white reflecting screen of modified reflecting pigment cells that gives the eyes a bright white color. The small, cylindrical A-segments penetrate the layers of white diffusing cells to become the axons of the optic nerve. In each case, the eye has lost its ability to form images and has become instead a very sensitive photon detector to utilize the dim light emitted by black smokers (Pelli and Chamberlain, 1989; Van Dover *et al.*, 1996).

The microvillar arrays in the R-segment are well organized in all five species of vent shrimp we have studied. We observed no evidence of rhabdom shedding—neither membrane whorls nor secondary lysosomal structures—in any of the retinas we examined. Indeed, the volume density of rhabdom in the R-segments is so high that any robust process of rhabdom shedding is virtually precluded for lack of space. Figure 7 shows semischematic tracings, not to a common scale, of micrographs of photoreceptors from five vent

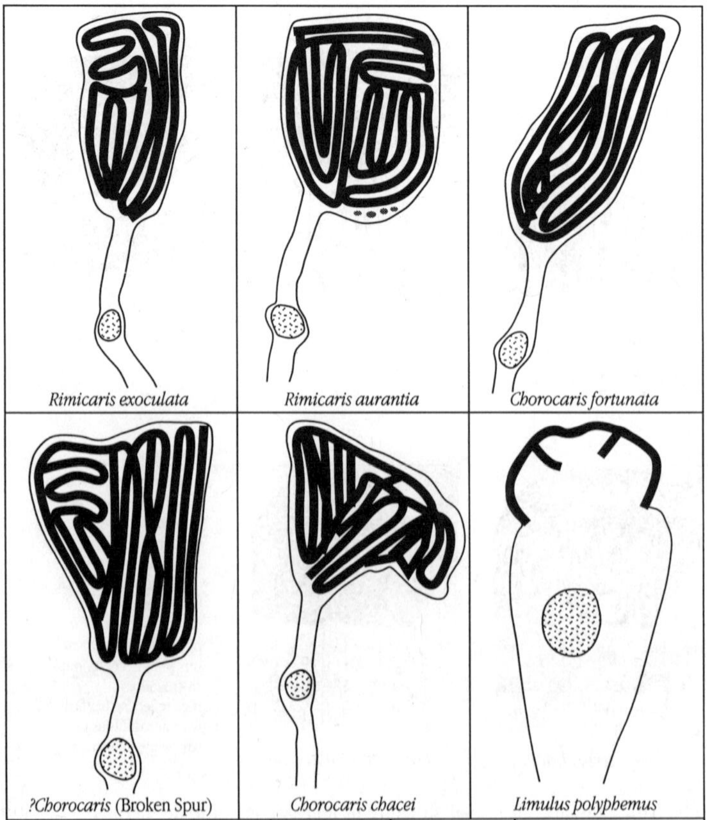

	R-Seg. Vol. (cu.μm)	Rh. Vol. Den.	Rh. Area/Vol. (sq. μm/cu. μm)	Rh. Membrane Area (sq. μm)
R. exoculata	48,589	75%	24.2	881,890
R. aurantia	26,456	83%	40.3	884,926
C. fortunata	37,390	80%	--	--
?C. (Broken Spur)	62,863	84%	39.8	2,101,636
C. chacei	35,118	60%	30.7	646,873
L. polyphemus	42,403	13%	41.0	226,007

Figure 7. Tracings of photoreceptors from longitudinal sections showing the rhabdom of the R-segment in black, the A-segment in white, and the nucleus. Five of the drawings show photoreceptors from hydrothermal-vent shrimp, which are adapted to dim constant light. The sixth drawing is a ventral photoreceptor from *Limulus*, a surface-dwelling species. The drawings are not to the same scale. Quantitative data for all six species are given below the drawings. (Adapted from Calman and Chamberlain, 1982; O'Neill *et al.*, 1995; Nuckley *et al.*, 1996; Kuenzler *et al.*, in press; and Lakin *et al.*, in press; R-segment volumes for hydrothermal-vent shrimp from Dan Cohen).

shrimp. A tracing of a ventral photoreceptor from *Limulus* is shown for comparison. Measured parameters for these photoreceptors are given at the bottom of Figure 7. The R-segments in vent shrimp photoreceptors are large; they have very high volume densities of rhabdom, and a large area of microvillar membrane is contained in each rhabdomere. The A-segments in vent shrimp photoreceptors are small, and the perikarya are thin spherical shells around the nuclei. In contrast, the *Limulus* ventral photoreceptor perikaryon is much larger than the whole R-segment, which itself has relatively little rhabdom. Furthermore, as shown in Figure 8, the perikaryon surrounding the nucleus in a vent shrimp photorecep-

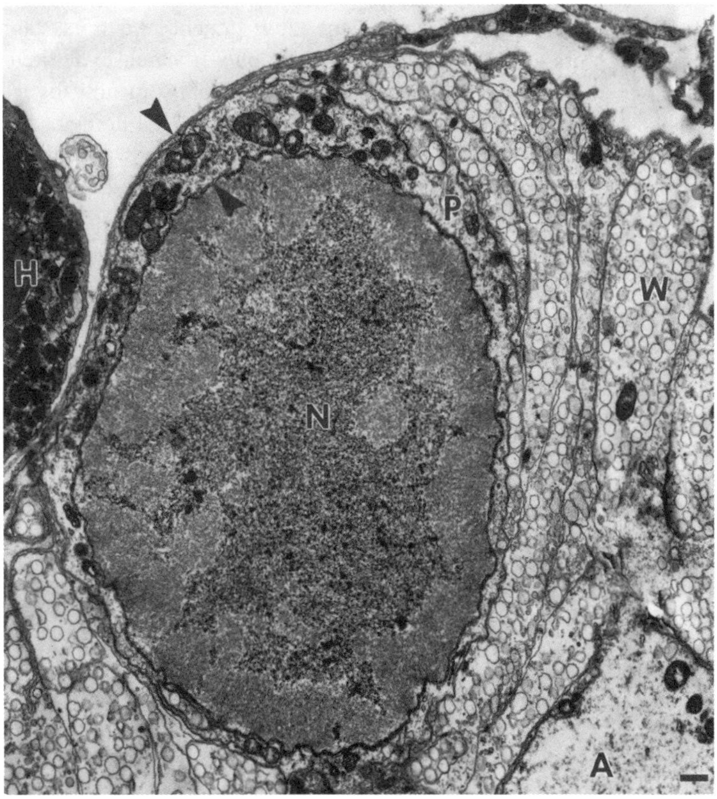

Figure 8. Ultrastructure of the cell body of photoreceptors in the retina of the hydrothermal vent shrimp, *Rimicaris exoculata.* The nucleus (N) is surrounded by a thin shell of perikaryon (P) indicated by the two arrow heads. Just above the nucleus, the A-segment cytoplasm is nearly devoid of organelles (A). Below the nucleus, there are screening pigment granules (not shown). There is no evidence of significant ongoing protein synthesis or membrane assembly in these photoreceptors. W, white diffusing cells; H, hemocyte. The bar represents 0.75 μm. (Micrograph courtesy of Patrick J. O'Neill.)

tor has relatively few synthesis organelles, suggesting that ongoing protein synthesis in these cells must be modest.

We conclude that there is no significant daily turnover of rhabdom membrane in these photoreceptors, at least not on a scale approaching that typical of photoreceptors in surface-dwelling invertebrates. This conclusion is based on the following observations: (1) there is no ultrastructural evidence of any kind in the R-segment or A-segment that rhabdom shedding is occurring (i.e., we found no secondary lysosomal structures); (2) the R-segment cytoplasm has insufficient volume to contain much shed rhabdom membrane; (3) the photoreceptor is completely surrounded by white diffusing cells whose cytoplasm also lacks any evidence of secondary lysosomal structures that might indicate an extracellular rhabdom shedding mechanism; (4) the cytoplasm of the stalk of the A-segment is essentially devoid of any membranous organelles or lysosomal structures; and (5) the perikaryon around the nucleus is extremely limited in extent and contains hardly any of the rough endoplasmic reticulum that would be expected if membrane synthesis were underway.

Given photostasis, one can interpret these data in the following way. Because there is no cyclic lighting in the hydrothermal vent environment and no seasonal variation therein, the photoreceptors never need to adjust their photon catch to achieve photostasis. Thus, the robust and metabolically expensive process of daily turning over a significant portion of the photosensitive membrane array is unnecessary as a homeostatic mechanism for photostasis. As a result there is no rhabdom shedding and renewal in these photoreceptors of the sort required in surface-dwelling species, which are faced with daily changes in day length and light intensity.

5. ONGOING INVESTIGATION OF PHOTOSTASIS IN *LIMULUS*

After a delay of nearly six years, we returned to investigating the nature of photostasis in *Limulus,* this time keeping the light intensity constant and varying the ratio of light to darkness (photoperiod) within the 24-hour cycle. We decided on this modification for two reasons: varying the photoperiod mimics the differences in day length that naturally occur during the seasons at northern latitudes, and it also permits the use of skeleton photoperiods—two pulses of light, one indicating dawn and the other dusk, separated by darkness—for further exploration of the photostasis hypothesis. On the basis of pilot studies that imply a circannual rhythm in retinal sensitivity, we believe that day length is an important parameter for *Limulus:* animals kept in constant darkness for many months have a period of increased sensitivity at night that progressively shortens in the fall and then progressively lengthens again in the spring (R. B. Barlow, Jr., unpublished data). Although animals living in northern latitudes are exposed to changes in both light intensity and day length, we chose to vary only day length in order to simplify the interpretation of our results.

In our current experiments, we also use the ERG to monitor retinal sensitivity. Our fundamental assumption is that the ERG is a reflection of the lateral eye's ability to capture photons. Evidence supporting this hypothesis comes from experiments showing that blocking spike generation in the second-order eccentric cells by injection of tetrodotoxin into the eye does not affect the shape or amplitude of the ERG (Kass and Barlow, 1984). Thus, it would appear, in *Limulus* at least, that the ERG results from the summed activity of the first-order photoreceptor cells. This fact allows us to use ERG's to assess photostasis qualitatively in *Limulus*.

In these experiments, two animals were initially entrained to a lighting cycle of 15 hours of light and 9 hours of darkness ("summer") for one week. After this entrainment, intensity-response curves were measured several times during the light portion of the light cycle and also on three successive days. Because it is extremely difficult to measure the saturated intensity-response curves in light-adapted horseshoe crabs, we allowed the animals to dark-adapt for one hour before measuring these functions. Our assumption is that any underlying anatomical changes (i.e., changes in aperture length or rhabdom distribution) resulting from the 1 hour of dark adaptation will be approximately the same for each lighting condition and will therefore not confound our results. These same two animals were then exposed to two subsequent lighting cycles following the same procedure outlined above: the animals were entrained for one week to the lighting cycles, and then the intensity-response curves were measured. The second lighting cycle used was 9 hours of light and 15 hours of darkness ("winter"), and the last lighting cycle used was a skeleton photoperiod (skeleton)—two 30-min pulses of light at subjective dawn and dusk separated by 8 hours of darkness—resulting in an implied day length of 9 hours of light and 15

Table 1. Saturated ERG responses from a single animal

	Saturated ERG level (μV)b			
Lighting condition[a]	Minimum	Maximum	Mean	Standard deviation
Summer	275	312	293	14.9
Winter	345	353	349	5.7
Skeleton	440	468	454	19.8

[a]"Summer" represents a lighting cycle of 15 hours of light and 9 hours of darkness; "Winter" represents a lighting cycle of 9 hours of light and 15 hours of darkness; and Skeleton represents a skeleton-photoperiod lighting cycle of 9 hours of implied light and 15 hours of darkness. The skeleton pulses consisted of 30 min of light.

[b]These data represent an average of all of the intensity-rêsponse functions that we measured for a given lighting condition.

hours of darkness. Because our initial results indicate that the intensity-response functions do not depend on the time during the light portion of the light cycle or the day in which they are measured, our reported curves represent an average of all of the functions that we measured for a given lighting cycle.

Our preliminary results from these experiments suggest that horseshoe crabs modulate photon catch in a manner consistent with photostasis: the saturation levels of the intensity-response functions are inversely proportional to photoperiod (Table 1). Inspection of the averaged intensity-response curves collected from a single animal (Fig. 9) shows that changing the photoperiod or using a skeleton photoperiod results in a vertical shift of these functions, indicating a change in retinal sensitivity that reflects a change in the eye's ability to capture photons. Of particular interest is the significant difference in response between the skeleton lighting condition (9L:15D skeleton) and the continuous lighting condition of matched day length (9L:15D). This result suggests that the implied photoperiod is not a critical parameter for the modulation of photon catch. It is also interesting to note that the winter skeleton condition produced a greater increase in sensitivity over the

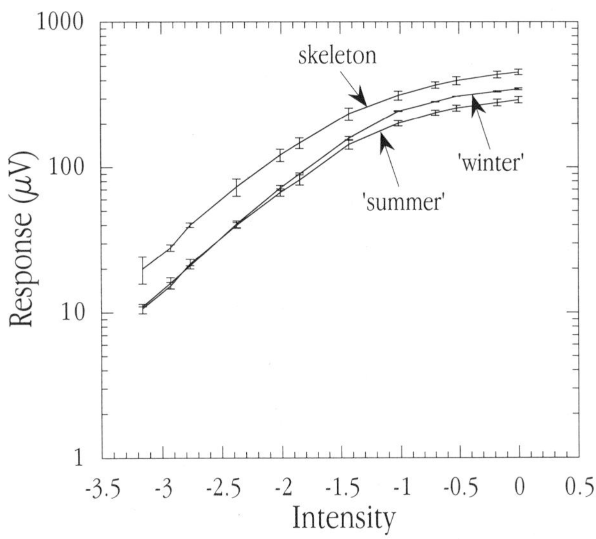

Figure 9. Averaged intensity-response functions measured from a single horseshoe crab that was subjected to three different lighting conditions. Standard deviations are indicated by bars. The maximum intensity of light incident on the eye (Log I = 0) corresponds to a photon flux of approximately 3 × 10^{15} photons/s at a wavelength of 595 nm. ("summer" represents a lighting cycle of 15 hours of light and 9 hours of darkness; "winter" represents a lighting cycle of 9 hours of light and 15 hours of darkness; and skeleton represents a skeleton-photoperiod lighting cycle of 9 hours of implied light and 15 hours of darkness. The skeleton pulses consisted of 30 min of light.)

summer condition than did the winter condition—just the prediction of the photostasis hypothesis.

Experiments are under way to determine which anatomical changes are responsible for the observed changes in sensitivity. Several potential candidates exist: aperture dimensions, rhabdom distribution, rhabdom volume, microvillar membrane packing, and rhodopsin packing could all affect the ability of the retina to capture photons. All of these changes, with the exception of aperture dimensions and rhabdom distribution, can be accounted for by the daily cycling of the photosensitive membrane. On the basis of our anatomical data, we will be able to quantify the photon catch in each of the lighting conditions to determine whether, in fact, this number stays approximately constant.

ACKNOWLEDGMENTS

We gratefully acknowledge the essential collaboration of our many colleagues who are co-authors on the papers discussed here. We thank Bob Barlow and Dan Cohen for sharing their unpublished data. We thank Bill Dossert and Chris Stathatos for technical assistance. Most of all, we thank Ted Williams for organizing and conducting the symposium that led to our writing this review. Parts of this research were supported by NIH, NSF, NSERC, and The Grass Foundation as listed in individual papers. Preparation of this review was supported by the Department of Bioengineering and Neuroscience at Syracuse University.

LITERATURE CITED

Barlow, R. B., Jr., Chamberlain, S. C., and Levinson, J. Z. (1980). *Limulus* brain modulates the structure and function of the lateral eye. *Science* 210:1037–1039.

Barlow, R. B., Jr., Chamberlain, S. C., and Lehman, H. K. (1989). Circadian rhythms in the invertebrate retina. In: Stavenga, D., and Hardie, R., eds. *Facets of Vision.* Springer-Verlag, Berlin. pp. 257–280.

Battelle, B.-A. (1991). Regulation of retinal functions by octopaminergic efferent neurons in *Limulus.* In: Osborne, N. N., and Chader, G. J., eds. *Progress in Retinal Research, vol. 10.* Pergamon Press, New York. pp. 333–355.

Blest, A. D., Stowe, S., and Price, D. G. (1980). The sources of acid hydrolases for photoreceptor membrane degradation in a grapsid crab. *Cell and Tissue Research* 205:229–244.

Calman, B. G., and Chamberlain, S. C. (1982). Distinct lobes of *Limulus* ventral photoreceptors. II. Structure and ultrastructure. *Journal of General Physiology* 80:839–862.

Calman, B. G., and Chamberlain, S. C. (1992). Localization of actin filaments and microtubules in the cells of the *Limulus* lateral and ventral eyes. *Visual Neuroscience* 9:365–375.

Chamberlain, S. C., and Barlow, R. B., Jr. (1979). Light and efferent activity control rhabdom turnover in *Limulus* photoreceptors. *Science* 206:361–363.

Chamberlain, S. C., and Barlow, R. B., Jr. (1984). Transient membrane shedding in *Limulus* photoreceptors: control mechanisms under natural lighting. *Journal of Neuroscience* 4:2792–2810.

Chamberlain, S. C., and Barlow, R. B., Jr. (1987). Control of structural rhythms in the lateral eye of *Limulus.* Interactions of diurnal lighting and circadian efferent activity. *Journal of Neuroscience* 7:2135–2144.

Chamberlain, S. C., and Hornstein, E. P. (1993). Photoreceptors, black smokers, and seasonal affective disorder: evidence for photostasis. In: Verrillo, R. T., ed. *Sensory Research, Multimodal Perspectives.* Lawrence Erlbaum Associates, Hillsdale, New Jersey. pp. 105–118.

Chamberlain, S. C., Meyer-Rochow, V. B., and Dossert, W. P. (1986). Morphology of the compound eye of the giant deep-sea isopod *Bathynomus giganteus. Journal of Morphology* 189:145–156.

Doughtie, D. G., and Rao, K. R. (1984). Ultrastructure of the eye of the grass shrimp, *Palaemonetes pugio.* General morphology, and light and dark adaptation at noon. *Cell and Tissue Research* 238:271–288.

Eguchi, E. (1976). Rhabdom structure and receptor potentials in single crayfish retinular cells. *Journal of Cellular and Comparative Physiology* 66:411–430.

Eguchi, E., and Waterman, T. H. (1967). Changes in retinal fine structure induced in the crab *Libinia* by light and dark adaptation. *Zeitschrift für Zellforschung* 79:209–229.

Fletcher, A., Murphy, T., and Young, A. (1954). Solutions of two optical problems. *Proceedings of the Royal Society A (London)* 223:216–225.

Hara, T. and Hara, R. (1972). Cephalopod retinochrome. In: H. J. A. Dartnall, ed. *Handbook of Sensory Physiology, Vol. VII/I.* Springer-Verlag, New York. pp. 720–746.

Hoenig, H., and Chamberlain, S. C. (1989). Testing the "photostasis" hypothesis in the *Limulus* lateral eye. *Investigative Ophthalmology and Visual Science Supplement* 30:291.

Home, E. M. (1975). Ultrastructure studies of development and light-dark adaptation of the eye of *Coccinella septempunctata* with particular reference to ciliary structures. *Tissue and Cell* 7:703–722.

Home, E. M. (1976). The fine structure of some carabid beetle eyes, with particular reference to ciliary structures in the retinula cells. *Tissue and Cell* 8:311–333.

Hornstein, E. P., and Chamberlain, S. C. (1991). Correlation of photoreceptor structure and lighting environment: implications for photostasis. *Society for Neuroscience Abstracts* 17:298.

Isono, K., Tanimura, T., Oda, Y., and Tuskahara, Y. (1988). Dependency on light and vitamin A derivatives of the biogenesis of 3-hydroxyretinal and visual pigment in the compound eyes of *Drosophila melanogaster. Journal of General Physiology* 92:587–600.

Itaya, S. K. (1976). Rhabdom changes in the shrimp, *Palaemonetes. Cell and Tissue Research* 166:265–273.

Jinks, R. N., White, R. H., and Chamberlain, S. C. (1996). Dawn, diacylglycerol, calcium, and protein kinase C—the retinal wrecking crew. A signal transduction cascade for rhabdom shedding in the *Limulus* eye. *Journal of Photochemistry and Photobiology B: Biology* 35:45–52.

Johnson, J. K., and Chamberlain, S. C. (1989). Membrane associated axial filaments in rhabdomeral microvilli of *Limulus* lateral eye photoreceptors. *Investigative Ophthalmology and Visual Science Supplement* 30:292.

Kass, L., and Barlow, R. B., Jr. (1984). Efferent neurotransmission of circadian rhythms in *Limulus* lateral eye. I. Octopamine-induced increases in retinal sensitivity. *The Journal of Neuroscience* 4:908–917.

Kier, C. K., and Chamberlain, S. C. (1990). Dual controls for screening pigment movement in photoreceptors of the *Limulus* lateral eye: circadian efferent input and light. *Visual Neuroscience* 4:237–255.

Kuenzler, R. O., Kwasniewski, J. T., Jinks, R. N., Lakin, R. C., Battelle, B.-A., Herzog, E. D., Kass, L., Renninger, G. H., and Chamberlain, S. C. (1997). Retinal anatomy of new bresiliid shrimp from Lucky Strike and Broken Spur on the Mid-Atlantic Ridge. *Journal of the Marine Biological Association of the United Kingdom* 77:707–725.

Lakin, R. C., Jinks, R. N., Battelle, B.-A., Herzog, E. D., Kass, L., Renninger, G. H., and Chamberlain, S. C. (1997). Retinal anatomy of *Chorocaris chacei,* a deep-sea hydrothermal vent shrimp from the Mid-Atlantic Ridge. *Journal of Comparative Neurology* 385:503–514.

Land, M. F. (1979). The optical mechanism of the eye of *Limulus. Nature* 280:396–397.

Loew, E. R. (1976). Light and photoreceptor degeneration in the Norway lobster, *Nephrops norvegicus* (L.). *Proceedings of the Royal Society, Series B* 193:31–44.

Martin, J. W., and Christiansen, J. C. (1995). A new species of the shrimp genus *Chorocaris* Martin & Hessler, 1990 (Crustacea: Decapoda: Bresiliidae) from hydrothermal vent fields along the Mid-Atlantic Ridge. *Proceedings of the Biological Society of Washington,* 108:220–227.

Martin, J. W., and Hessler, R. R. (1990). *Chorocaris vandoverae,* a new genus and species of hydrothermal vent shrimp (Crustacea, Decapoda, Bresiliidae) from the western Pacific. *Contributions in Science* 417:1–11.

Martin, J. W., Signorovitch, J., and Patel, H. (1997). A new species of *Rimicaris* (Crustacea, Decapoda, Bresiliidae) from the Snake Pit hydrothermal vent field of the Mid-Atlantic Ridge. *Proceedings of the Biological Society of Washington* 110:399–411

Meyer-Rochow, V. B. (1981). The eye of *Orchomene* sp. cf. *O. rossi,* an amphipod living under the Ross Ice Shelf (Antarctica). *Proceedings of the Royal Society, Series B* 212:93–111.

Meyer-Rochow, V. B., and Tiang, K. M. (1979). The effects of light and temperature on the structural organization of the eye of the Antarctic amphipod *Orchomene plebs* (Crustacea). *Proceedings of the Royal Society, Series B* 206:353–368.

Meyer-Rochow, V. B., and Walsh, S. (1977). The eyes of mesopelagic crustaceans. I. *Gennadas* sp. (Penaeidae). *Cell and Tissue Research* 184:87–101.

Meyer-Rochow, V. B., and Walsh, S. (1978). The eyes of mesopelagic crustaceans. III. *Thysanopoda tricuspidata* (Euphausiacea). *Cell and Tissue Research* 195:59–79.

Nilsson, H. L. (1982). Rhabdom breakdown in the eye of *Cirolana borealis* (Crustacea) caused by exposure to daylight. *Cell and Tissue Research* 227:633–639.

Nilsson, H. L., and Lindström, M. (1983). Retinal damage and sensitivity loss of a light-sensitive crustacean compound eye (*Cirolana borealis*). *Journal of Experimental Biology* 107:277–292.

Nuckley, D. J., Jinks, R. N., Battelle, B.-A., Herzog, E. D., Kass, L., Renninger, G. H., and Chamberlain, S. C. (1996). Retinal anatomy of a new species of bresiliid shrimp from a hydrothermal vent field on the Mid-Atlantic Ridge. *Biological Bulletin* 190:98–110.

O'Neill, P. J., Jinks, R. N., Herzog, E. D., Battelle, B.-A., Kass, L., Renninger, G. H., and Chamberlain, S. C. (1995). The morphology of the dorsal eye of the hydrothermal vent shrimp, *Rimicaris exoculata. Visual Neuroscience* 12:861–875.

Paulsen, R., and Schwemer, J. (1983). Biogenesis of blowfly photoreceptor membranes is regulated by 11-*cis* retinal. *European Journal of Biochemistry* 137:609–614.

Pelli, D. G., and Chamberlain, S. C. (1989). On the visibility of 350°C blackbody radiation by the shrimp *Rimicaris exoculata* and man. *Nature* 337:460–461.

Penn, J. S., and Williams, T. P. (1986). Photostasis: regulation of daily photon-catch by rat retinas in response to various cyclic illuminances. *Experimental Eye Research* 43:915–928.

Pepe, I. M., and Cugnoli, C. (1980). Isolation and characterization of a water-soluble photopigment from honeybee compound eye. *Vision Research* 20:97–102.

Pepe, I. M., Schwemer, J., and Paulsen, R. (1982). Characteristics of retinal-binding proteins from the honeybee retina. *Vision Research* 22:775–781.

Renninger, G. H., and Chamberlain, S. C. (1993). Modulation of photoreceptor function in *Limulus polyphemus* by a central circadian clock. In: Wiese, K., Gribakin, F. G., Popov, A. V., Renninger, G. H., eds. *Sensory Systems of Arthropods.* Birkhäuser, Basel. pp. 307–316.

Renninger, G. H., Kass, L., Gleeson, R. A., Van Dover, C. L., Battelle, B.-A., Jinks, R. N., Herzog, E. D., and Chamberlain, S. C. (1995). Sulfide as a chemical stimulus for deep-sea hydrothermal vent shrimp. *Biological Bulletin* 189:69–76.

Schönenberger, N. (1977). The fine structure of the compound eye of *Squilla mantis* (Crustacea, Stomatopoda). *Cell and Tissue Research* 176:205–233.

Schwemer, J. (1989). Visual pigments of compound eyes—structure, photochemistry, and regeneration. In: Stavenga, D. G., and Hardie, R. C., eds. *Facets of Vision.* Springer-Verlag, New York. pp. 112–133.

Schwemer, J., Pepe, I. M., Paulsen, R., and Cugnoli, C., 1984. Light-induced *trans-cis* isomerization of retinal by a protein from the honeybee retina. *Journal of Comparative Physiology A* 154:549–554.

Shelton, P. M. J., Gaten, E., and Chapman, C. J. (1985). Light and retinal damage in *Nephrops norvegicus* (L.) (Crustacea). *Proceedings of the Royal Society, Series B* 226:217–236.

Skalska-Rakowska, M., and Baumgartner, B. (1985). Longitudinal continuity of the subrhabdomeric cisternae in the photoreceptors of the compound eye of the drone, *Apis mellifera. Experientia* 41:43–45.

Smith, W. C., and Goldsmith, T. H. (1991a). The role of retinal photoisomerase in the visual cycle of the honeybee. *Journal of General Physiology* 97:143–165.

Smith, W. C., and Goldsmith, T. H. (1991b). Localization of retinal photoisomerase in the compound eye of the honeybee. *Visual Neuroscience* 7:237–249.

Smith, W. C., Friedman, M. A., and Goldsmith, T. H. (1992). Retinoids in the lateral eye of *Limulus:* evidence for a retinal photoisomerase. *Visual Neuroscience* 8:329–336.

Stein, P. J., Brammer, J. E., and Ostroy, S. E. (1979). Renewal of opsin in the photoreceptor cells of the mosquito. *Journal of General Physiology* 74:565–582.

Van Dover, C. L. (1995). Ecology of Mid-Atlantic Ridge hydrothermal vents. In: Parson, L. M., and Dixon, D. R., eds. *Hydrothermal Vents and Processes.* Geological Society Special Publication 87, pp. 257–294.

Van Dover, C. L., Szuts, E., Chamberlain, S. C., and Cann, J. R. (1989). A novel eye in "eyeless" shrimp from hydrothermal vents of the Mid-Atlantic Ridge. *Nature* 337:458–460.

Van Dover, C. L., Reynolds, G. T. , Chave, A. D., and Tyson, J. A. (1996). Light at deep-sea hydrothermal vents. *Geophysical Research Letters* 23:2049–2052.

Wharton, D. N., Jinks, R. N., Herzog, E. D., Battelle, B.-A., Kass, L., Renninger, G. H., and Chamberlain, S. C. (1997). Morphology of the eye of the hydrothermal vent shrimp, *Alvinocaris markensis. Journal of the Marine Biological Association of the United Kingdom* 77:1097–1108.

Williams, A. B. (1988). New marine decapod crustaceans from waters influenced by hydrothermal discharge, brine, and hydrocarbon seepage. *Fishery Bulletin* 86:263–287.

Williams, A. B., and Rona, P. A. (1986). Two new caridean shrimps (Bresiliidae) from a hydrothermal field on the Mid-Atlantic Ridge. *Journal of Crustacean Biology,* 6:446–462.

Williams, D. S. (1982). Ommatidial structure in relation to turnover of photoreceptor membrane in the locust. *Cell and Tissue Research* 225:595–617.

4

REGULATION OF RHODOPSIN EXPRESSION IN RHABDOMERAL PHOTORECEPTORS

Functional Analysis of the 5′ Flanking Region of the *Calliphora* Rh1 Opsin Gene

Armin Huber and Reinhard Paulsen

Institute of Zoology I
University of Karlsruhe
Kornblumenstrasse 13
76128 Karlsruhe, Germany

1. INTRODUCTION

Regulation of rhodopsin synthesis may occur at any of the biochemical steps required to transform the genetic information encoded in the opsin genes into the light-absorbing visual pigment, rhodopsin, contained in the photoreceptive membrane. These steps include the transcription of the opsin gene into mRNA, the translation of this mRNA into the opsin protein, chromophore binding, and correct protein folding, which may be aided by chaperones, e.g. by the *ninaA*-encoded cyclophilin homologue required for folding of *Drosophila* Rh1 and Rh2 opsin (Stamnes *et al.,* 1991; Baker *et al.,* 1994) or by the cyclophilin-related protein RanBP2 of vertebrate cone photoreceptor cells (Ferreira *et al.,* 1996). Co- and posttranslational modifications, rhodopsin glycosylation and binding of the retinal chromophore, which is 11-*cis* retinal in most vertebrate rhodopsins and 11-*cis* 3-hydroxyretinal in the visual pigments of dipteran flies (Vogt, 1983), are further prerequisites for the formation of mature rhodopsin (Fliesler and Basinger, 1985; Fliesler *et al.,* 1985; O'Tousa, 1992; Ozaki *et al.,* 1993; Huber *et al.,* 1994). Finally, in order to perform its function in photoreception, rhodopsin must be directed to and incorporated into the membranes constituting the photoreceptive cell compartment, the outer-segment membrane of vertebrate rod and cone cells or the microvillar membrane of rhabdomeral photoreceptors of invertebrates.

Two basic requirements for the regulation of rhodopsin synthesis can be distinguished. First, the expression of a distinct opsin gene must be activated in a specific type of photoreceptor, e.g. in one of the three types of cone cells of the human retina, which together express three distinct rhodopsin genes (Nathans *et al.,* 1986). The decision which rhodopsin is expressed in which cell is made during eye development. It is generally accepted that the regulation of cell-specific rhodopsin expression is performed at the transcriptional level by a defined set of transcription factors that bind to specific promoter sequences (*cis*-acting elements) of the opsin gene. Second, once the cell-specific transcription pathway has been activated, the mechanisms of receptor up- and down-regulation must become effective. This regulation may be achieved by control of the amount of rhodopsin that is newly synthesized and incorporated into the photoreceptive membrane during the process of rhodopsin turnover. It has been suggested that, apart from the regeneration of the photoreceptive compartment, a main function of rhodopsin turnover is the up- and down-regulation of the rhodopsin content of the photoreceptive membranes in order to adjust the light-absorbing properties of the photoreceptor cells to different environmental light conditions (Williams, 1984; Penn and Williams, 1986). Although the basic mechanism of cell-specific rhodopsin expression can be envisaged from the general principles of gene control at the transcriptional level, the level at which cellular rhodopsin renewal is regulated is not at all clear.

Other than in vertebrate model systems, cell-specific rhodopsin expression and the renewal of rhodopsin in differentiated cells are being studied primarily in fly photoreceptor cells, which provide a suitable model system because of the extensive genetic characterization of the visual system of *Drosophila.* In addition, a particularly attractive feature for studies investigating the control mechanism of rhodopsin renewal is the ease with which the rhodopsin content of the rhabdomeral fly photoreceptor membrane can be manipulated experimentally over a wide range. The present study focuses primarily on the question of how cell type–specific rhodopsin expression can be achieved in the visual system of dipteran flies. Our previous investigations intended to unravel the regulatory steps involved in the process of rhodopsin renewal in rhabdomeral photoreceptor cells will also be considered.

The visual system of dipteran flies is composed of the compound eyes and of small ocellar eye spots located on the vertex of the head (ocelli). Within the compound eyes, the rhodopsin-containing compartments, called rhabdomeres, of eight photoreceptor cells are arranged in a highly regular trapezoidal array. Photoreceptor cells R1–6 form the peripheral rhabdomeres of the open rhabdom. R7 and R8 cells form the apical and the basal portions, respectively, of a central rhabdomere. In these photoreceptor cells at least five different rhodopsins are expressed. The genes encoding three *Drosophila* rhodopsins of the compound eye and one of the ocelli were cloned about 10 years ago (O'Tousa *et al.*, 1985; Zuker *et al.*, 1985, 1987; Cowman *et al.*, 1986; Montell *et al.*, 1987; Fryxell and Meyerowitz, 1987). According to their sequence of cloning, these rhodopsins are referred to as Rh1–Rh4. Rh1, a blue-green-sensitive visual pigment, is expressed in the photoreceptor cells R1–6 and constitutes more than 90% of the total amount of the visual pigments present in the compound eye. Rh3 and Rh4 genes encode ultraviolet- and blue-sensitive rhodopsins of nonoverlapping sets of R7 cells. By reporter-gene analysis, Rh3 expression was also detected in a small population of R8 cells located at the extreme dorsal anterior margin of the compound eye (Fortini and Rubin, 1990). With the exception of the specialized R8 cells close to the anterior margin of the compound eye, physiological studies on larger flies (*Calliphora, Musca*) suggested that R7 and R8 cells are organized in two types of matched pairs, i.e. in R7 cells expressing Rh3, which are coupled with one of two subtypes of R8 cells, and in Rh4-expressing R7 cells coupled with the other R8 cell subtype. Recently, a rhodopsin gene expressed in one subtype of R8 cells (Rh5) was cloned (Chou *et al.,* 1996). Immunohistochemical localization of Rh5 rhodopsin demonstrated that it is present exclusively in those R8 cells that are located in the same ommatidia as Rh3-expressing R7 cells (Chou *et al.,* 1996). We have recently cloned the rhodopsin gene for the other subtype of R8 cells (Huber *et al.*, 1997). The Rh2 gene, originally suggested to encode a rhodopsin of R8 cells, was shown by *in situ* hybridization to be expressed in the ocelli (Pollock and Benzer, 1988). The only rhodopsin of larger flies for which the amino-acid sequence has been determined by sequencing of a corresponding cDNA clone is the *Calliphora* homologue of *Drosophila* Rh1 (Huber *et al.,* 1990).

The mechanism of terminal differentiation of the rhabdomeral photoreceptor cells is not understood in any detail. First insights have been gained by investigation of the 5′ flanking regions of the *Drosophila* rhodopsin genes Rh1 to Rh4. These studies comprised (i) sequence comparisons between the corresponding opsin genes of *Drosophila melanogaster* and *Drosophila virilis* and (ii) experiments in which sequentially deleted or mutated promoter fragments were tested for their ability to activate the transcription of reporter genes in the eyes of transgenic *Drosophila* (Mismer and Rubin, 1987, 1989; Mismer *et al.,* 1988; Fortini and Rubin, 1990). They revealed that a small fragment of the 5′ flanking region (base pairs −162 to +67 of the Rh1 opsin gene) is sufficient to determine cell-specific expression of Rh1 in R1–6 photoreceptor cells. Within this region a rhodopsin core sequence (RCS I) that seems to confer retina-specific expression of all *Drosophila* rhodopsin genes as well as two other putative *cis*-acting elements, termed rhodopsin upstream sequences A and B (RUS1A and RUS1B), were identified. RUS elements were defined for each of the four *Drosophila* opsin genes. They are located farther upstream than RCS I and show different sequences in Rh1, Rh2, Rh3, and Rh4 opsin genes. It has been suggested that RUS A and RUS B determine the cell-specific expression of the fly rhodopsins.

In order to obtain information about the evolutionary conservation of the mechanisms that determine cell-specific rhodopsin expression, we have analyzed the 5′ flanking region of the Rh1 opsin gene of *Calliphora* and have tested whether or not this promoter is able to interact *in vivo* with the transcription factors of *Drosophila* photoreceptor cells.

2. MATERIALS AND METHODS

2.1. Fly Stocks

Calliphora vicina Meig., chalky mutant, was reared on bovine liver, a procedure that yields a high rhodopsin content in the eyes. Larvae and pupae were reared in darkness at 25 °C. Adult flies were kept under a cycle of 12 hours of light and 12 hours of darkness each day (12L:12D). *Drosophila melanogaster* (strain Oregon R, white) was reared on a standard cornmeal diet at 22 °C under a 12L:12D cycle.

2.2. Northern Blot Analysis and Rhodopsin Measurements

Total RNA was isolated from *Calliphora* larvae, pupae, or heads of adults by the method of Chirgwin *et al.* (1979). Tissue was homogenized in a guanidinium thiocyanate solution, overlaid onto a cushion of $CsCl_2$ and ultracentrifuged in a Beckmann TLS-55 rotor. Total RNA obtained was electrophoresed through 1% agarose-formaldehyde gels and transferred onto a nylon membrane (Hybond N, Amersham Buchler) according to standard protocols (Sambrook *et al.,* 1989). RNA standards were purchased from Gibco BRL. The blots were hybridized with 50 ng/ml digoxigenin-labeled sense or antisense cRNA in 50% formamide, 5 × SSC (1 × SSC is 0.15 M NaCl, 0.015 M sodium citrate, pH 7.0), 0.1% laurylsarcosinate, 0.02% SDS, 4% blocking reagent (Boehringer Mannheim) at 65 °C for 15 hours. After hybridization, the blots were washed twice for 10 min in 2 × SSC, 0.1% SDS at 25 °C, and twice for 15 min in 0.1 × SSC, 0.1% SDS at 72 °C. Immunological detection of digoxigenin-labeled RNA was performed with a digoxigenin detection kit (Boehringer

Mannheim) with the luminescent dye AMPPD according to the manufacturer's instructions. Fluorographs documenting AMPPD luminescence were exposed 5 to 15 min using Kodak X-Omat AR films.

For determination of the rhodopsin content in the eyes of *Calliphora,* excised eyes were homogenized in 50 mM Na-phosphate buffer, pH 6.2. Eye membranes were sedimented by centrifugation (10 min, 13000 × g), and membrane proteins were extracted with 4% digitonin in 50 mM Na-phosphate buffer, pH 6.2. The rhodopsin concentration in these extracts was measured by difference spectrophotometry as has been described previously (Paulsen, 1984).

2.3. Library Screening, Sequencing, and Primer-Extension Analysis

Hybridization screening of a *Calliphora* genomic library (Durica *et al.,* 1987) with a cDNA encoding *Calliphora* Rh1 opsin (Huber *et al.,* 1990) was performed according to standard protocols (Sambrook *et al.,* 1989). A 1.4-kb EcoRI/NdeI restriction fragment of a genomic clone that hybridized with the Rh1 cDNA was subcloned into p-BluescriptKS+ and SK+ vectors (Stratagene). The nucleotide sequence of the subclone was determined for both strands by the dideoxy-chain-termination method (Sanger *et al.,* 1977) with templates generated by nested deletions. Elements homologous between the *Calliphora* and *Drosophila* 5′ flanking regions were identified by matrix analysis by the NMATPUS program of the PC-GENE software (Intelligenetics, Inc.).

Primer-extension analysis was performed as described by Sambrook *et al.* (1989) with 10 μg total RNA isolated from *Calliphora* eyes or thorax and a ^{32}P-labeled oligonucleotide (TTCCATTTTATAAAATTCAATTCAATTTATGTTTA) complementary to the translation start site as a primer. Primer-extension products were analyzed on 8 M urea–6% (w/v) polyacrylamide gels and visualized by autoradiography.

2.4. Generation of Transgenic *Drosophila* and Reporter-Gene Analysis

A *Drosophila* Rh1 promoter fragment (base pairs −833 to +67, Mismer and Rubin, 1987) and a *Calliphora* Rh1 promoter fragment (base pairs −1132 to +10, see Fig. 2) were cloned into the NotI site of the P-element vector pW-ATG-lac1 (Kuhn *et al.,* 1988). *Drosophila melanogaster* (strain Oregon R, white) was transformed with these constructs by P-element-mediated germ-line transformation according to the methods of Spradling and Rubin (1982).

10-μm cryosections were prepared from the heads of transgenic *Drosophila* as described by Wolfrum (1991) and were fixed in 2% glutaraldehyde in 50 mM cacodylate buffer, pH 7.3. Histochemical staining for β-galactosidase activity was carried out by covering of the sections for 2–20 h with 100 μl of staining gel consisting of 0.2% 5-bromo-4-chloro-3-indolyl-β-D-galactosid (X-Gal), 2% gelatin, 0.3% Triton X-100, 150 mM NaCl, 1 mM $MgCl_2$, 3.1 mM K_4 [Fe(II) $(CN)_6$], and 3.1 mM K_3 [Fe(III) $(CN)_6$] in 10 mM phosphate buffer, pH 8.0. Subsequently, sections were washed with water, mounted in Mowiol, and examined with a Zeiss microscope.

3. RESULTS

The decision which type of rhodopsin is expressed in which photoreceptor cell is made at the transcriptional level at a specific developmental stage. Figure 1 shows that ex-

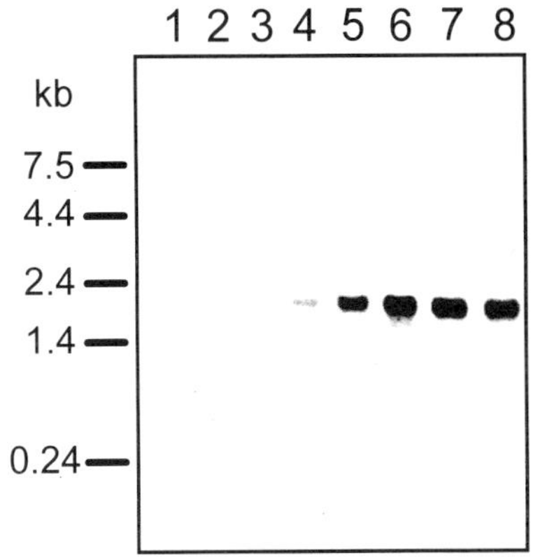

Figure 1. Developmental expression pattern of *Calliphora* Rh1 opsin mRNA. RNA was isolated from larvae (lane 1), total pupae 10 days before eclosion (lane 2), heads of pupae 3 days (lane 3) and 1 day (lane 4) before eclosion, and heads of adult flies at 1, 4, 11, and 18 days after eclosion (lanes 5–8, respectively). For each developmental stage, 10 individuals were used for RNA preparation, and 1/100 of the RNA obtained was size-fractionated through a 1% denaturing agarose gel and blotted onto a nylon membrane. Rh1 opsin mRNA was detected by hybridization with digoxigenin-labeled antisense cRNA transcribed from a *Calliphora* Rh1 opsin cDNA clone.

pression of the *Calliphora* Rh1 opsin gene is activated one day before eclosion from the pupa. It reaches its maximum level between days 1 and 4 after eclosion and then stays at a constant, high level for at least 18 days. These data reveal a developmental expression pattern similar to that previously described for the expression of the Rh1 opsin gene in *Drosophila* (Zuker *et al.,* 1985). Parallel to the opsin mRNA measurements, the rhodopsin content was determined spectrophotometrically in the already developed eyes of pupae 1 day before eclosion and in the eyes of adult flies in the first days after eclosion. Functional rhodopsin was first detected one day after eclosion (1.3 pmol/eye). At 2 and at 4 days after eclosion, 3.4 pmol/eye and 3.5 pmol/eye rhodopsin were measured, respectively. Thus, there is obviously a lag of about one day between opsin mRNA expression and the formation of functional rhodopsin.

As outlined above, regulatory elements in the 5′ flanking region of the *Drosophila* Rh1 opsin gene have been defined by sequence comparison of the opsin genes isolated from *Drosophila* as well as by mutation analyses. Characterization of an Rh1 opsin promoter from the more distantly related fly species *Calliphora* should yield additional information about conserved sequence elements important for Rh1 expression, because the gene control of *Drosophila* and *Calliphora* may have diverged through evolution more than can be expected for the genes of different *Drosophila* species.

In order to isolate the promoter region of the *Calliphora* Rh1 opsin gene, we screened a genomic DNA library in Lambda Embl 3 (Durica *et al.,* 1987) with a *Calliphora* Rh1 opsin cDNA clone (Huber *et al.,* 1990). 1.4 kilobases (kb) of a positive genomic clone were sequenced (Fig. 2). The sequence determined overlaps with the 5′ end of the Rh1 opsin cDNA sequence and comprises 60 base pairs (bp) of the coding sequence, a 97-bp intron, 205–211 bp of a 5′ untranslated leader sequence, and a 1.1-kb 5′ flanking region. The intron is located at exactly the same position (5 bp downstream of the ATG start codon) as the first intron in the *Drosophila* Rh1 opsin gene (Zuker *et al.,* 1985). We performed primer-extension analysis of Rh1 opsin mRNA to determine the transcription start site of the *Calliphora* Rh1 opsin gene. Three major extension products were obtained when retina RNAs were used as templates (Fig. 3), but none was obtained when RNAs isolated from thorax were used as a negative control. The sizes of the extension products indicated that the *Calliphora* Rh1 opsin gene contains three major transcription start sites located 211 bp, 205 bp, and 199 bp upstream of the ATG start codon. Thus, the 5′-untranslated leader sequence of each transcript of the *Calliphora* Rh1 opsin gene is longer than that of the *Drosophila* Rh1 opsin gene (171 bp), which contains a single transcription start site (Zuker *et al.,* 1985). However, the DNA sequence between bp −2 to +7 relative to the first transcription start site is identical in the Rh1 opsin genes of *Calliphora* and *Drosophila.* Multiple mRNA start sites were detected in none of the characterized *Drosophila* opsin genes, but two mRNA start sites are present in the human rhodopsin gene and in the human and murine phosphodiesterase β-subunit genes (Nathans and Hogness, 1984; Polo *et al.,* 1996).

Sequence comparison of the Rh1 opsin promoters of *Calliphora* and *Drosophila* revealed a weak overall homology, except for six short but highly conserved elements (Fig. 2). Among these elements are the TATA-box, which is located 32 nucleotides upstream of the first transcription start site, and an RCS I core element at position −90, the homologue of which has been proposed to determine tissue-specific expression of *Drosophila* Rh1 opsin (Fortini and Rubin, 1990). Thus, the regulation of tissue-specific expression of the Rh1 opsin genes of *Calliphora* and *Drosophila* may employ the same *cis*-acting element and probably homologous transcription factors. Elements homologous to *Drosophila* RUS1A and RUS1B, which have been implicated in the R1–6 cell-specific expression of Rh1 opsin (Fortini and Rubin, 1990), were not conserved in the 1.1-kb 5′ flanking region of the *Calliphora* Rh1 opsin gene. In the distal part of the *Calliphora* Rh1 opsin promoter (between bp −1132 and −200), four elements that show considerable homology to sequences in the *Drosophila* Rh1 opsin gene were identified (bp −979 to −965, bp −867 to 857, bp −675 to −666, and bp −303 to −280). It should be noted that, in the *Drosophila* Rh1 opsin gene, the corresponding sequences are always located closer to the mRNA start site than in *Calliphora;* i.e., the spacers between these elements are larger in the *Calliphora* Rh1 opsin promoter.

In order to investigate whether or not the tissue- and cell-specific expression of the *Calliphora* Rh1 opsin gene can be regulated by the transcription factors present in the *Drosophila* eye, we used an approach in which the expression of a bacterial reporter gene (LacZ) was driven by the *Calliphora* Rh1 promoter in *Drosophila* photoreceptor cells. *Drosophila* Oregon R, white, was transformed with P-element vectors (pW-ATG-lac1) that contained the 5′ flanking region of either the *Drosophila* (DmpRh1-pW-ATG-lac1) or the *Calliphora* (CvpRh1-pW-ATG-lac1) Rh1 opsin gene upstream of the *E. coli* LacZ gene. In these constructs, the *Drosophila white* gene was used as a selection marker. One transgenic *Drosophila* line carrying DmpRh1-pW-ATG-lac1 and three lines carrying CvpRh1-pW-ATG-lac1 were identified by their red eye color. Cryosections through the heads of flies carrying DmpRh1-pW-ATG-lac1 or CvpRh1-pW-ATG-lac1 were analyzed for β-galactosidase activity (Fig. 4). The positive control (DmpRh1-pW-ATG-lac1) showed intense β-galactosidase staining in the retina and in the first optic ganglion (the lamina) after 2 hours of incubation with the chromogen (Fig. 4A). A similar staining pattern was observed in two out of three transgenic lines that express the LacZ gene under the control of the *Calliphora* Rh1 promoter. These sections had to be incubated with the

Figure 2. Nucleotide sequence of the 5′ flanking region of *Calliphora* Rh1 opsin. Numbering at the left refers to the first transcription start site (arrow +1). Two other transcription start sites are indicated by arrows. The nucleotide sequence of seven conserved elements of the *Drosophila* Rh1 5′ flanking region is shown above the *Calliphora* sequence, and their position relative to the transcription start site is indicated. Stars denote putative capping signals. The TATA box and the ATG translation start codon are double-underlined. Single underlines indicate sequences that are identical with the corresponding portion of a *Calliphora* Rh1 cDNA clone (Huber *et al.,* 1990). (The DNA sequence of the *Calliphora* Rh1 opsin promoter is available at the EMBL data library under accession number Z23136.)

```
-1132 GAAATTGTCTTGATTCGTTCTTGGCCTCAAAAGATGAGCTGTTCTTTTGG
-1082 CTTGGAATCCATATGTTGCCAAAAAGATCGGAAAAGATCATAGCTAACAA
-1032 TGGTCAATTCTTTGAATAAGTTTATATTATACAAATGTTTCAAAATAAAA

      -610AGAAAagAAAcAAAA-596
 -982 GCTAGAAATTAAAAAAAAAATCGGCATTTTTAAGACATACACCCAATAAA
 -932 ACCGGTGGAATATATTCAGTATAACTGTACTAGATTACAGGGTCTCAGGT

                      -566AAAGAT-CaAT-554
 -882 AAAAATACATAGGGGAAAGATTCCATGATGAACTGGGGATAGGATCTAT
 -832 GTAGGTATTCTATGTAGGCTTTATCAGGGGAATAGGTTGGCTATTTTAGG
 -782 TAACAGGGCTTTGTGAATCCCAGTCTATACTTATTCTCTAAACTTTTAGA
 -732 CTGGGTTTTCACTTTATCTCTCTTTATATTACCCTACTTCTTTATTTACT

         -516TTCATaTCTT-507
 -682 TTCTAATTTCATTTCTTCTCTTCACATCTAGTTCTCCAATGGGTTTCCGA
 -632 GTTTTCCTTGATTTTCCAGCTTTTTATTGAACTTTTCTCATACGAATTAT
 -582 TGATTATTGTCTAATGATTTTCATGTCACATACTTTTAGGTTTTTTTTTT
 -532 AAAAGGAATAACAGCAGCATGTTAAATGTATTGCTGTTGTTAGGGGGTGT
 -482 TTGAAAAAAACATTTCCGAATACAACCAACAATTTTACATAAACTATAAT
 -432 TTCAAAGTGTTAAAGAACAGAAGCATTGGCCACGTTTTGGTTTTTTTTT
 -382 TCTTTTAAATTTTCTTTGTTTTGGGCAAAGACATCATCATCATCTTTAAG

                                  -199CCTTTAACT-----ATTGACA
 -332 AACGAATTTAAAAGAAAGCAAAAGGAAACCCTTTAACTGAATTATTGACA

      TTG-181
 -282 TTGAAACGTTTGAAAAAATCATCACCCTGAGACAGAAATATGAAAATAAA
 -232 GTGAAAAGCTTAAAAATAAAATAATACGACTGGGTATTTCTGTAAGCAAG
 -182 TATATAAAGGTAAGCAGTATATATGCATTTAAATACAAATCTTTATATAT

                                                -76CT-AATTg
 -132 TATGTCTGTCTGCCACTTATTTAAACACTTTAAACACAAGTACTTAATTT

      AATT-66                                      -36AAAAC
  -82 AATTTTCTTGTTAAAACCTAATTTTCTGTTTAAACACCTCAAACCAAAAC

                                        ↱+1
      --TATAAA-26         ********  -2TTGCAGGTT+7
  -32 TATATAAACGCGGGTATTTTCAGTTTCCGCTTGCAGGTTGATATCGAGCA
                                        ↳+1   ↳     ↳

         ********
  +19 CCAAATCAGTATCTAGTATCTCACCGGTCAAAAAAATAAAAAAAGGAATT
  +69 GAAATTTAAAAAAAAAAAAATTAACTTTGTGCAACTAAAATTTACGGATT
 +119 TTCTGGGTGATTGTGAAATGAGATTTTGAAAAGTTTCCATTTTCAAAACA
 +169 GATAAACAATTTTATAAACATAAATTGAATTGAATTTTATAAAATGGAAA
 +219 GGTATTAATGATTTAAATTAAATTTAAAAAAAAAATGTGAAATTGTGTTG
 +269 GAAAAGTTTAAAGTGATTTTATAATGGTTTGTTTTTGTTTAATAATAGAT
 +319 ACTCGACACCGCTAATAGGGCCAAGCTTTGCCGCCCTAACAAATGGAT
```

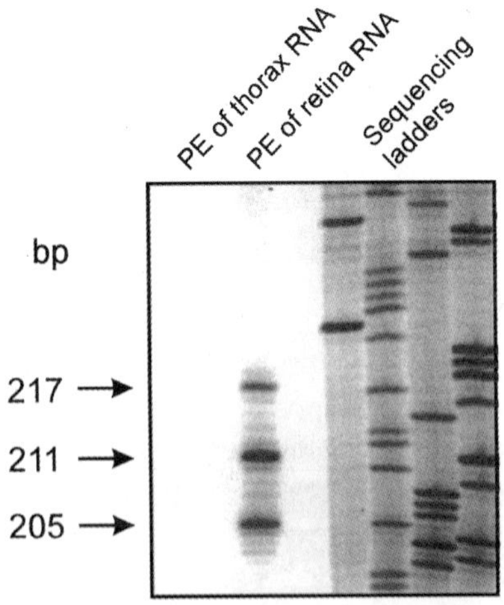

Figure 3. Primer extension analysis (PE) of *Calliphora* Rh1 opsin mRNA. A ^{32}P 5′ end-labeled oligonucleotide complementary to nucleotides −29 to +6 relative to the translation start site of the *Calliphora* Rh1 opsin gene was annealed to 10 μg of total RNA isolated from *Calliphora* thorax or retina. After reverse transcription of the templates, the primer extension products were separated on a 6% denaturing polyacrylamide gel and autoradiographed. The sizes of the major primer-extension products obtained from retina RNA (205, 211, and 217 bp), which were determined by use of sequencing reactions as size markers, are indicated at the left.

chromogen for 20 hours to show β-galactosidase staining, and the staining intensity was much weaker than that in the positive control, indicating that the *Calliphora* Rh1 promoter cannot fully substitute for the *Drosophila* Rh1 promoter. The possibility that the staining observed in these sections results from internal β-galactosidase activity is ruled out by the fact that sections of only two out of three transgenic lines were stained. Also, wild-type

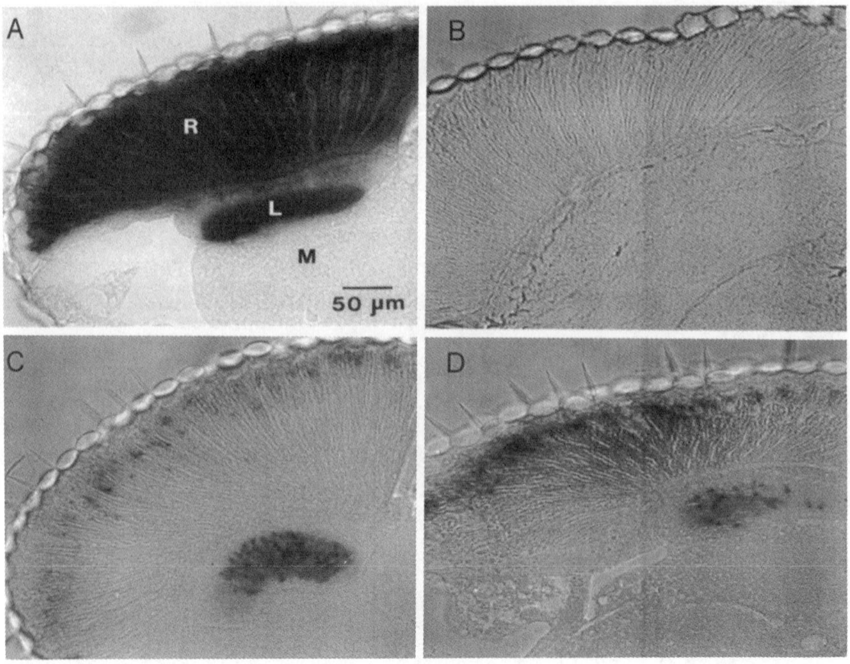

Figure 4. Spatial expression pattern of the LacZ reporter gene under the control of the *Drosophila* (A) and *Calliphora* (B–D) Rh1 opsin promoter. Longitudinal cryosections through the heads of transgenic *Drosophila* lines were assayed for β-galactosidase activity (dark staining). A shows a section through the head of a fly from a transgenic line carrying the construct DmpRh1-pW-ATG-lac1. In B–D three independent transgenic lines carrying CvpRh1-pW-ATG-lac1 were analyzed. *R*, retina; *L*, lamina; *M*, medulla.

controls incubated under the same conditions showed no detectable β-galactosidase staining in the retina (data not shown). The negative result obtained with one of the three investigated CvpRh1-pW-ATG-lac1 lines is probably due to position effects, i.e. to the localization of the transgene in the genome. In the positive CvpRh1-pW-ATG-lac1 lines, the β-galactosidase activity is not distributed evenly in the eye. The retina is preferentially stained in the distal part, close to the cornea, where the cell nuclei of R1–6 and of R7 cells are located. As is the case in flies carrying DmpRh1-pW-ATG-lac1, β-galactosidase activity is also detected in the lamina of transgenic lines carrying CvpRh1-pW-ATG-lac1. It was shown previously that staining of the lamina is due to diffusion of β-galactosidase along the axons of the peripheral photoreceptor cells R1–6, which project into the first optic ganglion (Mismer and Rubin, 1987). Most of the central photoreceptor cells R7 and R8 project into the second optic ganglion (the medulla), and expression of LacZ under the control of an R7 opsin promoter has been shown to result in β-galactosidase activity in the retina and in the second optic ganglion (Fortini and Rubin, 1990). Staining of the medulla is not detected in flies carrying CvpRh1-pW-ATG-lac1. Although it cannot be ruled out that expression of the reporter gene in R7 or in R8 cells, which should result in a staining signal in the medulla, is below the detection limit of our method, the current data suggest that the *Calliphora* Rh1 promoter drives gene expression specifically in R1–6 cells of transgenic *Drosophila.* This finding is unexpected because the *Calliphora* Rh1 promoter does not contain the putative *cis*-acting elements RUS1A and RUS1B, which are thought to determine R1–6 cell-specific expression in *Drosophila.*

4. DISCUSSION

In the present study we have characterized the 5′ flanking region of the *Calliphora* Rh1 rhodopsin gene. We have demonstrated that the promoter contained in this region is able to drive the photoreceptor-specific expression of a reporter gene in the *Drosophila* eye, although the 5′ flanking regions of the Rh1 opsin genes of these two species show little overall sequence homology.

In the core region of the *Calliphora* and *Drosophila* Rh1 opsin promoter, an RCS I element is conserved. This putative *cis*-acting element is also present in the 5′ flanking region of other *Drosophila* genes involved in phototransduction, e.g. in the genes encoding arrestin 1 (Hyde *et al.,* 1990), a retina-specific cyclophilin (Schneuwly *et al.,* 1989), chaoptin (Reinke *et al.,* 1988), the *ninaC* gene product (Montell and Rubin, 1988), and the transient receptor potential protein (Montell and Rubin, 1989). Moreover, a similar sequence, located either on the same DNA strand as in the dipteran rhodopsin promoters or on the opposite DNA strand, is found in the promoter regions of vertebrate rhodopsins (see Fig. 5 and Table 1), arrestin, interphotoreceptor retinoid-binding protein, and phosphodiesterase β-subunit genes (Kikuchi *et al.,* 1993; Polo *et al.,* 1996). In the photoreceptor-specifically expressed genes of vertebrates, this sequence is referred to as photoreceptor-conserved element I (PCE I). It has been shown by gel-mobility–shift analyses that putative transcription factors isolated from vertebrate retinas bind to the PCE I element and also to the RCS I element of the *Drosophila* Rh2 opsin gene (Kikuchi *et al.,* 1993). It has therefore been suggested that the basic mechanisms for the photoreceptor-specific gene expression are evolutionarily conserved from fly to man. It is therefore striking that the RCS I elements of the *Calliphora* and *Drosophila* Rh1 opsin genes are not perfectly conserved. The RCS I element analyzed here (CTTAATTTAATT) differs from the consensus sequence determined for *Drosophila* RCS I elements

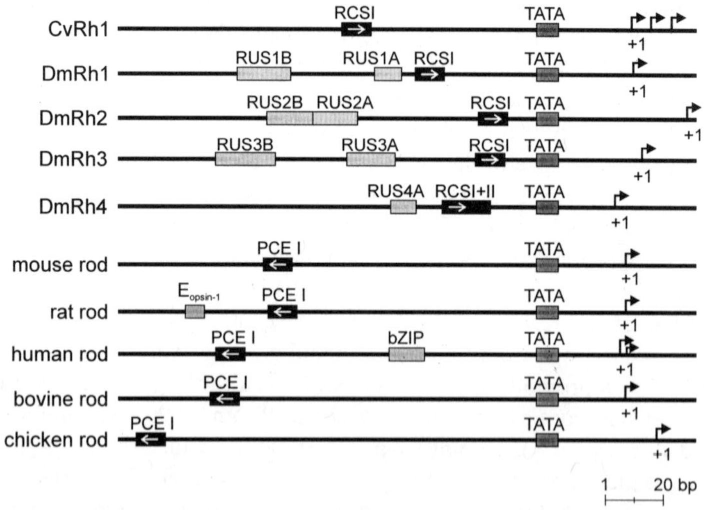

Figure 5. Schematic structure of invertebrate and vertebrate opsin promoters. The mRNA start sites are depicted by broken arrows. Putative *cis*-acting elements, including TATA-box (TATA), rhodopsin core sequences (RCS I) of *Calliphora vicina* Rh1 (CvRh1) and *Drosophila melanogaster* Rh1 to Rh4 (DmRh1–DmRh4) opsin promoters, and the RCS I homologue photoreceptor conserved element I (PCE I) of vertebrate rhodopsin promoters are indicated by boxes. Arrows within these boxes illustrate the orientation of RCS I and PCE I elements on the coding or reverse strand, respectively. Also shown are elements specific for *Drosophila* and vertebrate opsin promoters: rhodopsin upstream sequences A and B (RUS1A–RUS4A, RUS1B–RUS3B) of *Drosophila* Rh1 to Rh4, the E opsin-1 box of the rat rhodopsin gene, and the bZIP-BE box of the human rhodopsin gene.

(CTAATTG(G/A)(G/A)TT) by the insertion of a T at position 2 and by a G → T exchange at position 7. Whether or not this variation is species-specific must be determined by analysis of the promoter regions of other photoreceptor-specifically expressed *Calliphora* genes. Despite these differences in the RCS I element and the weak conservation of other proximal promoter elements, the *Calliphora* Rh1 opsin promoter is able to induce a retina-specific and, within the detection limits of our method, an R1–6 cell-specific expression of the LacZ reporter gene in the *Drosophila* eye. One or more of the four upstream elements

Table 1. Nucleotide sequence of the RCS I and PCE I elements of invertebrate and vertebrate opsin promoters of Figure 5. Underlined nucleotides are identical in seven or more elements

Gene	RCS I/PCE I element		
CvRh1	-92	CTTAATTTAATT	-81
DmRh1	-76	CT–AATTGAATT	-66
DmRh2	-69	CT–AATTGGGTT	-59
DmRh3	-56	CT–AATCCAATT	!46
DmRh4	-61	CT–AATTGAATT	!51
Mouse rhodopsin	-117	CT–AATTGGCTT	-127
Rat rhodopsin	-115	CT–AATTAGCTT	-125
Human rhodopsin	-124	CT–AATTGGCTT	-134
Bovine rhodopsin	-126	TT–AATTGGCTT	-136
Chicken rhodopsin	-171	GT–AATTGGCTT	-181

(located between bp −979 and bp −280) that are conserved between the *Calliphora* and *Drosophila* Rh1 opsin genes may contribute to an R1–6 cell-specific opsin expression. The results illustrate that the transcriptional machinery involved in photoreceptor-specific gene expression is similar in the two distantly related fly species. The hypothesis that this machinery is conserved over much greater evolutionary distances could be tested by screening of suitable vertebrate promoters for their ability to drive gene expression in *Drosophila.*

Besides the core region of the *Drosophila* Rh1 opsin promoter (bp −162 to +67), which contains the RCS I and the RUS1A and RUS1B elements, two enhancer elements that serve partially redundant functions have been identified between bp −215 and −162 and between bp −501 and −219. Removal of both enhancer elements resulted in a 30-fold decrease in expression of a reporter gene (Mismer and Rubin, 1987). A similar organization into a core region and an enhancer region has recently been proposed for the regulatory sequences of vertebrate rhodopsin genes (Nie *et al.,* 1996). In bovine rhodopsin the core region comprises bp −176 to +70, and a rhodopsin enhancer region (RER) is located between bp −2045 and −1923. As in *Drosophila,* the core region is sufficient to direct photoreceptor cell-specific expression, but RER increases the level of expression about 10-fold. RER sequences are also present in the regulatory regions of the human, mouse, and rat rhodopsin genes but the location of this element relative to the mRNA start site varies significantly, from 1.5 to 2 kb upstream (Nie *et al.,* 1996).

The expression of the lacZ gene was much weaker under the control of the 1.1-kb 5′ flanking region of the *Calliphora* Rh1 opsin gene than under the control of the *Drosophila* Rh1 opsin promoter. This finding can be interpreted in two ways. The putative *cis*-acting elements like RCS I are not perfectly conserved between *Calliphora* and *Drosophila.* Transcription factors of *Drosophila* photoreceptor cells may therefore bind to the *Calliphora* Rh1 opsin promotor much more inefficiently than to the *Drosophila* Rh1 opsin homologue. Alternatively, the low level of expression of the reporter gene may be explained by the lack of enhancer-like elements in the 1.1-kb 5′ flanking region of the *Calliphora* Rh1 opsin gene, which may be located farther upstream than in the *Drosophila* Rh1 rhodopsin promoter.

A fundamental question with respect to rhodopsin renewal in differentiated fly photoreceptor cells is whether or not up- and down-regulation of the rhodopsin content involves alterations in rhodopsin gene transcription. This question has been addressed by several laboratories for Rh1 rhodopsin expressed in the major fly photoreceptor cells R1–6. These photoreceptor cells are a particularly attractive model system, because the Rh1 rhodopsin content can easily be manipulated over a wide range (between 2% and 100%) without photoreceptor cell degeneration or severe alterations of the structure of the photoreceptive membrane. Methods to manipulate the Rh1 rhodopsin content include carotenoid/retinoid deprivation and replacement (Goldsmith *et al.,* 1964; Paulsen and Schwemer, 1979, 1983; Sapp *et al.,* 1991), as well as specific illumination programs that deplete and regenerate specifically the 11-*cis* form of the retinal chromophore (Schwemer, 1984). Each manipulation of the chromophore supply of the photoreceptor cells was shown to affect not only the concentration of functional rhodopsin, which can be determined spectrophotometrically or electrophysiologically, but also the amount of the apoprotein opsin present in the photoreceptive membrane. For example, a down-regulation of rhodopsin from 100 to 13%, induced simply by maintenance of the flies for one week under permanent green light illumination, is accompanied by the loss of an equivalent amount of opsin (Huber *et al.,* 1994). Whether these alterations of the opsin content are due to a regulation of gene transcription or of regulatory steps at the posttranscriptional level has been a subject of controversy in the literature. Using reporter-gene assays, Sun *et al.* (1993) found that the expression of a reporter gene under the control of the Rh1 opsin-gene promoter is reduced in carotenoid/retinoid-deprived *Drosophila* and is increased after the deprived flies are fed a diet containing carotenoids and retinoids. Accordingly, it was concluded that Rh1 opsin-gene transcription depends on a sufficient supply of carotenoids and retinoids in the photoreceptor cells. We and others have found that the Rh1 opsin mRNA level is not changed under conditions that dramatically change the amount of opsin protein (Ozaki *et al.,* 1993; Huber *et al.,* 1994). In the moth *Manduca sexta*, carotenoid/retinoid deprivation seemed even to elevate the level of opsin mRNA's in the eye (Chase *et al.,* 1996). For *Drosophila* and *Calliphora,* it was further shown that photoreceptor cells depleted of the 11-*cis* form of the retinal chromophore do synthesize nascent opsin at a level similar to that in the presence of chromophore (Ozaki *et al.,* 1993; Huber *et al.,* 1994), but in the absence of chromophore, the nascent opsin is not efficiently processed to the mature form and is not incorporated into the photoreceptive membrane. Therefore, the presence of retinal chromophore, and probably its attachment to nascent opsin, is required for rhodopsin maturation into a conformation that is not retained and degraded in cytoplasmic membrane systems, e.g. the endoplasmic reticulum and Golgi cysternae, but it is properly transported to the photoreceptive membrane. Recent studies by Picking *et al.* (1996) may resolve these discrepancies by providing evidence that retinoids, most probably retinoic acid, are required in the photoreceptor cells to maintain a high level of opsin gene transcription. This putative activator of opsin-gene transcription was probably depleted in the studies of Sun *et al.* (1993) but not under the conditions applied by Ozaki *et al.* (1993) and Huber *et al.* (1994). When the flies were supplied with retinoic acid in addition to an otherwise carotenoid/retinoid-deficient diet, opsin-gene transcription was not down-regulated, but rhodopsin renewal was still blocked, indicating a posttranscriptional regulation step (Picking *et al.,* 1996). Taken together, alterations of rhodopsin synthesis at the transcriptional level seem to involve retinoids other than the 11-*cis* hydroxyretinal chromophore that is required later to ensure proper rhodopsin maturation.

ACKNOWLEDGMENTS

The authors are grateful to M. Bähner and J. Bentrop for their help in the generation of transgenic *Drosophila.* U. Wolfrum gave helpful advice on the preparation of *Drosophila* eye slices, and I. Huber assisted in the preparation of the manuscript. The pW-ATG-lac1 P-element vector was kindly provided by J. Urban (University of Mainz). The *Drosophila* Rh1 opsin promoter was a gift from C. S. Zuker (University of California, San Diego). The work was supported by grants from the Deutsche Forschungsgemeinschaft.

REFERENCES

Baker, E. K., Colley, N. J., and Zuker, C. S. (1994). The cyclophilin homolog NinaA functions as a chaperone, forming a stable complex *in vivo* with its protein target rhodopsin. *EMBO Journal* 13:4886–4895.

Chase, M. R., Bennett, R. R., and White, R. H. (1996). Expression of opsin mRNA in normal and Vitamin A deficient retinas of the sphingid moth *Manduca sexta. Visual Neuroscience* 13:353–358.

Chirgwin, J. M., Pryzbyla, A. E., MacDonald, R. J., Rutter, W. J. (1979). Isolation of biologically active ribonucleic acid from sources enriched in ribonuclease. *Biochemistry* 18, 5294–5299.

Chou, W.-H., Hall, K. J., Wilson, D. B., Wideman, C. L., Townson, S. M., Chadwell, L. V., and Britt, S. G. (1996). Identification of a novel *Drosophila* opsin reveals specific patterning of the R7 and R8 photoreceptor cells. *Neuron* 17:1101–1115.

Cowman, A. F., Zuker, C. S., and Rubin, G. M. (1986). An opsin gene expressed in only one photoreceptor cell type of the *Drosophila* eye. *Cell* 44:705–710.

Durica, D. S., Restrepo, M. A., Thomas, T. L., and Beckingham, K. (1987). Isolation and characterization of *abl* gene sequences in *Calliphora erythrocephala. Gene* 59:63–76.

Ferreira, P. A., Nakayama, T. A., Pak, W. L., and Travis, G. H. (1996) Cyclophilin-related protein RanBP2 acts as chaperone for red/green opsin. *Nature* 383:637–640.

Fliesler, S. J., and Basinger, S. F. (1985). Tunicamycin blocks the incorporation of opsin into rod outer segment membranes. *Proceedings of the National Academy of Sciences of the USA* 82:1116–1120.

Fliesler, S. J., Rayborn, M. E., and Hollyfield, J. G. (1985). Membrane morphogenesis in retinal rod outer segments: inhibition by tunicamycin. *Journal of Cell Biology* 100:574–587.

Fortini, M. E., and Rubin, G. M. (1990). Analysis of *cis*-acting requirements of the Rh3 and Rh4 genes reveals a bipartite organization to rhodopsin promoters in *Drosophila melanogaster. Genes and Development* 4:444–463.

Fryxell, K. J., and Meyerowitz, E. M. (1987). An opsin gene that is expressed in the R7 photoreceptor cell of *Drosophila. EMBO Journal* 6:443–451.

Goldsmith, T. H., Barker, R. J., and Cohen, C. F. (1964). Sensitivity of visual receptors of carotenoid/retinoid-depleted flies: a vitamin A deficiency in an invertebrate. *Science* 146:65–67.

Huber, A., Smith, D. P., Zuker, C. S., and Paulsen, R. (1990). Opsin of *Calliphora* peripheral photoreceptors R1–6: homology with *Drosophila* Rh1 and post-translational processing. *Journal of Biological Chemistry* 265:17906–17910.

Huber, A., Wolfrum, U., and Paulsen, R. (1994). Opsin maturation and targeting to rhabdomeral photoreceptor membranes requires the retinal chromophore. *European Journal of Cell Biology* 63:219–229.

Huber, A., Schulz, S., Bentrop, J., Groell, C., Wolfrum, U., and Paulsen, R. (1997). Molecular cloning of *Drosophila* Rh6 rhodopsin: the visual pigment of a subset of R8 photoreceptor cells. *FEBS Letters* 406:6–10.

Hyde, D. R., Mecklenburg, K. L., Pollock, J. A., Vithelic, T. S., and Benzer, S. (1990). Twenty *Drosophila* visual system cDNA clones: one is a homolog of human arrestin. *Proceedings of the National Academy of Sciences of the USA* 87:1008–1012.

Kikuchi, T., Raju, K., Breitman, M. L., and Shinohara, T. (1993). The proximal promoter of the mouse arrestin gene directs gene expression in photoreceptor cells and contains an evolutionarily conserved retinal factor-binding site. *Molecular and Cellular Biology* 13:4400–4408.

Kuhn, R., Schäfer, U. and Schäfer, M. (1988). pW-ATG-lac, P-element vectors for lacZ transcriptional gene fusions in *Drosophila. Nucleic Acids Research* 16:4163.

Mismer, D., and Rubin, G. M. (1987). Analysis of the promoter of the *nina*E opsin gene in *Drosophila melanogaster. Genetics* 116:565–578.

Mismer, D., and Rubin, G. M. (1989). Definition of *cis*-acting elements regulating expression of the *Drosophila melanogaster nina*E opsin gene by oligonucleotide-directed mutagenesis. *Genetics* 121:77–87.

Mismer, D., Michael, W. M., Laverty, T. R., and Rubin, G. M. (1988). Analysis of the promoter of the Rh2 Opsin Gene in *Drosophila melanogaster. Genetics* 120:173–180.

Montell, C., and Rubin, G. M. (1988). The *Drosophila nina*C locus encodes two photoreceptor cell specific proteins with domains homologous to protein kinases and myosin heavy chain head. *Cell* 52:757–772.

Montell, C., and Rubin, G. M. (1989). Molecular characterization of the *Drosophila trp* locus: a putative integral membrane protein required for phototransduction. *Neuron* 2:1313–1323.

Montell, C., Jones, K., Zuker, C. S., and Rubin, G. (1987). A second opsin gene expressed in the ultraviolet-sensitive R7 photoreceptor cells of *Drosophila melanogaster. Journal of Neuroscience* 7:1558–1566.

Nathans, J., and Hogness, D. S. (1984). Isolation and nucleotide sequence of the gene encoding human rhodopsin. *Proceedings of the National Academy of Sciences of the USA* 81:4851–4855.

Nathans, J., Thomas, D., and Hogness, D. S. (1986). Molecular genetics of human color vision: the genes encoding blue, green, and red pigments. *Science* 232:193–202.

Nie, Z., Chen, S., Kumar, R., and Zack, D. J. (1996). RER, an evolutionarily conserved sequence upstream of the rhodopsin gene, has enhancer activity. *Journal of Biological Chemistry* 271:2667–2675.

O'Tousa, J. E. (1992) Requirement of N-linked glycosylation site in *Drosophila* rhodopsin. *Visual Neuroscience* 8:385–390.

O'Tousa, J. E., Baehr, W., Martin, R. L., Hirsch, J., Pak, W. L., and Applebury, M. L. (1985). The *Drosophila ninaE* gene encodes an opsin. *Cell* 40:839–850.

Ozaki, K. H., Nagatani, H., Ozaki, M., and Tokunaga, F. (1993). Maturation of major *Drosophila* rhodopsin, *ninaE,* requires chromophore 3-hydroxyretinal. *Neuron* 10:1113–1119.

Paulsen, R. (1984). Spectral characteristics of isolated blowfly rhabdoms. Journal of Comparative Physiology A 155:47–55.

Paulsen, R., and Schwemer, J. (1979). Vitamin A deficiency reduces the concentration of visual pigment protein within blowfly photoreceptor membranes. *Biochimica et Biophysica Acta* 577:385–390.

Paulsen, R., and Schwemer, J. (1983). Biogenesis of blowfly photoreceptor membranes is regulated by 11-*cis* retinal. *European Journal of Biochemistry* 137:609–614.

Penn, J. S., and Williams, T. P. (1986). Photostasis: regulation of daily photon-catch by rat retinas in response to various cyclic illuminances. *Experimental Eye Research* 43:915–928.

Picking, W. L., Chen, D. M., Lee, R. D., Vogt, M. E., Polizzi, O. L., Marietta, R. G., and Stark, W. S. (1996). Control of *Drosophila* opsin gene expression by carotenoids and retinoic acid: Northern and western analyses. *Experimental Eye Research* 63:493–500.

Pollock, J. A., and Benzer, S. (1988). Transcript localisation of four opsin genes in three visual organs of *Drosophila*; Rh2 is ocellus specific. *Nature* 333:779–782.

Polo, A. D., Rickman, C. B., and Farber, D. B. (1996). Isolation and initial characterization of the 5′ flanking region of the human and murine cyclic guanosine monophosphate-phosphodiesterase β-subunit genes. *Investigative Ophthalmology and Visual Science* 37:551–560.

Reinke, R., Krantz, D. E., Yen, D., and Zipursky, S. L. (1988). Chaoptin, a cell surface glycoprotein required for *Drosophila* photoreceptor cell morphogenesis, contains a repeat motiv found in yeast and human. *Cell* 52:291–301.

Sambrook, J., Fritsch, E. F., and Maniatis, T. (1989). *Molecular Cloning (A Laboratory Manual),* 2nd edition. Cold Spring Harbor Laboratory Press, Cold Spring Harbor, New York.

Sanger, F., Nicklen, S., and Coulson, A. R. (1977). DNA sequencing with chainterminating inhibitors. *Proceedings of the National Academy of Sciences of the USA* 74:5463–5467.

Sapp, R. J., Christianson, J. S., Maier, L., Studer, K., and Stark, W. S. (1991). Carotenoid replacement therapy in *Drosophila:* recovery of membrane, opsin and visual pigment. *Experimental Eye Research* 53:73–79.

Schneuwly, S., Shortridge, R. D., Larrivee, D. C., Ono, T., Ozaki, M., and Pak, W. L. (1989). *Drosophila ninaA* gene encodes an eye-specific cyclophilin (cyclosporin A-binding protein). *Proceedings of the National Academy of Sciences of the USA* 86:5390–5394.

Schwemer, J. (1984). Renewal of visual pigment in photoreceptors of blowfly. *Journal of Comparative Physiology* 154:535–547.

Spradling, A. C., and Rubin, G. M. (1982). Transposition of cloned P elements into *Drosophila* germ line chromosomes. *Science* 218:341–347.

Stamnes, M. A., Shieh, B.-H., Chuman, L., Harris, G. L., and Zuker, C. S. (1991). The cyclophilin homolog *ninaA* is a tissue-specific integral membrane protein required for the proper synthesis of a subset of *Drosophila* rhodopsins. *Cell* 65:219–227.

Sun, D., Chen, D.-M., Harrelson, A., and Stark, W. S. (1993). Increased expression of chloramphenicol acetyltransferase by carotenoid and retinol replacement in *Drosophila* opsin promoter fusion stocks. *Experimental Eye Research* 57:177–187.

Vogt, K. (1983). Is fly visual pigment a rhodopsin? *Zeitschrift für Naturforschung Teil C* 38:329–333.

Williams, T. P. (1984). Some properties of old and new rhodopsin in single *Bufo* rods. *Journal of General Physiology* 83:841–852.

Wolfrum, U. (1991). Distribution of F-actin in the compound eye of *Calliphora erythrocephala* (Diptera, Insecta). *Cell and Tissue Research* 263:399–403.

Zuker, C. S., Cowman, A. F., and Rubin, G. M. (1985). Isolation and structure of a rhodopsin gene from *Drosophila melanogaster. Cell* 40:851–858.

Zuker, C. S., Montell, C., Jones, K., Laverty, T., and Rubin, G. M. (1987). A rhodopsin gene expressed in photoreceptor cell R7 of the *Drosophila* eye: homology with other signal-transducing molecules. *Journal of Neuroscience* 7:1550–1557.

5

ROD OUTER SEGMENT (ROS) RENEWAL

An Underlying Mechanism for Maintaining Photostasis

Jeri-Lynn Schremser-Berlin

Howe Laboratory
Harvard School of Medicine
Massachusetts Eye and Ear Infirmary
243 Charles Street
Boston, Massachusetts 02114

1. INTRODUCTION

The method by which the vertebrate rod cell replaces integral membrane proteins in the outer segment is unique to this type of photoreceptor. The process is termed rod outer-segment (ROS) turnover, and one turnover period for rats is 10 days (Bairati and Orzalesi, 1963; Droz, 1963; Young, 1967, 1971a, b; Hall *et al.*, 1968; Anderson and Fisher, 1976). The vertebrate ROS, the light-catching apparatus of the cell, is composed of a stack of membranous discs surrounded by a plasma membrane (Sjöstrand, 1953; Moody and Robertson, 1960; Cohen, 1968). Newly synthesized rhodopsin, an integral membrane protein, is inserted into discs at the base of the outer segment. Once rhodopsin molecules are incorporated into a disc, they remain there for the lifetime of the disc (Young, 1967; Daemen, 1973; Steinberg *et al.*, 1980; Papermaster and Schneider, 1982). As new discs are formed, they progressively displace the previously synthesized discs toward the tip of the outer segment. When the discs reach the tip, they are shed in groups surrounded by a plasma membrane. The packets of discs, ingested by pigment epithelial cells, are called phagosomes, and they are degraded by lysosomal enzymes within these cells (Young and Bok, 1969; Ishikawa and Yamada, 1970; Spitnas and Hogan, 1970; Young, 1971b; Roof *et al.*, 1991). A burst of disc shedding occurs daily, 1–2 hours after light onset (Hollyfield *et al.*, 1976; LaVail, 1976; Besharse *et al.*, 1977a, b; Kinney and Fisher, 1978). For mammals, the shedding burst is under circadian control (Goldman *et al.*, 1980; LaVail, 1980; Teirstein *et al.*, 1980).

Young proposed that ROS turnover, also called ROS renewal, functioned as preventive maintenance, replacing molecules "before they wear out" (Droz, 1963; Young, 1967, 1971a, b; Anderson and Fisher, 1976), and this hypothesis was supported by several subsequent studies (Kaplan *et al.*, 1978; Baylor *et al.*, 1979; Andrews and Cohen, 1983; Williams, 1984). The renewal process plays a significant part in preserving the integrity of the

Photostasis and Related Phenomena, edited by Williams and Thistle
Plenum Press, New York, 1998

fragile rod cell and could in addition serve as a potential mechanism for adaptive change (reviewed by Penn and Anderson, 1992). This concept, also called plasticity, arose from the belief that membrane turnover enables a cell to respond rapidly to its environment (Bairati and Orzalesi, 1963; Finean *et al.,* 1978). Demonstration of the dynamism of the photoreceptor membrane structure in both vertebrates and invertebrates in studies in which environmental lighting was varied made the hypothesis testable (White and Lord, 1975; Battelle and LaVail, 1978; Blest, 1978; Penn and Williams, 1986; Hoenig and Chamberlain, 1989; Hornstein and Chamberlain, 1991; Organisciak *et al.,* 1991; Penn *et al.,* 1992; Schremser and Williams, 1992; reviewed by Penn and Anderson, 1992).

Photostasis, the regulation of photon-catching ability, is a relatively new idea that derives its name from the observation that groups of albino rats reared in different light intensities all absorb the same number of photons per retina per day (Penn and Williams, 1986). If animals truly regulate the number of photons their eyes absorb, ROS turnover is a candidate for the mechanism underlying such a process.

Pigment concentration in the rod cell can be altered by changes in rates of outer segment membrane synthesis and breakdown. Both rates influence ROS length, which in turn contributes to whole-retina rhodopsin levels. Changes in synthesis of rhodopsin and its incorporation into the discs contribute as well, and visual pigment level in the retina directly regulates light absorption (White and Lord, 1975). Animals could adjust the photon-catching ability of their retinas by rebuilding photoreceptive membranes in response to environmental light intensity changes. Continuous turnover of ROS membrane, which lends ROS length and phospholipid:opsin ratio a dynamic quality, is a potential mechanism for regulation of photon-catching ability (Penn and Anderson, 1992).

Levels of other photoreceptor-specific proteins, including arrestin and transducin, have been shown to change with lighting environment (Broekhuyse *et al.,* 1985; Philp *et al.,* 1987; Whelan and McGinnis, 1988; Farber *et al.,* 1991; Organisciak *et al.,* 1991; McGinnis *et al.,* 1992). In most studies, however, the light-mediated changes in the levels of visual transduction proteins were studied in relation to light damage, not photon catch. It has been suggested that the susceptibility of the retina to damage by light is influenced by the relative levels of visual-cell proteins at the time of intense light exposure (Noell and Albrecht, 1971; Rapp and Williams, 1980; Penn and Williams, 1986; Organisciak *et al.,* 1991; Rapp *et al.,* 1993). Regulation by the rod cell of transduction proteins, including rhodopsin, according to environmental light may be the common element underlying both normal adaptation and its extreme case, cell death. For example, regulation or, more accurately, deregulation of photon catch may be one reason for light damage to rod cells, but more information is needed before we can be certain this is true.

Long-term adaptation is defined as a modification of the rod cell at the cellular or molecular level requiring days to weeks to develop fully, as distinct from a change over seconds or minutes, such as recovery of bleached pigment after light exposure. Determining the time course of cellular changes during the renewal process after a change in light intensity is the first step to understanding the potential role of ROS renewal as the mechanism for adaptation. A few studies of frogs and pigmented mice have examined the renewal process in detail, but their focus was effects of temperature changes or high-intensity light conditions on the renewal process (Besharse *et al.,* 1977a, b; Besharse and Hollyfield, 1979; Hollyfield, 1979; Hollyfield and Rayborn, 1982; Hollyfield *et al.,* 1982).

For years, investigators have known that ROS renewal occurs and have used it as an explanation for remodeling of rod cells. In experiments involving adaptation of the retina to different light environments, previous investigators have documented the state of the rod before and after a change in light intensity and have followed some but not all of the cellular changes that occur as a result (Organisciak *et al.,* 1991; Penn and Anderson, 1987, 1992; Penn *et al.,* 1992). In the present study, we have investigated these changes in detail and have shown that, by exploiting the mechanisms of the renewal process, rod cells do indeed remodel themselves in accordance with their light environment.

Other points that must be demonstrated before we can conclude that photostasis results from ROS renewal are (a) an effect on synthesis of rhodopsin by changes in environmental light intensity, (b) measurable differences in rhodopsin packing density between basal and tip discs in single ROSs from animals that have been moved from one lighting intensity to another, and (c) absorbance, after adaptation to a new light intensity, of the same number of photons per retina per day that were absorbed under the old.

2. METHODS

2.1. Animals

Sprague-Dawley albino rats (breeder pairs from Charles River, Wilmington, MA, USA) were born and maintained in controlled cyclic lighting (12 hours light, 12 hours dark, lights on at 0800 hours). Two intensities were used: a "low"-intensity condition of 3 lux and a "high"-intensity condition of 200 lux. NIH guidelines for care and use of animals were followed.

"Natives" were those animals that remained in the same cyclic intensity throughout their lives, and "adventives" were those that were switched to a new cyclic intensity for a period of 1 to 28 days. For example, a "day-7, 200-lux adventive" was born in 3 lux, was switched at 12 weeks to 200 lux, and experienced the new cyclic light intensity for 7 days. All switched animals were moved at exactly 12 weeks of age. All native animals were assayed at 12 weeks of age, and adventives were assayed at 12 weeks plus the time in the new light. All animals were switched during their dark period in black boxes and experienced the new intensity when lights came on the next morning. Rats described as "dark-adapted" received the full day's complement of light (12 hours) and were then put into a separate dark room at the end of their light period until the time of the assay the next morning.

2.2. Dark-Adapted Rhodopsin Assay

Rats were dark-adapted for 11 hours before the assay, and all subsequent procedures were performed in dim red light (which produced no measurable bleaching). For full details see Schremser and Williams (1995a). Briefly, rhodopsin was extracted in 1% Emulphogene, and rhodopsin concentration in the detergent extract was calculated and converted to units of rhodopsin molecules per retina.

2.3. Morphometric Analysis of Retinal Tissue

Eyes were enucleated and placed in a preparation of 1% paraformaldehyde and 2.5% glutaraldehyde buffered to pH 7.2 with 0.2 M PIPES buffer. Fixation and processing were as previously described by Schremser and Williams (1995a). ROS length measurements were made 1 mm and 3 mm from the optic nerve head in both the superior and inferior hemispheres of each retina. The four measurements, from the base to the tip of the outer segments, were averaged to yield ROS length for each retina. In addition, the outer nuclear layer (ONL) area, which is proportional to the number of photoreceptor cells, was calculated from measurements taken every 250 μm along the lengths of both the superior and inferior hemispheres, starting from the optic nerve head. For quantification of phagosomes, all inclusion bodies in the pigment epithelial cells that were greater than 0.75 μm in any dimension were defined as large phagosomes and were counted as an index of disc shedding (LaVail, 1980). Five consecutive 180-μm lengths were examined in the posterior retina on each side of the optic nerve head starting at 400 μm from the optic nerve head. The mean number of phagosomes found in ten 180-μm lengths of retinal pigment epithelium (RPE) in a single eye was calculated.

2.4. Intraocular Isotope Injection

According to the method of O'Brien (1993), each eye was injected with a volume of 2 μl containing 20 μCi of a ^{3}H-leucine and ^{3}H-phenylalanine mixture. All injections were done between 1000 and 1030 hours (Schremser and Williams, 1995b).

2.5. Protein Purification, SDS-PAGE, and Immunoblotting Analysis

Briefly, whole-retina Emulphogene extracts were run over a Con A-Sepharose column and eluted with α-methylmannoside. This eluant, containing peak fractions of rhodopsin, was run on a 12% gel after the samples were adjusted to 2% SDS. Aliquots of the pre- and postcolumn samples were used for scintillation counting, which yielded counts incorporated into total protein and counts incorporated into rhodopsin, respectively. For details see Schremser and Williams (1995b). A monoclonal antibody to opsin (MAb k16–155C, a generous gift from P. Hargrave) (Adamus *et al.,* 1991) was used to detect opsin on the Western blot.

2.6. Microspectrophotometry (MSP)

The microspectrophotometer used was fashioned after the one built by E. F. MacNichol, Jr. (MacNichol, 1978; Williams, 1984). It is a single-photon-counting computer-assisted machine that can scan the visible spectrum in less than a second. Transverse illumination was used, and absorbance spectra were determined at the base and the tip of single rat ROSs (Schremser and Williams, 1995b).

2.7. RNA Isolation and Northern Blot

Total RNA was isolated from all groups at 1000 hours, the same time of day at which the intraocular injections were performed. The opsin mRNA level found at this time was used for comparisons among groups of rats under different lighting conditions (Schremser and Williams, 1995b). The level of opsin mRNA was determined by hybridization of a ^{32}P-labeled antisense RNA probe to total retinal RNA, immobilized on S&S BA-S nitrocellulose. The probe was made from the pBBO4 EcoR1/Pst 1 fragment of bovine opsin cDNA (Nathans and Hogness, 1983). The blot was air dried, and the levels of labeled probe hybridized to opsin mRNA were quantified with a Betascope scanner.

2.8. Statistical Analysis

Student's t-tests were performed, and the level of significance was set at $p < 0.05$.

3. RESULTS

3.1. ROS Length

The 3-lux native average ROS length was 34.3 ± 1.8 μm (average ± SD), and the 200-lux native average length was significantly shorter, 27.7 ± 0.9 μm, a difference of 30% ($p < 0.01$). Animals moved to a new intensity gradually acquired a ROS length similar to that of natives of that intensity (Fig. 1). The day-7 measurement of the 200-lux adventives showed that ROS length had decreased to 17.9 μm at a calculated rate of 2.3 μm day^{-1}—after the switch to bright light, the ROSs became shorter. The 3-lux adventives followed a different pattern. On day 1 at 3 lux, ROSs were slightly shorter, and by day 3, ROS length had decreased significantly, to 19.6 μm ($p < 0.01$), at a calculated rate of 1.7 μm day^{-1}. From day 3 to day 7 at 3 lux, however, ROS length increased at a calculated rate of 2.1 μm day^{-1}. The net result of the switch to dim light was that ROSs became longer.

3.2. Phagosome Counts and Shedding Patterns

Shedding patterns, based on phagosome counts, were the same for 3-lux and 200-lux natives, but changes in shedding patterns were observed in adventives. After the first day at 200 lux, adventives reestablished the usual shedding pattern, a burst 1–2 hours after light onset. On the first day at 200 lux, significantly fewer ($p < 0.01$) phagosomes were observed in the pigment epithelium, but thereafter, 200-lux adventives and 200-lux natives did not differ in phagosome counts (Fig. 2). A change from low to high intensity did not produce major changes in the shedding pattern, yet the ROSs became shorter.

The 3-lux adventives showed a major change in shedding pattern (Fig. 3). Shedding pattern remained unchanged on day 1, but by day 3, shedding had decreased and continued to decrease through day 7. The normal shedding pattern was reestablished by day 21, at which time the average ROS length of the 3-lux adventives was similar to that of the 3-lux natives. Changes in shedding pattern paralleled ROS length changes in the 3-lux adventives. From day 1 to day 3, length decreased slightly, but from day 3 through day 7,

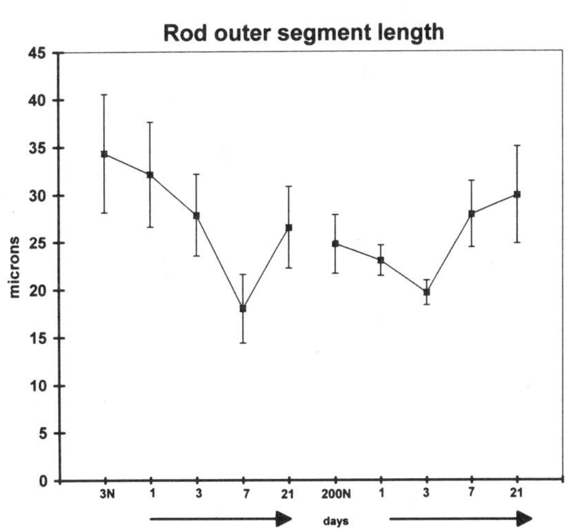

Figure 1. Changes in ROS length with time at new light intensities. Arrows indicate the progression of time at the new intensity. Each point is the average of measurements from several eyes; error bars are ± SD. 3N and 200N are the 3-lux native and 200-lux native values, respectively. Over time, all adventives whether switched into higher or lower intensities, gradually acquire ROS lengths like those of animals born and reared at those intensities.

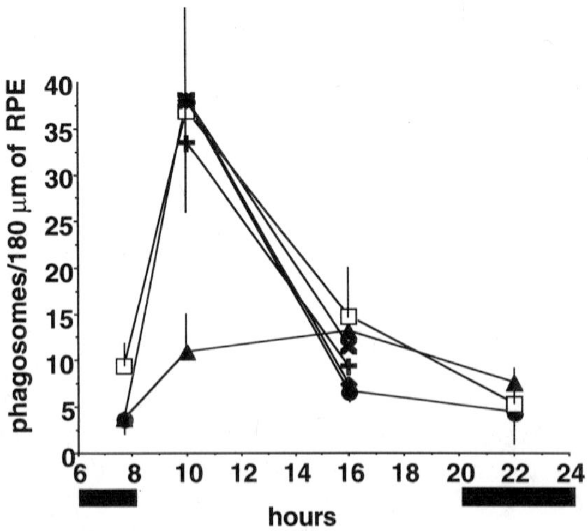

Figure 2. Phagosome counts in retinal pigment epithelium for 200-lux adventives, days 1 to 21. Except for day 1 (filled triangles) shedding pattern did not change in timing or in number of phagosomes. Filled circles, 3-lux natives; open squares, 200-lux natives; filled triangles, day-1 200-lux adventives; diamonds, day-3 200-lux adventives; +'s, day-7 200-lux adventives; x's, day-21 200-lux adventives. Dark bars at bottom represent the dark portion of the light-dark cycle. Error bars are ± SD.

ROS average lengths continued to increase as shedding decreased. The decrease in shedding in the 3-lux adventives resulted in longer ROSs.

3.3. Rhodopsin Levels

The 3-lux native animals had $1.44 \pm 0.10 \times 10^{15}$ rhodopsin molecules per retina (2.39 nmoles), and the 200-lux native animals had $0.75 \pm 0.04 \times 10^{15}$ rhodopsin molecules per retina (1.25 nmoles), a difference of 48% ($p < 0.01$). Because the difference in ROS length between the two groups was 30%, differences in ROS length alone could not account for the difference in rhodopsin levels.

After day 1, the rhodopsin levels of 200-lux adventives dropped in a linear fashion through day 7 (Fig. 4), at -0.13×10^{15} rhodopsin molecules lost per day, equivalent to 0.05×10^{15} molecules lost per micron of ROS between day 1 and day 7. There was a tran-

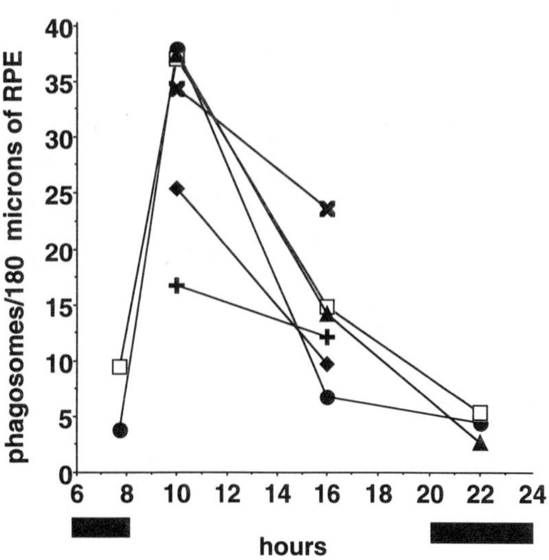

Figure 3. Phagosome counts in retinal pigmental epithelium for 3-lux adventives, days 1 to 21. Filled circles, 3-lux natives; open squares, 200-lux natives; filled triangles, day-1 3-lux adventives; diamonds, day-3 3-lux adventives; +'s, day-7 3-lux adventives; x's, day-21 3-lux adventives. Day-1 shedding was indistinguishable from the native pattern, but thereafter shedding decreased until day 7. By day 21, the 3-lux adventive pattern again resembled the native pattern. Dark bars at bottom represent the dark portion of the light-dark cycle.

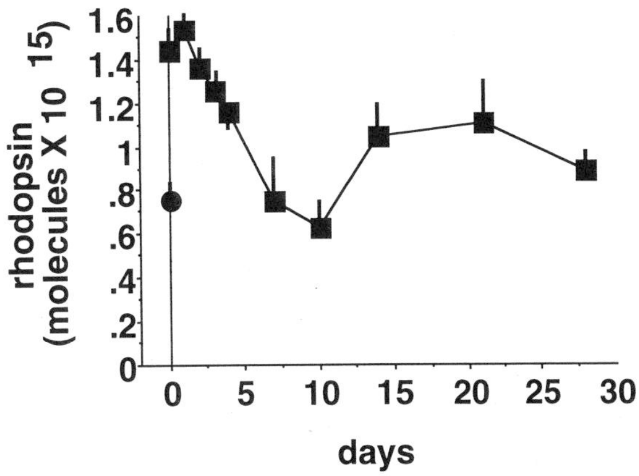

Figure 4. Dark-adapted rhodopsin levels in 200-lux adventives as a function of time at the new, higher light intensity. During the first 7 days, rhodopsin levels fell linearly with time. Thereafter, the levels oscillated around the 200-lux native value and eventually approached the 200-lux native level (filled circle). Error bars are ± SD.

sient increase in rhodopsin values after day 10, but by day 28, rhodopsin values had dropped significantly to 0.89×10^{15} rhodopsin molecules per retina ($p < 0.01$).

The 3-lux adventives showed an increase in rhodopsin levels but not a linear one (Fig. 5). The rhodopsin level did not change significantly from day 1 to day 3, but by day 7, a significant increase in rhodopsin level had occurred, and values began to approach those of 3-lux natives by day 28.

3.4. ONL Measurements

There was no significant difference in ONL area between 3-lux natives (0.341 ± 0.005 mm^2) and 200-lux natives (0.314 ± 0.023 mm^2). To verify that the decrease in shedding in the 3-lux adventives was due to shedding of less ROS material and not to loss of rod cells, we measured ONL area and found no significant difference between 200-lux natives and day-7, 3-lux adventives. The fewer phagosomes found in the pigment epithelium of 3-lux adventives was therefore not due to loss of rod cells.

Changes did occur in ONL area with age (Penn *et al.,* 1985, 1992; Curcio *et al.,* 1993). Animals reared at 3 lux and switched, at 12 weeks, to 200 lux had lost 17% of ONL area by day 21 (when they were 15 weeks old) compared with the 3-lux native group

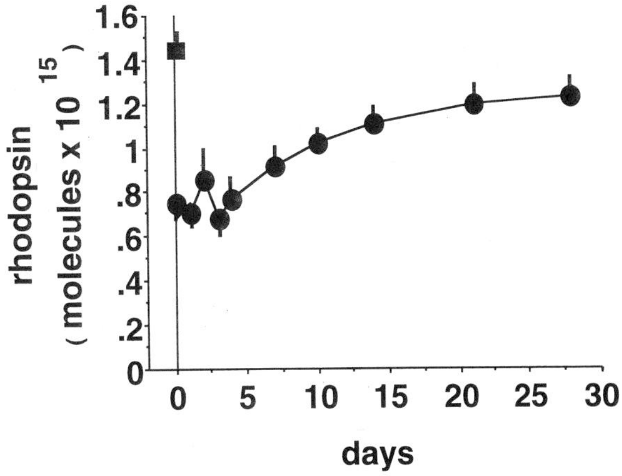

Figure 5. Dark-adapted rhodopsin levels in 3-lux adventives as a function of time at the new, lower light intensity. During the first 5 days, the rhodopsin levels oscillated around the 200-lux native value. Thereafter, the levels increased and approached the 3-lux native value (filled square). Error bars are ± SD.

(when they were 12 weeks old). Animals reared at 200 lux and switched at 12 weeks of age to 3 lux had lost 25% of the ONL area by day 21 compared with the 200-lux native group. In both adventive groups, the decrease in ONL area by day 21 was significant as compared to the ONL area of the native groups. The 3-lux and 200-lux adventives lost cells at different times after the switch, however. For example, 200-lux adventives lost 11% of the cells by day 3, and by day 7 they had lost 13%. From day 7 to day 21, another 4% was lost, for a final loss of 17%. The 3-lux adventives did not lose any cells in the first 7 days, but 25% of the cells were lost between day 7 and day 21.

3.5. Net Synthesis of Rhodopsin

By measuring incorporation of a pulse of ^{3}H-amino acids into newly synthesized rhodopsin, we determined that the net synthesis of rhodopsin depends on light intensity. The amount of incorporation per pulse of amino acid, in units of μCi per two retinas, was proportional to the amount of newly synthesized rhodopsin accumulated for any given day.

The 200-lux adventives began with an initial amino acid incorporation value of 0.55 μCi per two retinas (Fig. 6). On day 1 at 200 lux, incorporation dropped by 82% ($p < 0.01$) to a value of 0.10 μCi per two retinas. By day 3, the incorporation value of 200-lux adventives (0.30 μCi per two retinas) closely matched that of the 200-lux natives (0.28 μCi per two retinas). There was a transient increase in incorporation on day 10, but by day 28 the 200-lux adventive incorporation value was 0.18 μCi per two retinas. Thus, at higher light intensity, the net synthesis of rhodopsin was reduced.

The 3-lux adventives began with an incorporation value of 0.28 μCi per two retinas (Fig. 7). On day 1 at 3 lux, incorporation increased to 0.43 μCi per two retinas, reflecting an increase in net synthesis of rhodopsin. On day 6, the 3-lux adventives attained an incorporation value comparable to that of the 3-lux natives. By day 10 at 3 lux, however, incorporation had decreased to 0.30 μCi per two retinas. The incorporation value again increased to 0.38 μCi per two retinas, and the last value recorded, on day 28 in the new light, was 0.33 μCi per two retinas. Thus, at lower light intensity, the net synthesis of rhodopsin was increased.

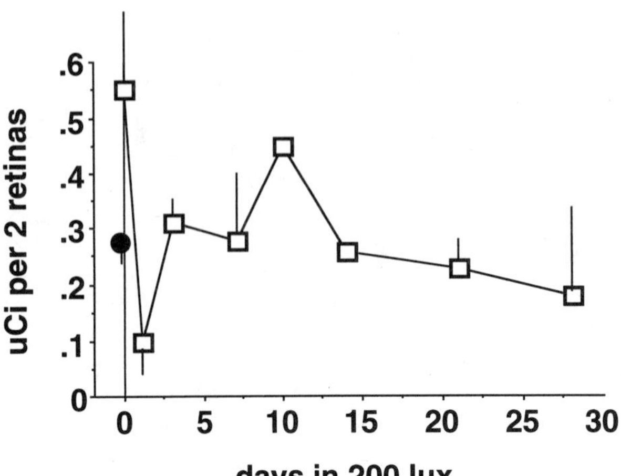

Figure 6. Incorporation of radiolabeled amino acid into newly synthesized rhodopsin (net synthesis) as a function of time in the new 200-lux environment. The square at day 0 represents the incorporation by 3-lux native rats ($n = 10$) before they were switched to 200 lux. Note that the net synthesis of rhodopsin drops on day 1 in the more intense light. Thereafter, aside from the day-10 point, the level of incorporation is indistinguishable from that of 200-lux natives (filled circle on the day-0 line). Each point represents the mean of 4–20 retinas. Error bars are ± SD.

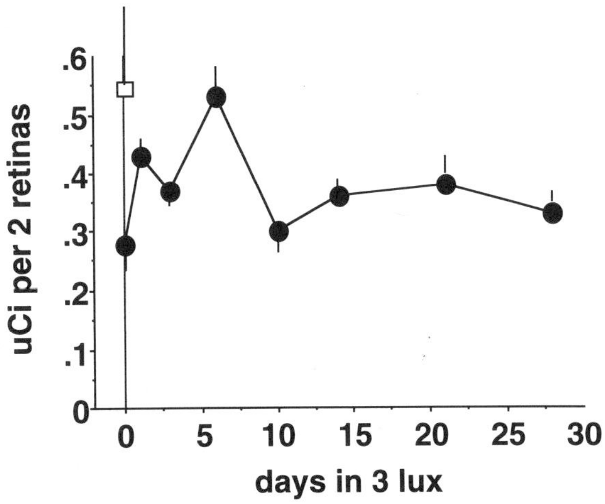

Figure 7. Incorporation of radiolabeled amino acid into newly synthesized rhodopsin (net synthesis) as a function of time in the new 3-lux environment. The filled circle at day 0 represents the incorporation by 200-lux native rats (n = 9) before they were switched to 3 lux. Note that it is about half the incorporation found in 3-lux natives (open square on the day-0 line), consistent with the observation that dark-adapted rhodopsin levels in the 3-lux natives and 200-lux natives stand in the ratio of 2:1. Incorporation rises to the native 3-lux value by day 7 but then drops. This unexpected fall may be due to ONL cell loss, which occurs during this time (see text). Each point represents the mean of 4–18 retinas. Error bars are ± SD.

3.6. SDS-Page and Western Blot

The membrane proteins in the eluant from the Con-A Sepharose column purification assay were resolved on an SDS 12% polyacrylamide gel. From the gel it was determined, on the basis of apparent molecular weight, that the major protein in the eluant was opsin. This conclusion was confirmed by Western blot analysis using bovine opsin antibody, MAb k16–155C. These results verified the assumption that the majority of counts from the column eluant were derived from tritiated amino acids incorporated into rhodopsin.

3.7. Microspectrophotometry

Microspectrophotometric measurements on single rod cells demonstrated that differences reflected in whole-retina rhodopsin extractions were present at the single-cell level. The difference in retinal rhodopsin levels between the two light-intensity groups was 48%. In agreement with that result, the difference in transverse absorbance of single ROSs from the two light intensity groups was 45% ($p < 0.01$). The diameter of the ROSs appeared the same in all lighting conditions. These changes in packing density of rhodopsin per disc suggest a difference in net synthesis of rhodopsin in animals reared in different lighting intensities, corroborating our incorporation results. Recall that the difference in incorporation values between 3-lux and 200-lux animals was 49%.

Figure 8A shows a spectrum obtained from a 3-lux native ROS; the transverse absorbance for this ROS was 0.0092 ± 0.0009. The average absorbance based on values from 24 similar 3-lux native ROSs was 0.0101 ± 0.0021. Figure 8B shows the absorbance spectrum of a 200-lux native ROS, in which the transverse absorbance was 0.0058 ± 0.0014. The average transverse absorbance for a 200-lux native single ROS was 0.0056 ± 0.0018 (n = 16).

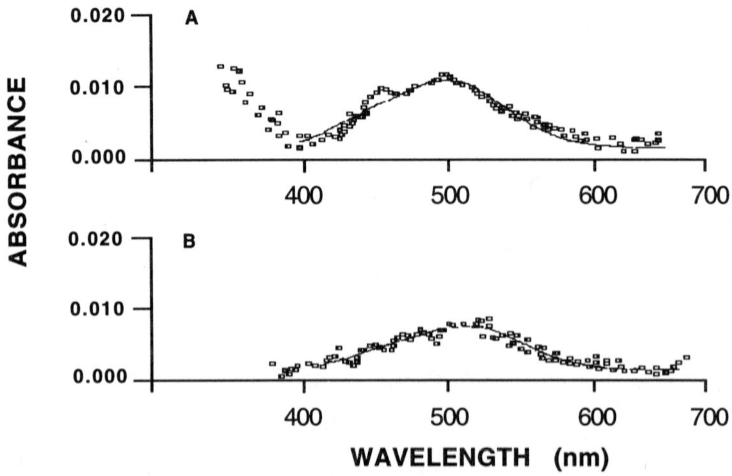

Figure 8. (A) Absorption spectrum of a single ROS taken from a 3-lux native. The transverse absorbance of this ROS was 0.0092 and was typical of a native 3-lux ROS. The absorbance value of this and all spectra in the present paper were obtained at λ_{max} as determined by a curve-fitting routine that subtracts out baseline shifts. (B) Absorption spectrum of a single ROS taken from a 200-lux native. The transverse absorbance of this ROS was 0.0058 and was typical of a native 200-lux ROS. The average difference in absorbance of 3 and 200 lux ROSs is 45% and is in agreement with the difference in whole-retina rhodopsin results, 48%.

Several individual rods of adventives were also measured. For this experiment, the animals were kept at the new cyclic intensity for 5 days (approximately one-half of a turnover period), so discs at the ROS tips should represent the initial intensity of light, and those at the ROS bases should represent discs formed after the switch to the new light intensity. In fact, the absorbance at the ROS tip of a day-5 200-lux adventive ($A = 0.0095 \pm 0.0015$) closely matched that of a 3-lux native (Figs. 8A, 9A). The absorbance at the base of the same ROS ($A = 0.0055 \pm 0.0014$) resembled that of a 200-lux native (Figs. 8B, 9B). The difference in absorbance between the base and the tip of the same ROS may reflect the changes in packing density that occur when an animal is moved to a new lighting environment.

3.8. mRNA Levels

The level of opsin mRNA was determined by Northern blot analysis. The results show that the same amount of opsin mRNA was present in day-1 200-lux adventives as in 200-lux natives, so the abrupt decrease in rhodopsin net synthesis on day 1 in the new, more intense light (Fig. 6) is not due to a decline in message level on day 1.

4. DISCUSSION

Several lines of evidence suggest that ROS renewal is the mechanism that confers plasticity on the rod cell (reviewed by Penn and Anderson, 1992). To demonstrate that it does so, we followed the step-by-step transformation of the rod cells after animals were switched to a new cyclic lighting intensity, and we quantified the daily changes that oc-

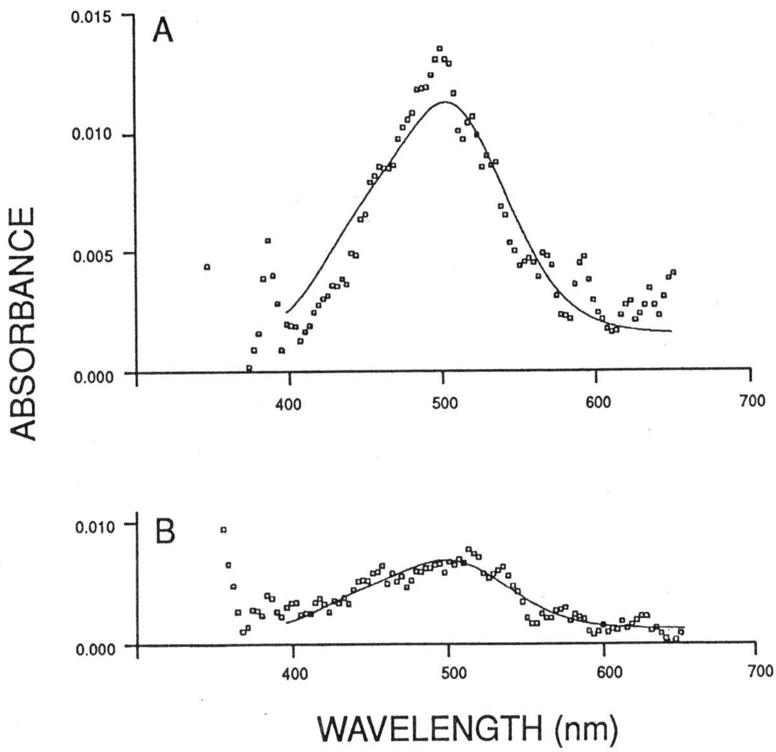

Figure 9. Absorption spectra of rhodopsin taken from opposite ends of the same individual ROS from a day-5 200-lux adventive. The absorbance is greater at the tip of the ROS, top panel (A), than at the base, bottom panel (B), reflecting the greater packing density in 3-lùx discs (tip) than in 200-lux discs (base).

curred in the renewal process during the change. It had been previously established that ROS length changes when animals are exposed to a light intensity different from that of their native environment (Battelle and LaVail, 1978). In the present study, we investigated how changes in ROS length were achieved by the rod cell, both when the animal was moved from low- to high-intensity light and when it was moved from high- to low-intensity light.

We found that the pattern of ROS disc shedding is modified in the latter group but not in the former; shedding decreased only on the first day in the more intense light. In a previous study, in rats switched to more intense light, phagosome degradation increased on day 1 (Baker *et al.,* 1986). Our results may therefore reflect not a decrease in shedding but rather an increase in the rate of phagosome degradation in the RPE. After day 1, the 200-lux adventives showed their usual shedding pattern, yet ROS length decreased, suggesting that disc addition had diminished. Future autoradiographic studies may reveal whether this is the case.

In animals moved to less intense light, shedding remained unchanged during the first 3 days, but ROS length decreased, perhaps because discs were added at a lower rate. After day 3, however, the ROSs began to lengthen, and shedding decreased. Lower numbers of phagosomes in the RPE of these animals resulted from decreased shedding per rod and not from a decrease in the number of cells, because no cell loss was evident within the first 7 days in the dim light.

Disc addition and disc shedding are not necessarily coupled and can be separately regulated in the rat. The same is true in mice during the development of the rod cells (LaVail, 1973) and in amphibians held in different light conditions (Besharse *et al.*, 1977a, b). The present study demonstrates that changes in shedding occur in adult albino rats in response to changes in environmental light intensity and shows indirectly that disc addition may do so as well.

Changes in ROS length do not fully account for the measured difference in rhodopsin levels between animals reared at 3 lux and at 200 lux, so it has been suggested that packing density of rhodopsin per disc also depends on the light environment (Organisciak and Noell, 1977; Battelle and LaVail, 1978; Penn and Williams, 1986; Penn and Anderson, 1992). Results from the present study verify that indeed rhodopsin packing is different in rods of animals reared at 3 lux and at 200 lux. First, animals reared at the two different intensities synthesize different amounts of rhodopsin per day (0.28 μCi per two retinas at 200 lux, 0.55 μCi per two retinas at 3 lux, a difference of 49%). Second, our microspectrophotometric data showed that ROSs constructed at 200 lux contained 45% less rhodopsin than ROSs constructed at 3 lux.

The rapid drop in synthesis, to 0.38 μCi per two retinas, on day 10 in animals moved to the less intense light is largely explained by the 25% loss of rod cells between day 7 and day 21. If the remaining cells continued their new rate of rhodopsin synthesis, cell loss alone could lower synthesis to 0.40 μCi per two retinas. After only 3 days, the incorporation values of animals moved to more intense light had decreased to a level equivalent to that of animals reared there. By day 21, it had decreased to 0.23 μCi per two retinas. This value is far lower than the calculated value of 0.46 μCi per two retinas that could be accounted for by the 17% cell loss, indicating that net synthesis of rhodopsin in the remaining cells had decreased.

By day 28 in the new light, the 200-lux adventive value was 0.18 μCi per two retinas, and the 3-lux adventive value was 0.33 μCi per two retinas. Although these values are lower than those of 200- and 3-lux natives, they differ by about the same percentage (45%) as do the two native groups (49%), so animals are able to adjust rhodopsin net synthesis when they are moved to a new light intensity.

Perhaps our most striking result was direct demonstration that packing density of rhodopsin in individual ROS discs changes when animals are moved from one light intensity to another. Microspectrophotometry revealed that, in animals moved from low-intensity to higher-intensity light and tested after 5 days (one half of a turnover period), discs at an ROS tip differed in transverse absorbance from those at the base of the same individual ROS. Further, the absorbance value at the base matched that of an animal reared at the lower intensity, and the value at the tip matched that of an animal reared at the higher one. Differences in ROS diameter cannot explain these results; the diameters at the tip and base would have to differ by at least a factor of two, and no such change in diameter was seen.

Preliminary studies using 8 lux and 400 lux (Williams *et al.*, 1988) indicated that rats moved from one light regime to another, once adapted to the new light, catch the same number of photons per animal per day (approximately 1×10^{16}) as those native to the new light regime. The present study has shown that all aspects of the renewal process, including membrane synthesis and degradation (disc addition and disc shedding), rhodopsin synthesis, and rhodopsin incorporation into discs contribute to modification of the rod cell in response to the light environment. ROS renewal is the ongoing mechanism that allows rod cells to vary their ability to catch photons, so we can say with confidence that renewal is one of the underlying mechanisms for maintaining photostasis.

Photoreceptor renewal is not the only process regulated by the light environment. Changes in antioxidant levels and polyunsaturated fatty acid content of ROS membranes indicate that an intricate combination of retinal processes are affected and regulated by the animals' light environment (Penn and Anderson, 1992). Photoreceptor physiology may also be governed by the light environment (Penn *et al.,* 1989). Studies using pigmented rats (Williams *et al.,* 1994) and invertebrates (Hoenig and Chamberlain, 1989) will determine whether photostasis is a general phenomenon. Because results so far indicate that photostasis occurs in both vertebrates and invertebrates, it may play a significant role at a fundamental level of photoreceptor physiology and perhaps influence the physiology of the animal as a whole.

REFERENCES

Adamus G., Zam, Z. S., Arendt, A., Palczewski, K., McDowell, J. H., and Hargrave, P. A. (1991). Anti-rhodopsin monoclonal antibodies of defined specificity: characterization and application. *Vision Research* 31:17–31.

Anderson, D. H., and Fisher, S. K. (1976). The photoreceptors of diurnal squirrels: outer segment structure, disc shedding and protein renewal. *Journal of Ultrastructure Research* 55:119–141.

Andrews, L. D., and Cohen, A. I. (1983). Freeze-fracture studies of photoreceptor membranes: new observations bearing upon the distribution of cholesterol. *Journal of Cell Biology* 97:749–755.

Bairati, A., Jr., and Orzalesi, N. (1963). The ultrastructure of the pigment epithelium and of the photoreceptor-pigment epithelium junction in the human retina. *Journal of Ultrastructure Research* 9:484–496.

Baker, B. N., Moriya, M., and Williams, T. P. (1986). Alteration of disk-shedding patterns by light-onset of higher than normal intensity. *Experimental Eye Research* 42:535–546.

Battelle, B.-A., and LaVail, M. M. (1978). Rhodopsin content and rod outer segment length in albino rat eyes: modification by dark adaptation. *Experimental Eye Research* 26:487–497.

Baylor, D. A., Lamb, T. D., and Yau, K.-W. (1979). Responses of retinal rods to single photons. *Journal of Physiology (London)* 288:613–634.

Besharse, J. C., and Hollyfield, J. G. (1979). Turnover of mouse photoreceptor outer segments in constant light and darkness. *Investigative Ophthalmology and Visual Science* 18:1019–1024.

Besharse, J. C., Hollyfield, J. G., and Rayborn, M. E. (1977a). Turnover of rod photoreceptor outer segments. II. Membrane addition and loss in relationship to light. *Journal of Cell Biology* 75:507–527.

Besharse, J. C., Hollyfield, J. G., and Rayborn, M. E. (1977b). Photoreceptor outer segments: accelerated membrane renewal in rods after exposure to light. *Science* 196:536–538.

Blest, A. D. (1978). The rapid synthesis and destruction of photoreceptor membrane by a dinopid spider: a daily cycle. *Proceedings of the Royal Society of London, Series B, Biological Sciences* 200:463–483.

Broekhuyse, R. M., Tolhuizen, E. F. J., Janssen, A. P. M., and Winkes, H. J. (1985). Light induced shift and binding of S-antigen in retinal rods. *Current Eye Research* 4:613–618.

Cohen, A. I. (1968). The ultrastructure of the rods of the mouse retina. *American Journal of Anatomy* 107:23–48.

Curcio, C. A., Millican, C. L., Allen, K. A., and Kalina, R. E. (1993). Aging of the human photoreceptor mosaic: evidence for selective vulnerability of rods in central retina. *Investigative Ophthalmology and Visual Science* 34:3278–3296.

Daemen, F. J. M. (1973). Vertebrate rod outer segment membranes. *Biochimica et Biophysica Acta* 300:255–288.

Droz, B. (1963). Dynamic condition of protein in the visual cells of rats and mice as shown by radiography with labeled amino acids. *Anatomical Record* 145:157–167.

Farber, D. B., Seager-Danciger, J. S., and Organisciak, D. T. (1991). Levels of mRNA encoding proteins of the cGMP cascade as a function of light environment. *Experimental Eye Research* 53:781–786.

Finean, J. B., Coleman, R., and Michell, R. H. (1978). *Membranes and Their Cellular Functions,* 2nd edition. Halstead Press, New York.

Goldman, A. I., Teirstein, P. S., and O'Brien, P. J. (1980). The role of ambient lighting in circadian disc shedding in the rod outer segment of the rat retina. *Investigative Ophthalmology and Visual Science* 19:1257–1267.

Hall, M. O., Bok, D., and Bacharach, A. D. E. (1968). Visual pigment renewal in the mature frog retina. *Science* 161:787–789.

Hoenig, H., and Chamberlain, S. C. (1989). Testing the "photostasis" hypothesis in the *Limulus* lateral eye. *Investigative Ophthalmology and Visual Science Supplement* 30:291.

Hollyfield, J. G. (1979). Membrane addition to photoreceptor outer segments: progressive reduction in the stimulatory effect of light with increased temperature. *Investigative Ophthalmology and Visual Science* 18:977–981.

Hollyfield, J. G., and Rayborn, M. E. (1982). Membrane assembly in photoreceptor outer segments: progressive increase in "open" basal discs with increased temperature. *Experimental Eye Research* 34:115–119.

Hollyfield, J. G., Besharse, J. C., and Rayborn, M. E. (1976). The effect of light on the quantity of phagosomes in the pigment epithelium. *Experimental Eye Research* 23:623–635.

Hollyfield, J. G., Rayborn, M. E., Verner, G. E., Maude, M. B., and Anderson, R. E. (1982). Membrane addition to rod photoreceptor outer segments: light stimulates membrane assembly in the absence of increased membrane biosynthesis. *Investigative Ophthalmology and Visual Science* 22:417–427.

Hornstein, E. P., and Chamberlain, S. C. (1991). Correlation of photoreceptor structure and lighting environment: implications for photostasis. *Society for Neuroscience Abstracts* 17:298.

Ishikawa, T., and Yamada, E. (1970). The degradation of the photoreceptor outer segments within the pigment epithelial cell of the rat retina. *Journal of Electron Microscopy* 19:85–91.

Kaplan, M. W., Deffebach, M. E., and Liebman, P. A. (1978). Birefringence measurements of structural inhomogeneities in *Rana pipiens* rod outer segments. *Biophysical Journal* 23:59–69.

Kinney, M. S., and Fisher, S. K. (1978). The photoreceptors and the pigment epithelium of adult *Xenopus* retina: morphology and outer segment renewal. *Proceedings of the Royal Society of London* 201:131–147.

LaVail, M. M. (1973). Kinetics of rod outer segment renewal in the developing mouse retina. *Journal of Cell Biology* 58:650–661.

LaVail, M. M. (1976). Rod outer segment disc shedding in relation to cyclic lighting. *Experimental Eye Research* 23:277–280.

LaVail, M. M. (1980). Circadian nature of rod outer segment disc shedding in the rat. *Investigative Ophthalmology and Visual Science* 19:407–411.

McGinnis, J. F., Whelan, J. P., and Donoso, L. A. (1992). Transient, cyclic changes in mouse visual cell gene products during the light-dark cycle. *Journal of Neuroscience Research* 31:584–590.

MacNichol, E. F., Jr. (1978). A photon-counting MSP for the study of single vertebrate photoreceptor cells. In: Cool, S. J., and Smith, E. L., eds. *Frontiers of Visual Science.* Springer-Verlag, Berlin. pp. 194–208.

Moody, M. F., and Robertson, J. D. (1960). The fine structure of some retinal photoreceptors. *Journal of Biophysical and Biochemical Cytology* 7:87–91.

Nathans, J., and Hogness, D. S. (1983). Isolation, sequence analysis and intron-exon arrangement of the gene encoding bovine rhodopsin. *Cell* 34:807–814.

Noell, W. K., and Albrecht, R. (1971). Irreversible effects of visible light on the retina: role of vitamin A. *Science* 192:76–80.

O'Brien, P. J. (1993). *In vivo* labeling of retinal components by intraocular injection. *Methods in Neuroscience* 15:75–85.

Organisciak, D. T., and Noell, W. K. (1977). The rod outer segment phospholipid/opsin ratio of rats maintained in darkness or cyclic light. *Investigative Ophthalmology and Visual Science* 16:188–190.

Organisciak, D. T., Xie, A., Wang, H.-M., Jiang, Y.-L., Darrow, R. M., and Donoso, L. A. (1991). Adaptive changes in visual cell transduction protein levels: effect of light. *Experimental Eye Research* 53:773–779.

Papermaster, D. S., and Schneider, B. G. (1982). Biosynthesis and morphogenesis of outer segment membranes in vertebrate photoreceptor cells. In: McDevitt, D. S., ed. *Cell Biology of the Eye.* Academic Press, New York. pp. 475–531.

Penn, J. S., and Anderson, R. E. (1987). Effect of light history on rod outer segment membrane composition in the rat. *Experimental Eye Research* 44:767–778.

Penn, J. S., and Anderson, R. E. (1992). Effects of light history on the rat retina. In: Osborne, N. N., and Chader, G. J., eds. *Progress in Retinal Research, Volume 11.* Pergamon Press, New York. pp. 75–98.

Penn, J. S., and Williams, T. P. (1986). Photostasis: regulation of daily photon-catch by rat retinas in response to various cyclic illuminances. *Experimental Eye Research* 43:915–928.

Penn, J. S., Baker, B. N., Howard, A. G., and Williams, T. P. (1985). Retinal light-damage in albino rats: lysosomal enzymes, rhodopsin and age. *Experimental Eye Research* 41:275–284.

Penn, J. S., Thum, L. A., and Naash, M. I. (1989). Photoreceptor physiology in the rat is governed by the light environment. *Experimental Eye Research* 49:205–215.

Penn, J. S., Tolman, B. L., Thum, L. A., and Koutz, C. A. (1992). Effect of light history on the rat retina: time course of morphological adaptation and readaptation. *Neurochemical Research* 17:91–99.

Philp, N. J., Chang, W., and Long, K. (1987). Light-stimulated movement in rod photoreceptor cells of the rat retina. *FEBS Letters* 225:127–131.

Rapp, L. M., and Williams, T. P. (1980). A parametric study of retinal light damage in albino and pigmented rats. In: Williams, T. P., and Baker, B. N., eds. *The Effects of Constant Light on Visual Processes.* Plenum Press, New York. pp. 135–159.

Rapp, L. M., Fisher, P. L., and Dhindsa, H. S. (1993). Role of outer segment renewal in photoreceptor cell recovery from light damage. *Investigative Ophthalmology and Visual Science* 34 (Suppl):1434.

Roof, D., Adamian, M., Jacobs, D., and Hayes, A. (1991). Cytoskeletal specializations at the rod photoreceptor distal tip. *Journal of Comparative Neurology* 305:289–303.

Schremser, J.-L., and Williams, T. P. (1992). Photoreceptor plasticity in the albino rat retina following unilateral optic nerve section. *Experimental Eye Research* 55:393–399.

Schremser, J.-L., and Williams, T. P. (1995a). Rod outer segment (ROS) renewal as a mechanism for adaptation to a new intensity environment. I. Rhodopsin levels and ROS length. *Experimental Eye Research* 61:17–24.

Schremser, J.-L., and Williams, T. P. (1995b). Rod outer segment (ROS) renewal as a mechanism for adaptation to a new intensity environment. II. Rhodopsin synthesis and packing density. *Experimental Eye Research* 61:25–32.

Sjöstrand, F. S. (1953). The ultrastructure of the outer segment of rods and cones of the eye as revealed by the electron microscope. *Journal of Cellular and Comparative Physiology* 42:15–70.

Spitnas, M., and Hogan, M. J. (1970). Outer segments of photoreceptors and the pigment epithelium. Archives of Ophthalmology 4:810–819.

Steinberg, R. H., Fisher, S. K., and Anderson, D. H. (1980). Disk morphogenesis in vertebrate photoreceptors. *Journal of Comparative Neurology* 190:501–518.

Teirstein, P. S., Goldman, A. I., and O'Brien, P. J. (1980). Evidence for both local and central regulation of rat rod outer segment disc shedding. *Investigative Ophthalmology and Visual Science* 19:1268–1273.

Whelan, J. P., and McGinnis, J. P. (1988). Light-dependent subcellular movement of photoreceptor proteins. *Journal of Neuroscience Research* 20:263–270.

White, R. H., and Lord, E. (1975). Diminution and enlargement of the mosquito rhabdom in light and darkness. *Journal of General Physiology* 65:583–598.

Williams, T. P. (1984). Some properties of old and new rhodopsin in single *Bufo* rods. *Journal of General Physiology* 83:841–852.

Williams, T. P., Penn, J. S., Bush, R. A., and Makino, C. L. (1988). Renewal of rod outer segment and regulation of daily photon catch by the rat retina. In: *Proceedings of the Yamada Conference XXI.* Yamada Science Foundation, Kyoto, Japan. pp. 255–260.

Williams, T. P., Baker, B. N., and Dodge, J. (1994). Pigmented rats, pupillary reflex and photostasis. *Investigative Ophthalmology and Visual Science* 35 (Suppl.):1608.

Young, R. W. (1967). The renewal of photoreceptor outer segments. *Journal of Cell Biology* 33:61–72.

Young, R. W. (1971a). The renewal of rod and cone outer segments in the rhesus monkey. *Journal of Cell Biology* 49:303–318.

Young, R. W. (1971b). Shedding of discs from rod outer segments in the rhesus monkey. *Journal of Ultrastructure Research* 34:190–203.

Young, R. W., and Bok, D. (1969). Participation of the retinal pigment epithelium in the rod outer segment renewal process. *Journal of Cell Biology* 42:392–403.

6

ENVIRONMENTAL LIGHT AND AGE-RELATED CHANGES IN RETINAL PROTEINS

D. T. Organisciak, R. M. Darrow, R. A. Darrow, and L. A. Lininger*

The Petticrew Research Laboratory
Department of Biochemistry and Molecular Biology
Wright State University
Dayton, Ohio 45435

1. INTRODUCTION

Light-induced photoreceptor-cell degeneration is now widely studied as an *in vivo* model of cellular injury, hereditary and age-related visual-cell dysfunction, and cellular death. Like other multifactorial tissue pathologies, retinal light damage involves a complex series of events modulated by environmental light history, genetic inheritance, age, and the nature of the light-exposure protocol (Noell *et al.,* 1966; Noell and Albrecht, 1971; O'Steen *et al.,* 1974; Williams and Howell, 1983; LaVail *et al.,* 1987; Penn and Anderson, 1987; for reviews, see Organisciak and Winkler, 1994; Rapp, 1995). Although death of photoreceptor cells appears to involve both necrotic and apoptotic processes (Edward *et al.,* 1991; Shahinfar *et al.,* 1991; Organisciak *et al.,* 1995; Remé *et al.,* 1995; Abler *et al.,* 1996), the mechanism of retinal light damage is still unknown. Cellular oxidation and adaptation have each been implicated as the underlying reason for retinal light damage.

1.1. Oxidation as a Causative Factor in Light Damage

Much of the evidence suggesting that oxidation occurs in light damage comes from studies with natural or synthetic antioxidants. In 1974, Hayes described macular degeneration in monkeys fed a vitamin E deficient diet for more than 2 years. He attributed the disruption of rod outer segment (ROS) membranes to lipid peroxidation (Hayes, 1974). Subsequently, Farnsworth and Dratz (1976) showed that vitamin E did in fact prevent the oxidative loss of ROS polyunsaturated fatty acids *in vitro,* but studies with vitamin-E- and selenium-deficient rats were less conclusive in that light damage was not enhanced in the deficient animals (Stone *et al.,* 1979b). This finding has been both confirmed (Katz and Eldred, 1989) and disputed (Kagan *et al.,* 1981).

* Present address: Naval Training Center, Great Lakes, Chicago, Illinois 60088.

Photostasis and Related Phenomena, edited by Williams and Thistle
Plenum Press, New York, 1998

The first definitive reports of an antioxidative protective effect in acute intense-light retinal damage involved rats given ascorbic acid before light treatment. Ascorbate-treated animals exhibited greater preservation of the visual-cell protein rhodopsin, as well as overall cell morphology, than did rats not supplemented with ascorbate (Organisciak *et al.*, 1984, 1985; Li *et al.*, 1985). Furthermore, vitamin C may reduce oxidized vitamin E during intense light exposure, thereby preventing its loss from retina (Hunt *et al.*, 1984; Wiegand *et al.*, 1986; Penn *et al.*, 1987; Stoyanovsky *et al.*, 1995). Using similar approaches, other studies with synthetic antioxidants, including dimethylthiourea (DMTU) (Lam *et al.*, 1990; Organisciak *et al.*, 1992b) and WR-77913, a radioprotective dye (Remé *et al.*, 1991), or drugs such as flunarizine, a calcium overload blocker (Edward *et al.*, 1991), and the unnatural D-stereoisomer of L-ascorbic acid (Organisciak *et al.*, 1991a, 1992a) all showed protection against retinal light damage.

Stimulated, perhaps, by the unusually high concentration of polyunsaturates in ROS membrane lipids (Anderson and Maude, 1972; Daemen, 1973; Stone *et al.*, 1979a; Fliesler and Anderson, 1983), a parallel line of investigations involving docosahexaenoic acid (DHA) was conducted. Wiegand *et al.* (1983) first showed that DHA levels decreased in ROS from rats exposed to intense light for extended periods, while retinal peroxide levels increased. Similarly, Penn and Anderson (1987) showed that ROS DHA levels were lower in rats reared in bright light than in others reared in dim light. In addition, rats fed diets deficient in the essential fatty acid linolenic acid exhibited a decrease in ROS DHA levels (Anderson and Maude, 1972) and a decrease in the extent of retinal light damage (Organisciak *et al.*, 1987; Bush *et al.*, 1991). Furthermore, when rats are treated with antioxidants, the loss of ROS DHA is less than in light-exposed rats not given the antioxidants (Organisciak et al., 1985, 1987, 1992b).

De La Paz and Anderson (1992) reported that hydrophobic radicals, derived from compounds such as linoleic acid hydroperoxide, but not hydroxyl radicals, were effective in promoting lipid peroxidation in isolated ROS. Weigand *et al.* (1983) found an increase in UV absorbance in ROS isolated from long-term light-exposed rats. Organisciak *et al.* (1983) estimated that nanomolar quantities of organic hydroperoxides were present in rat retinas following short light-exposure periods, but recent work using a purified phospholipid hydroperoxide glutathione peroxidase indicates that lipid hydroperoxides probably never exceed 1.0 nmol/retina (Darrow *et al.*, 1994). Glycolytic enzymes, as potential markers of oxidation in ROS, have been shown to be partially inhibited after intense light exposure (Organisciak and Winkler, 1994). Recently, Darrow *et al.* (1997) demonstrated that the activity of all-*trans* retinol dehydrogenase (*t*-RDH), an oxidatively sensitive membrane-bound ROS enzyme (Zimmerman *et al.*, 1975), was preserved in light-exposed rats pretreated with DMTU, but glutamine synthetase (GS), another oxidatively sensitive enzyme located in Müller cells (Burns and Robles, 1990), was unaffected by intense light. Accordingly, the evidence strongly suggests that light-induced oxidation occurs in the photoreceptor cell, especially in the ROS organelle, although the type of oxidant (or free radical) and quantitative measures remain elusive.

1.2. Adaptive Mechanisms as a Basis for Retinal Damage

Other evidence points to a series of photoreceptor-cell adaptive changes or responses as the important variable in determining the outcome of acute intense light exposure. Noell and Albrecht (1971) first suggested that long-term rearing environment was a critical factor contributing to the greater susceptibility to light damage of dark-reared rats than of those reared in weak cyclic light. Birch and Jacobs (1980) found a similar relationship between rearing environment and light damage in rats ascribable to photopic and scotopic mechanisms. In earlier studies the rhodopsin content of the retina, rhodopsin packing in ROS, and ROS length were found to be greater in rats reared in darkness than in those reared in weak cyclic light (Organisciak and Noell, 1977). Battelle and LaVail (1978) also found an increase in rhodopsin in the retinas of dark-reared rats, attributable to an increase in ROS length, and in rats changed from cyclic light to darkness, to an increase in rhodopsin packing density in nascent ROS disks. More recently, the levels of other visual-cell transduction proteins, and their mRNAs, have been found to change in response to changes in rearing environments. In rats transferred from a weak cyclic light environment to darkness, α-transducin as well as rhodopsin levels increased and S-antigen (arrestin) declined; the converse was true in rats changed from darkness to weak cyclic light (Farber *et al.*, 1991; Organisciak *et al.*, 1991b).

In their now classic paper, Penn and Williams (1986) described the ability of retinal photoreceptors to adapt metabolically to environmental light so as to regulate the total number of photons absorbed each day. According to the concept of "photostasis," rhodopsin content can be altered by changes in the number of ROS disk membranes, in ROS length, in rhodopsin packing, or in the rate of regeneration of the bleached visual pigment (Penn and Williams, 1986). Penn and Anderson (1987) examined ROS membranes in rats reared from birth at various intensities of cyclic light. Animals reared in dim (5-lux) light had a higher rhodopsin level, a higher content of polyunsaturated fatty acids in their ROS, and a greater rhodopsin packing density in ROS membranes than did others reared in brighter (800-lux) light (Penn and Anderson 1987). More recently, Schremser and Williams (1995a, b) reported that ROS length increases in rats moved from a 200-lux rearing environment to a 3-lux environment and decreases in those undergoing the opposite change. Using intraocular injections of [^{3}H] amino acids, phagosome counts in retinal pigment epithelial cells, and microspectrophotometry, they concluded that basal disk addition and the shedding and phagocytosis of apical ROS tips can be uncoupled. This finding then could explain the alterations in ROS length and rhodopsin packing density they found (Schremser and Williams, 1995b) and supports the concept of photostasis.

In another study Penn *et al.* (1987) found that retinal light damage was greater in rats reared in a low-light environment than in others reared in a higher-light environment. In these rats damage was inversely related to the level of the antioxidants α-tocopherol and ascorbic acid and to the activities of glutathione-based antioxidative enzymes, both of which were elevated in the high-light-environment rats. Organisciak *et al.* (1991b) found that susceptibility to light damage increased after rats originally reared in weak cyclic light spent additional time in darkness. The time course of damage was correlated with increases in rhodopsin and α-transducin and a simultaneous decrease in S-antigen levels. Similarly, Rapp *et al.* (1990) found an age-dependent increase in light damage in rats, based on greater rhodopsin levels/ROS length.

The purpose of the present study was to examine changes in retinal proteins or enzymes and susceptibility to light damage in rats as a function of light environment and age. To address the question of whether older animals, like younger ones, are able to adapt to their light environment, we changed rats from weak cyclic light to darkness and vice versa. We also studied ROS membrane lipid synthesis in rats reared in weak cyclic light or darkness.

2. MATERIALS AND METHODS

2.1. Animal Maintenance, Light Environment, and Exposure

The animals used in these studies were male albino Sprague Dawley rats (Harlan, Inc., Indianapolis, IN), purchased as weanlings and kept in their respective rearing environments until the age of 2 or 8 months. The animals were obtained at different times so that rats of the two ages could be studied simultaneously. All rats were maintained in either a dim cyclic light environment (hereafter called light-reared) or darkness (dark-reared), fed *ad libitum,* and given free access to water. The dim light environment consisted of 12 hours of white light per day (on 08:00, off 20:00), maintained at 20–30 lux by multiple 7-W night lights run at less than line voltage. Dark-reared rats were exposed to red light >600 nm for less than 30 minutes/day for cage cleaning and feeding.

Some of the 2- or 8-month-old rats were changed from their original rearing environment to the opposite environment and maintained there for as long as 16 days. During this period some rats were sacrificed for rhodopsin analysis or for gel electrophoresis of retinal proteins. After 16 days other rats were exposed to intense visible light, for determination of their susceptibility to retinal light damage, then placed in darkness for 2 weeks before sacrifice for rhodopsin measurements. All rats were killed in a CO_2-saturated chamber. Animals in this study were used in accord with the ARVO resolution on the use of animals in research.

The intense light exposure consisted of continuous light for periods of 4 (for dark-reared rats) or 8 hours (for light-reared rats) in green Plexiglas chambers, which transmit 490- to 580-nm light (Noell *et al.,* 1966). Light intensity was 1200–1500 lux during exposures. The light exposure periods were chosen to induce about 50% photoreceptor cell loss (dark-reared) and no cell loss (light-reared). Accordingly, the degree to which a dark-reared rat adapted to its new cyclic-light environment could be estimated by the degree of its resistance to an 8-hour intense-light exposure. Conversely, a light-reared rat moved to darkness would be expected to acquire an increased susceptibility to retinal damage from only a 4-hour light exposure. In each case, rats from the original rearing environment were also exposed for comparable periods. Others were placed into darkness for 2 weeks to serve as unexposed controls.

2.2. Rhodopsin, S-Antigen and α-Transducin Analysis

Rhodopsin in the excised eyes of rats was determined by spectrophotometric analysis of Emulphogene BC-720 detergent extracts. The visual-cell protein was quantitated from the difference between absorbances at 500 nm before and after bleaching of the extracts (Delmelle *et al.,* 1975).

The levels of S-antigen and α-transducin were detected by Western analysis of retinal homogenates as previously described (Organisciak *et al.,* 1991b). In these studies anti S-antigen (C 10 C 10), a gift from L. A. Donoso, and anti α-transducin (TF-15) from B. Fung were used simultaneously on the same transblot. Control experiments established that neither antibody cross reacted with proteins other than its specific antigen. For staining, an alkaline-phosphatase-conjugated second antibody was used with 5-bromo-4-chloro-3 indolyl phosphate and nitro-blue tetrazolium. Quantitation was accomplished by scanning with a Fotodyne CCD camera and use of Collage Image Analysis Software.

2.3. ROS Membrane Synthesis

Two-month-old rats, previously maintained in either dim cyclic light or darkness, were injected intravitreally in one eye with 10 μCi [1,2–^{14}C]sodium acetate (116 mCi/mmol; ICN Pharmaceuticals, Irvine, CA) for labeling of retinal lipids. After injection, these animals were returned to their respective rearing environments until their sacrifice 4 or 8 days later. Their eyes were dissected under dim red light, and ROS were prepared from homogenates of four retinas from each group of animals. Purified band I ROS were isolated and removed after discontinuous sucrose gradient centrifugation (Organisciak *et al.,* 1991b). Lipids were extracted into chloroform:methanol (2:1), and specific activities were determined for ROS membrane lipids by means of a lipid phosphorus spectrophotometric assay. For determination of adaptive changes, some cyclic-light-reared animals were injected intravitreally with isotope and then transferred to darkness for periods of 4 or 8 days.

2.4. Oxidatively Sensitive Retinal Enzymes

Individual retinas, previously stored in liquid N_2 or at $-80°C$, were homogenized at 0°C in 0.5 ml of an isotonic salt mixture buffered with 10 mM HEPES at pH 7.4, containing protease inhibitors. We determined *t*-RDH activity by converting all-*trans* retinal (*t*-retinal) to [^{3}H]all-*trans* retinol (*t*-retinol), using [^{3}H]NADPH, as previously described (Saari *et al.,* 1993; Darrow *et al.,* 1997). The [^{3}H]*t*-retinol was partitioned into petroleum ether for determination of radioactivity, and picomoles of product calculated on the assumption that the specific radioactivity of the [^{3}H]*t*-retinol formed was the same as that of the [^{3}H]NADPH added.

GS activity was determined by the γ-glutamyl transferase reaction (Chader, 1971; Thorndike and Reif-Lehrer, 1971; Darrow *et al.,* 1997); product was quantitated from the absorbance at 500 nm of authentic γ-glutamyl hydroxamate.

2.5. Antioxidative Enzymes

For catalase and glutathione peroxidase (GPX), single retinas were disrupted at 0°C in 0.5 ml of 0.04 M Tris HCl, pH 7.8, containing 0.2 mM sodium diethylenetriamine-pentaacetate (DTPA) and protease inhibitors, by intermittent probe sonication for three 10-second periods. Fifty µl of 10% octyl glucoside was then added, and the clarified sonicate kept at 0–4°C for the determination of enzymatic activity. Catalase activity was determined by the decrease in absorbance of H_2O_2 at 240 nm at 25°C (Aebi, 1984) over that of a control where catalase was inhibited with 0.01% sodium azide. For comparison with other enzymes, we expressed the activity in nanomoles per minute, by multiplying the experimentally determined rate constant by the concentration of H_2O_2 used in the GPX assay.

GPX was determined at 37°C from the decrease in absorbance at 340 nm resulting from the net oxidation of NADPH, using 0.175 mM H_2O_2 as substrate, in the presence of excess glutathione reductase (Flohé and Günzler, 1984), and quantitated from the extinction coefficient of NADPH. Because H_2O_2 oxidizes glutathione nonenzymatically at a significant rate under our assay conditions, the decrease in absorbance in the absence of enzyme was measured and subtracted from that observed in the complete system.

We also localized the activities of the various enzymes in the retina by measuring their activities in isolated ROS and in whole-retina homogenates from 2-month-old rats and in retinal homogenates from 9-month-old normal and Royal College of Surgeons (RCS) dystrophic rats, which lack photoreceptor cells (Delmelle *et al.,* 1975).

3. RESULTS

3.1. Rhodopsin as a Function of Rearing Environment and Age

To determine whether 8-month-old (older) rats could adapt to a new rearing environment as well as 2-month-old (young) animals could, we changed young and older light-reared rats to darkness and vice versa. After various periods in their new environments,

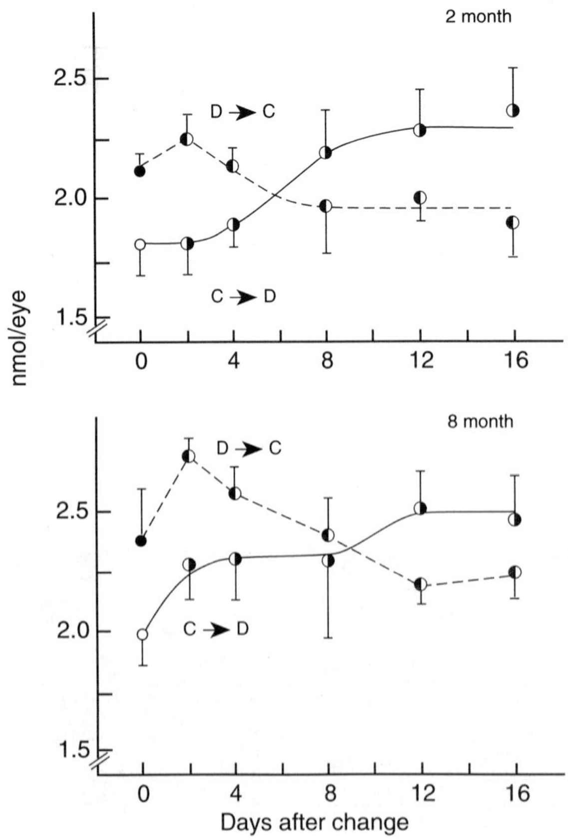

Figure 1. Rhodopsin in rats changed from one rearing environment to another. Rats reared in weak cyclic light (○) or darkness (●) were changed to the opposite condition for up to 16 days. (◐) dark to cyclic light; (◑) cyclic light to darkness. Results are the mean ± SD for n = 4 pairs of eyes from 2-month- or 8-month-old rats.

animals were killed for rhodopsin measurements. As expected, young light-reared rats made a smooth transition to darkness by increasing their rhodopsin content (Fig. 1). In these rats, rhodopsin increased from 1.85 to 2.30 nmol/eye within 16 days. The half-time of the change appeared to be 5 days, or one-half of the 10-day time period normally required for complete ROS turnover. Similarly, young dark-reared rats changed to the cyclic-light environment decreased their rhodopsin content from 2.15 to 1.95 nmol/eye over the 16-day period. Once again the half-time of the change appeared to be 5 days.

The transitions to new rearing environments in older rats were different from those in the young animals. Light-reared rats changed to darkness increased their rhodopsin content from 2.0 to 2.5 nmol/eye within 16 days. The increase in rhodopsin, although almost the same as that in young rats, appeared to occur with a half-time of 8 days. In older rats changed from darkness to weak cyclic light, the change in rhodopsin was only 0.1 nmol, or 2.35 to 2.25 nmol/eye within 16 days. Furthermore, a transient increase in rhodopsin occurred 2 days after the change to cyclic light, suggesting that the normally synchronous process of ROS synthesis and degradation in older rats was disrupted.

3.2. Changes in Rearing Environment and Retinal Light Damage

On the basis of known differences in light-damage susceptibility between dark- and light-reared rats, we exposed young and older rats to intense visible light. Young dark-

Table 1. Rhodopsin in rats changed from one environment to another (nmol/eye)[a]

Age (months)	Light exposure	Cyclic-light reared	Dark reared	Moved from dark to light	Moved from light to dark
2	0 hr	2.06 ± 0.11 (4)	2.25 ± 0.14 (4)	1.96 ± 0.18 (9)	2.26 ± 0.15 (9)
	4 hr	—	0.91 ± 0.31 (8)	—	0.94 ± 0.28 (5)
	8 hr	2.03 ± 0.17 (9)	—	1.57 ± 0.55 (5)	—
% of control		98.5	40.4	80.1	41.6
8	0 hr	2.16 ± 0.03 (4)	2.36 ± 0.22 (4)	2.27 ± 0.19 (9)	2.45 ± 0.19 (9)
	4 hr	—	0.86 ± 0.26 (8)	—	1.21 ± 0.30 (5)
	8 hr	1.69 ± 0.20[b] (9)	—	1.48 ± 0.56 (5)	—
% of control		78.2	36.4	65.2	49.4

[a]Results are the mean ± SD for (n) pairs of eyes from 2- and 8-month-old rats maintained in their original rearing environment or changed to the opposite environment 16 days before intense light exposure. Rhodopsin measured 2 weeks after intense light treatment.
[b]Value significantly lower than for 2-month-old cyclic-light-reared rats $p < 0.005$.

reared rats lost approximately 50% of their retinal rhodopsin after 4 hours of intense light; light-reared rats are resistant to damage from up to 8 hours of continuous exposure. Accordingly, the ability of a light-reared rat to adapt to darkness can be determined by measurement of rhodopsin recovery after a 4-hour light exposure. Similarly dark to cyclic-light adaptation can be determined by exposure of rats to 8 hours of intense light.

As shown in Table 1, young light-reared rats exposed to intense light retained 98% of the rhodopsin content found in unexposed animals, whereas dark-reared rats retained only 40% of the rhodopsin found in their unexposed controls. When dark-reared rats were changed to cyclic light for 16 days before an 8-hour intense light exposure, they recovered an average of 80% of the control rhodopsin level, but the range of values seems to show that the transition was complete for some rats but incomplete for others. Alternatively, rats changed from cyclic light to darkness for 16 days exhibited nearly the same rhodopsin recovery (42%) as rats always reared in darkness. In young rats, therefore, the change from dark to cyclic light yielded variable and incomplete adaptation, whereas light-reared rats appeared to make a complete transition to the dark environment.

Older light-reared rats retained only 78% rhodopsin after an 8-hour light exposure; young rats retained 98% ($p < 0.005$). This result indicates that older light-reared rats are more susceptible to light damage than are young ones. Conversely, older dark-reared rats appear to be no more sensitive to light damage than are young dark-reared rats; the older rats retained 36% rhodopsin, compared to 40% for the young ones. Older rats changed from cyclic light to darkness recovered 49% of rhodopsin, a value 7% greater than predicted from the 42% found in young ones. Similarly, older rats changed from dark to cyclic light recovered 65% rhodopsin, or about 15% less than did young ones.

3.3. Changes in Other Visual-Cell Transduction Proteins

Figure 2 contains data from Western transblots of retinal proteins probed with anti S-antigen and anti α-transducin monoclonal antibodies. For young light-reared rats changed to darkness, the level of S-antigen decreased while α-transducin levels increased. After 16 days in the new environment, retinas from these rats exhibited 42% lower S-antigen staining than did day-0 rats. Transducin levels increased by 12% during the same 16-day period. The transition for young dark-reared rats changed to dim cyclic light was

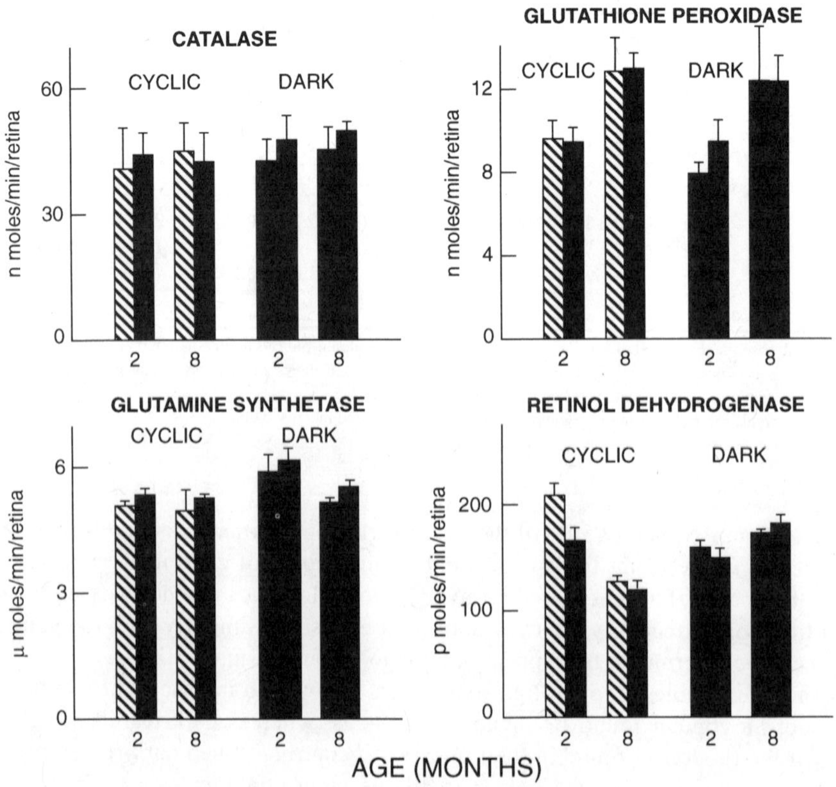

Figure 2. Retinal S-antigen and α-transducin in rats changed from one rearing environment to another. Western blots of retinal proteins from 2- and 8-month-old rats changed from cyclic light to darkness (C → D) or darkness to cyclic light (D → C) for 16 days before sacrifice. Upper panel, 2-month-old rats: cyclic-light-reared (0 days), C → D (16 days); dark-reared (0 days), D → C (16 days). Lower panel, 8-month-old rats. Upper band, S-antigen; lower band, α-transducin.

different. Although S-antigen levels were lower than in rats from the cyclic-light environment, they increased by 40%, while α-transducin levels declined by 41% from the day-0 control values. In older rats, the transition from darkness to weak cyclic light was associated with a 13% increase in S-antigen and a 14% decrease in α-transducin. Older light-reared rats changed to darkness exhibited no change in α-transducin but a 39% decrease in S-antigen levels.

3.4. Rod Outer Segment Membrane Synthesis and Rearing Environment

To determine whether the differences in rearing environment also affected ROS lipids, we isolated ROS from young rats given intraocular injections of [^{14}C]acetate. The specific activities of ROS membrane lipids 4 and 8 days after injection appear in Table 2. In one experiment, ROS isolated from light-reared rats 4 days after injection contained 23% more label than those isolated from dark-reared animals ($p < 0.005$). A second series of experiments resulted in similar differences for the light- and dark-reared rats (line 2, Table 2), but when rats were changed from dim cyclic light to darkness, the specific activity of

Table 2. ROS lipid-specific activities in cyclic-light- and dark-reared rats (DPM/nmol PO_4)[a]

Time after injection	Light reared	Moved from light to dark	Dark reared
	90.3 ± 14.7 (7)	—	69.2 ± 14.2[b] (7)
4 days	82.8 (1)	117.6 ± 25.9 (4)	68.8 ± 11.7[b] (5)
8 days	107.2 (2)	65.8 ± 23.2 (4)	56.3 ± 15.1 (6)

[a]Results are the mean ± SD for (n) experiments using ROS isolated from 2-month-old rats.
[b]Significantly lower than in cyclic-light-reared rats, $p < 0.005$, or, for light-to-dark rats, $p < 0.01$.

ROS lipids had not decreased after 4 days. On the other hand, ROS lipid-specific activity in rats changed from cyclic light to darkness for 8 days did show the expected decline. In this experiment the decrease from the light-reared level was almost 40%, and the specific activities of the dark-reared control and experimental animals were similar.

3.5. Retinal Enzymes and Rearing Environment

Table 3 gives the relative activities and distribution of the four retinal enzymes we studied. *t*-RDH, a specialized single-function enzyme, has a catalytic capacity of approximately 200 pmol/minute/retina. Catalase and GPX have 190 and 45 times the activity of *t*-RDH, underscoring the importance of antioxidation in the retina. It is also of interest that catalase has about four times the potential of GPX for detoxifying H_2O_2. GS, at 5 μmol/minute/retina, is the most active enzyme of the group. Also shown in Table 3 are values for the relative distribution of these activities between photoreceptor cells and inner retinal layers. Our findings confirm that *t*-RDH is essentially confined to ROS (Zimmerman *et al.,* 1975), whereas GS is located in the Müller cells (Riepe and Norenberg, 1977). Similarly, both catalase and GPX appear to be primarily (90%) located in the inner retinal layers. Upon direct assay of concentrated band I ROS, no catalase or GPX activity was found. Atalla *et al.* (1987) found no immunohistochemical staining for catalase in the outer segments of rat photoreceptor cells but did find immunoreactivity in the inner segments. This pattern may account for our finding of 12% of the total retinal catalase in the photoreceptor cells (Table 3).

Table 3. Distribution of retinal enzymes in albino rats[a]

	Enzyme (activity/retina)			
	t-RDH (212 ± 11 pmol/minute)	GS (5.1 ± 0.1 μmol/minute)	Catalase (41.0 ± 0.1 nmol/minute)	GPX (9.5 ± 1 nmol/minute)
Rod outer segment	—	—	0%	0%
Photoreceptor cells	93%	0%	12%	7.5%
Inner retina layers	7%	100%	88.5%	92.5%

[a]Activity/retina represents the average ± SD from three or four homogenates of whole retinas from two-month-old rats. Activities in ROS represent direct determinations on concentrated isolated band I ROS from two-month-old rats. The distributions of enzyme activities between photoreceptor cells and inner retinal layers were obtained by comparison of values from whole homogenates of retinas from 9-month-old RCS dystrophic rats with those from normal rats of the same age.

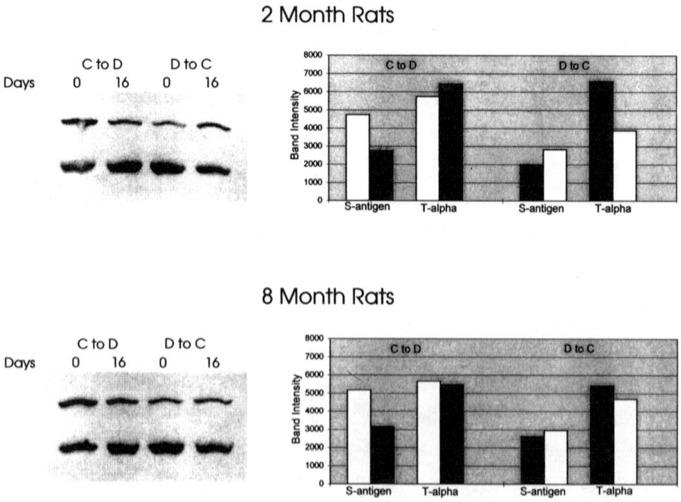

Figure 3. Retinal enzyme activities in rats. Retinas were excised immediately after light exposure of 24 hours for cyclic-light-reared or 3–4 hours for dark-reared rats. Enzyme activities were determined as described in Methods. Values represent the average ± SD of three or four retinas. Cross-hatched bars represent cyclic-light-reared rats unexposed to intense light. Filled bars represent dark-reared rats without light exposure. Adjacent open bars represent intense-light-exposed rats of the same type.

Figure 3 shows the effect of age, light exposure, and rearing conditions on catalase, GPX, GS, and *t*-RDH. Intense light had little effect on any of the activities, except for *t*-RDH in young light-reared rats, where light reduced the activity by 21% ($p < 0.005$). Other studies (Darrow *et al.,* 1997), however, have shown a consistent decrease in *t*-RDH-activity in young dark-reared rats exposed to intense light for 4 hours, when retinas were harvested 20 hours after exposure. These studies compared the effect of light in the presence and in the absence of the antioxidant DMTU, suggesting that the decrease in both types of animals was an oxidative phenomenon. *t*-RDH was also 41% lower ($p < 0.001$) in older rats than in younger ones, when neither were exposed to damaging light. In similar rats exposed to damaging light for 24 hours the difference was 29% ($p < 0.001$).

The activities of catalase and GS did not vary with age or intense light exposure, but the antioxidant enzyme GPX increased significantly between 2 and 8 months, in both light- and dark-reared rats. When we combined the data for light-exposed and unexposed animals, which were similar, GPX increased by 44% ($p < 0.001$) in light-reared and by 41% ($p < 0.001$) in dark-reared animals.

4. DISCUSSION

Our results demonstrate that young rats are able to adapt to changes in rearing environment, but the transition from dim cyclic light to darkness appeared to be more complete than that from darkness to cyclic light. In young light-reared rats, the transition to darkness was complete within 16 days (Fig. 1), and their light-damage susceptibility was identical to that of rats always reared in darkness (Table 1). As judged by the 80% average recovery of rhodopsin for intense-light-exposed rats previously changed from darkness to dim cyclic light, some of these rats did not acquire the protection typical of light-reared rats. This result might be related to the lower endogenous S-antigen levels in long-term dark-reared rats (Fig. 2), because high S-antigen levels are correlated with reduced light damage (Organisciak *et al.,* 1991b). Thus the capacity of rats to adapt to changes between dim-cyclic-light and dark rearing environments depends on the direction of the transition. This result differs from the case in which rats are transferred from one cyclic light intensity to another (Penn and Williams, 1986; Penn *et al.,* 1987) and may be related to other changes associated with total dark maintenance.

In older rats, the changes based on rearing environment were similar to but less complete than those in the young animals. The relatively smooth transitions seen in young rats were not present in the older animals, and the half-times of the changes were different (Fig. 1). In addition, there was a substantial, although transient, increase in rhodopsin levels in the older rats 2 days after a change in rearing conditions. This increase suggests that the normally coupled processes of ROS disk addition and shedding or phagocytosis are more easily uncoupled in older rats by changes in rearing environment. Using younger rats changed from one light intensity to another, Schremser and Williams (1995a, b) concluded that these events were indeed uncoupled by a change in light levels. In their study they found an increase in [^{3}H]amino acid incorporation into rhodopsin that was correlated with its synthesis in rats changed from a high-light to a low-light environment. Our study of ROS membrane lipids shows that cyclic-light-reared rats incorporated significantly more label than did dark-reared animals and that about 8 days were required to approach the lower specific activity in rats changed from cyclic light to darkness (Table 2). Taken together, the results of Schremser and Williams (1995a, b) and our findings indicate that ROS protein and lipid metabolism may also be uncoupled. Thus, the increases found in rhodopsin packing in ROS disks upon rearing-environment change may arise from both increased opsin synthesis and a decrease in ROS membrane lipid synthesis.

The older light-reared rats were also more susceptible to light damage than the young rats (Table 1). When older rats were changed to the opposite rearing condition, they did not acquire the expected sensitivity or resistance to light damage, and the changes in S-antigen and α-transducin were generally less than in young rats. Older dark-reared rats, however, were not more susceptible to damage than young dark-reared rats. This result suggests that long-term darkness affects the expression of genes that normally increase susceptibility to retinal light damage.

We compared the expression of four enzymes in young and older rat retinas by measuring the activities of *t*-RDH and GS, as markers of oxidation, and catalase and GPX, enzymes thought to serve an antioxidative role in the retina. *t*-RDH activity was reduced by intense light exposure and age in cyclic-light-reared rats (Fig. 3). The loss of activity with age in rats reared at low cyclic light levels may represent a chronic deterioration of photoreceptor function not seen when rats are reared in total darkness. On the basis of its location in ROS membranes and the prevention of its inactivation by DMTU (Darrow *et al.,* 1997), the loss of activity also appears to be oxidative in nature. Although GPX was not affected by intense light, its activity was greater in older rat retinas than in those from young animals. This increase may arise from oxidative stresses, which are increased in older animals, and may represent a protective adaptation (Stadtman, 1992). Interestingly, both catalase and GPX appear to be localized in the inner retinal layers (Table 3). This pattern may help to explain why the inner retinal neurons are not damaged by intense light while the photoreceptors undergo massive structural and functional changes (Noell *et al.,* 1966). If oxidation of *t*-RDH leads to its inactivation in ROS, then the lack of an effect on GS, a Müller cell enzyme, suggests that light damage is largely confined to the photoreceptors of the rat retina. These are the same cells that exhibit remarkable adaptive changes in protein levels, based on light rearing environment (Batelle and LaVail, 1978; Organisciak *et al.,* 1991b) and photostasis (Penn and Williams, 1986). Accordingly, retinal light damage appears to involve oxidative processes that can be modulated by adaptive mechanisms related to photon catch and to subsequent steps in the visual transduction cascade.

ACKNOWLEDGMENTS

Special thanks to L. Barsalou for excellent technical assistance. Supported by NIH grant EY-01959, M. Petticrew, a T. Pruett seed grant on aging, and Ohio Lions.

LITERATURE CITED

Abler, A. S., Chang, C.-J., Fu, J., Tso, M. O. M., and Lam, T. T. (1996). Photic injury triggers apoptosis of photoreceptor cells. *Research Communications in Molecular Pathology and Pharmacology* 92:177–189.

Aebi, H. (1984). Catalase *in vitro. Methods in Enzymology* 105:121–161.

Anderson, R. E., and Maude, B. B. (1972). Lipids of ocular tissues. VIII. The effects of essential fatty acid deficiency on the phospholipids of the photoreceptor membranes of rat retina. *Archives of Biochemistry and Biophysics* 151:270–276.

Atalla, L., Fernandez, M. A., and Rao, N. A. (1987). Immunohistochemical localization of catalase in ocular tissue. *Current Eye Research* 6:1181–1187.

Battelle, B. A., and LaVail, M. M. (1978). Rhodopsin content and rod outer segment length in albino rats: modification by dark adaptation. *Experimental Eye Research* 26:487–497.

Birch, D. G., and Jacobs, G. H. (1980). Light-induced damage to photopic and scotopic mechanisms in the rat depends on rearing conditions. *Experimental Neurology* 68:269–286.

Burns, M. S., and Robles, M. (1990). Müller cell GFAP expression exhibits a gradient from the focus of photoreceptor light damage. *Current Eye Research* 9:479–486.

Bush, R. A., Remé, C. E., and Malnoë, A. (1991). Light damage in the rat retina: the effects of dietary deprivation of N-3 fatty acids on acute structural alterations. *Experimental Eye Research* 53:741–752.

Chader, G. J. (1971). Hormonal effects in the neural retina. 1. Glutamine synthetase development in the retina and liver of the normal and triiodothyronine-treated rat. *Archives of Biochemistry and Biophysics* 144:657–662.

Daemen, F. J. M. (1973). Vertebrate rod outer segment membranes. *Biochimica et Biophysica Acta* 300:255–288.

Darrow, R. A., Darrow, R. M., Lininger, L. A., and Organisciak, D. T. (1994). Lipid hydroperoxides in rat retina: enzymatically determined upper limits *in vitro* and *in vivo. Investigative Ophthalmology and Visual Science* 35(Suppl.):2136.

Darrow, R. A., Darrow, R. M., and Organisciak, D. T. (1997). Biochemical characterization of cell-specific enzymes in light-exposed rat retinas: oxidative loss of all-*trans* retinol dehydrogenase activity. *Current Eye Research* 16:144–151.

De La Paz, M. A., and Anderson, R. E. (1992). Lipid peroxidation in rod outer segments: role of hydroxyl radical and lipid hydroperoxides. *Investigative Ophthalmology and Visual Science* 33:2091–2096.

Delmelle, M., Noell, W. K., and Organisciak, D. T. (1975). Hereditary retinal dystrophy in the rat: rhodopsin, retinol and vitamin A deficiency. *Experimental Eye Research* 21:369–380.

Edward, D. P., Lam, T.-T., Shahinfar, S., Li, J., and Tso, M. O. M. (1991). Amelioration of light-induced retinal degeneration by a calcium overload blocker, flunarizine. *Archives of Ophthalmology* 109:554–562.

Farber, D. B., Seager-Danciger, J., and Organisciak, D. T. (1991). Levels of mRNA encoding proteins of the cGMP cascade as a function of light environment. *Experimental Eye Research* 53:781–786.

Farnsworth, C. C., and Dratz, E. A. (1976). Oxidative damage of retinal rod outer segment membranes and the role of vitamin E. *Biochimica et Biophysica Acta* 443:556–570.

Fliesler, S. J., and Anderson, R. E. (1983). Chemistry and metabolism of lipids in the vertebrate retina. *Progress in Lipid Research* 22:79–131.

Flohé, L., and Günzler, W. A. (1984). Assays of glutathione peroxidase. *Methods in Enzymology* 105:114–121.

Hayes, K. C. (1974). Retinal degeneration in monkeys induced by deficiencies of vitamin E or A. *Investigative Ophthalmology* 13:499–510.

Hunt, D. F., Organisciak, D. T., Wang, H. M., and Wu, R. L. C. (1984). α-Tocopherol in the developing rat retina: a high pressure liquid chromatographic analysis. *Current Eye Research* 11:1281–1288.

Kagan, V. E., Kuliev, I. Y., Spirichev, V. B., Shvedova, A. A., and Kozlow, Y. P. (1981). Accumulation of lipid peroxidation products and depression of retinal electrical activity in vitamin E deficient rats exposed to high-intensity light. *Bulletin of Experimental Biology and Medicine* 91:144–147.

Katz, M. L., and Eldred, G. E. (1989). Failure of vitamin E to protect the retina against damage resulting from bright light exposure. *Investigative Ophthalmology and Visual Science* 30:29–36.

Lam, S., Tso, M. O. M., and Gurne, D. H. (1990). Amelioration of retinal photic injury in albino rats by dimethylthiourea. *Archives of Ophthalmology* 108:1751–1757.

LaVail, M. M., Gorrin, G. M., Repaci, M. A., Thomas, L. A., and Ginsberg, H. M. (1987). Genetic regulation of light damage to photoreceptors. *Investigative Ophthalmology and Visual Science* 28:1043–1048.

Li, Z.-Y., Tso, M. O. M., Wang, H.-M., and Organisciak, D. T. (1985). Amelioration of photic injury in rat retina by ascorbic acid: a histopathologic study. *Investigative Ophthalmology and Visual Science* 26:1589–1598.

Noell, W. K., and Albrecht, R. (1971). Irreversible effects of visible light on the retina: role of vitamin A. *Science* 172:76–80.

Noell, W. K., Walker, V. S., Kang, B. S., and Berman, S. (1966). Retinal damage by light in rats. *Investigative Ophthalmology* 5:450–473.

Organisciak, D. T., and Noell, W. K. (1977). The rod outer segment phospholipid/opsin ratio of rats maintained in darkness or cyclic light. *Investigative Ophthalmology and Visual Science* 16:188–190.

Organisciak, D. T., and Winkler, B. S. (1994). Retinal light damage: practical and theoretical considerations. In: Osborn, N. N., and Chader, G. J., eds. *Progress in Retinal and Eye Research, Vol. 13.* Pergamon Press, New York. pp. 1–29.

Organisciak, D. T., Favreau, P., and Wang, H.-M. (1983). The enzymatic estimation of organic hydroperoxides in the rat retina. *Experimental Eye Research* 36:337–349.

Organisciak, D. T., Wang, H.-M., and Kou, A. L. (1984). Ascorbate and glutathione levels in the developing normal and dystrophic rat retina: effect of intense light exposure. *Current Eye Research* 3:257–267.

Organisciak, D. T., Wang, H.-M., Li, Z.-Y., and Tso, M. O. M. (1985). The protective effect of ascorbate in retinal light damage of rats. *Investigative Ophthalmology and Visual Research* 26:1580–1588.

Organisciak, D. T., Wang, H.-M., and Noell, W. K. (1987). Aspects of the ascorbate protective mechanism in retinal light damage of rats with reduced ROS docosahexaenoic acid. In: Hollyfield, J. G., Anderson, R. E., and LaVail, M. M., eds. *Degenerative Retinal Disorders: Clinical and Laboratory Investigations.* Liss, New York. pp. 455–468.

Organisciak, D. T., Darrow, R. M., Bicknell, I. R., Jiang, Y.-L., Pickford, M., and Blanks, J. C. (1991a). Protection against retinal light damage by natural and synthetic antioxidants. In: Anderson, R. E., Hollyfield, J. G., and LaVail, M. M., eds. *Retinal Degenerations.* CRC Press, Boca Raton. pp. 190–201.

Organisciak, D. T., Xie, A., Wang, H.-M., Jiang, Y.-L., Darrow, R. M., and Donoso, L. A. (1991b). Adaptive changes in visual cell transduction protein levels: effect of light. *Experimental Eye Research* 53:773–779.

Organisciak, D. T., Bicknell, I. R., and Darrow, R. M. (1992a). The effects of L- and D-ascorbic acid administration on retinal tissue levels and light damage in rats. *Current Eye Research* 11:231–241.

Organisciak, D. T., Darrow, R. M., Jiang, Y.-L., Marak, G. E., and Blanks, J. C. (1992b). Protection by dimethylthiourea against retinal light damage in rats. *Investigative Ophthalmology and Visual Science* 33:1599–1609.

Organisciak, D. T., Kutty, R. K., Leffak, M., Wong, P., Messing, S., Wiggert, B., Darrow, R. M., and Chader, G. J. (1995). Oxidative damage and responses in retinal nuclei arising from intense light exposure. In: Anderson, R. E., LaVail, M. M., and Hollyfield, J. G., eds. *Degenerative Diseases of the Retina.* Plenum Press, New York. pp. 9–17.

O'Steen, W. K., Anderson, K. V., and Shear, C. R. (1974). Photoreceptor degeneration in albino rats: dependency on age. *Investigative Ophthalmology* 13:334–339.

Penn, J. S., and Williams, T. P. (1986). Photostasis: regulation of daily photon-catch by rat retinas in response to various cyclic illuminances. *Experimental Eye Research* 43:915–928.

Penn, J. S., and Anderson, R. E. (1987). Effect of light history on rod outer segment membrane composition in the rat. *Experimental Eye Research* 44:767–778.

Penn, J. S., Naash, M. I., and Anderson, R. E. (1987). Effect of light history on retinal antioxidants and light damage susceptibility in the rat. *Experimental Eye Research* 44:779–788.

Rapp, L. M. (1995). Retinal Phototoxicity, Chapter 32. In: Chang, L. W., and Dyer, R. S., eds. *Handbook of Neurotoxicology.* Marcel Dekker, New York. pp. 963–1003.

Rapp, L. M., Tolman, B. L., Koutz, C. A., and Thum, L. A. (1990). Predisposing factors to light-induced photoreceptor cell damage: retinal changes in maturing rats. *Experimental Eye Research* 51:177–184.

Remé, C. E., Braschler, U. F., Roberts, J., and Dillon, J. (1991). Light damage in the rat retina: effects of a radioprotective agent (WR-77913) on acute rod outer segment disk disruptions. *Photochemistry and Photobiology* 54:137–142.

Remé, C. E., Weller, M., Szczesny, P., Munz, K., Hafezi, F., Reinboth, J.-J., and Clausen, M. (1995). Light induced apoptosis in the rat retina *in vivo:* morphological features, threshold on a time course. In: Anderson, R. E., LaVail, M. M., and Hollyfield, J. G., eds. *Degenerative Diseases of the Retina.* Plenum Press, New York. pp. 19–25.

Riepe, R. E., and Norenberg, M. D. (1977). Müller cell localization of glutamine synthetase in rat retina. *Nature* 268:654–655.

Saari, J. C., Bredberg, D. L., Garwin, G. G., Buczlko, J., Wheeler, T., and Palczewski, K. (1993). Assays of retinoid dehydrogenases by phase partition. *Analytical Biochemistry* 213:128–132.

Schremser, J.-L., and Williams, T. P. (1995a). Rod outer segment (ROS) renewal as a mechanism for adaptation to a new intensity environment. I. Rhodopsin levels and ROS length. *Experimental Eye Research* 61:17–24.

Schremser, J.-L., and Williams, T. P. (1995b). Rod outer segment (ROS) renewal as a mechanism for adaptation to a new intensity environment. II. Rhodopsin synthesis and packing density. *Experimental Eye Research* 61:25–32.

Shahinfar, S., Edward, D. P., and Tso, M. O. M. (1991). A pathologic study of photoreceptor cell death in retinal photic injury. *Current Eye Research* 10:47–59.

Stadtman, E. R. (1992). Protein oxidation and aging. *Science* 257:1220–1224.

Stone, W. L., Farnsworth, C. C., and Dratz, E. A. (1979a). A re-investigation of the fatty acid content of bovine, rat and frog retinal rod outer segments. *Experimental Eye Research* 28:387–397.

Stone, W. L., Katz, M. L., Lurie, M., Marmor, M. F., and Dratz, E. A. (1979b). Effects of dietary vitamin E and selenium on light damage to the rat retina. *Photochemistry and Photobiology* 29:725–730.

Stoyanovsky, D. A., Goldman, R., Darrow, R. M., Organisciak, D. T., and Kagan, V. E. (1995). Endogenous ascorbate regenerates vitamin E in the retina directly and in combination with exogenous dihydrolipoic acid. *Current Eye Research* 14:181–189.

Thorndike, J., and Reif-Lehrer, L. (1971). A sensitive assay for glutamyl transferase. *Enzyme* 12:235–241.

Wiegand, R. D., Giusto, N. M., Rapp, L. M., and Anderson, R. E. (1983). Evidence for rod outer segment lipid peroxidation following constant illumination of the rat retina. *Investigative Ophthalmology and Visual Science* 24:1433–1435.

Wiegand, R. D., Joel, C. O., Rapp, L. M., Nielsen, J. C., Maude, M. B., and Anderson, R. E. (1986). Polyunsaturated fatty acids and vitamin E in rat rod outer segments during light damage. *Investigative Ophthalmology and Visual Science* 27: 727–733.

Williams, T. P., and Howell, W. L. (1983). Action spectrum of retinal light-damage in albino rats. *Investigative Ophthalmology and Visual Science* 24:285–287.

Zimmerman, W. F., Lion, F., Daemen, F. J. M., and Bonting, S. L. (1975). Biochemical aspects of the visual process. XXX. Distribution of stereospecific retinol dehydrogenase activities in subcellular fractions of bovine retina and pigment epithelium. *Experimental Eye Research* 21:325–332.

7

THE ORIGIN OF THE MAJOR ROD- AND CONE-DRIVEN COMPONENTS OF THE RODENT ELECTRORETINOGRAM AND THE EFFECT OF AGE AND LIGHT-REARING HISTORY ON THE MAGNITUDE OF THESE COMPONENTS

E. N. Pugh, Jr. ,[1] B. Falsini,[2] and A. L. Lyubarsky[1]

[1]Departments of Psychology and Ophthalmology
University of Pennsylvania
3815 Walnut Street
Philadelphia, Pennsylvania 19104-6196
[2]Department of Ophthalmology
Catholic University of Rome
Italy

1. INTRODUCTION

The electroretinogram (ERG) is a complex field potential evoked by light. The ERG response to brief flashes has long provided an important assessment tool in basic and clinical research (see, e.g., Berson, 1992; Jacobson *et al.,* 1994). In the past decade, several rapidly growing and overlapping areas of retinal research have led to extensive basic and clinical applications of the ERG.

One is phototransduction. This rich area of research on the biochemical and physiological mechanisms underlying vertebrate photoresponses has yielded a nearly complete molecular description of the phototransduction cascade of rods and a generally consistent though less advanced description of the cone cascade (for reviews, see Koutalos and Yau, 1993; Pugh and Lamb, 1993). Research in phototransduction has led to the formulation of a model of the activation phase of the transduction cascade (Lamb and Pugh, 1992) that is directly applicable to analysis of the a-wave of the ERG (Pugh and Lamb, 1993; Breton *et al.,* 1994; Cideciyan and Jacobson, 1996; Lyubarsky and Pugh, 1996).

A second burgeoning area of research that has led to increased application of the ERG is that of the molecular biology of hereditary eye disease. The genotypes of a great number of human retinal disorders can now be identified (e.g., Daiger *et al.,* 1995; Shastry, 1997). The explosive growth of genotype information about human eye disease has spurred application of the ERG for retinal phenotyping and for ongoing clinical assessment of disease progression.

Photostasis and Related Phenomena, edited by Williams and Thistle
Plenum Press, New York, 1998

A third, relatively new area of research now propelling increased use of the ERG is also molecular biological: the development of lines of rodents with transgenes for retinally expressed proteins (Lem and Makino, 1996). These animal lines include mice with specific transgenes known to cause hereditary eye disease in humans (e.g., Wong, 1997). The ERG can provide an efficient and effective noninvasive screen of retinal function in these animals. Because animal models of hereditary disease provide a major resource for the evaluation of pharmacological, molecular-biological, and other interventions in human hereditary disease, and because electroretinographic assessment in human patients is straightforward, the value and use of the ERG in transgenic animals is rapidly expanding.

1.1. Goals of the Present Chapter

This chapter has three goals. The first is to provide a tutorial on the principal components of the flash-evoked ERG, namely, the rod-dominated a-wave and the rod- and cone-dominated b-waves. To achieve this goal we will re-present material from recently published papers, along with schematics that illustrate how certain retinal cell types are thought to give rise to the specific field potentials or components of the ERG known as the a-waves and b-waves. Our second goal is to illustrate how rod- and cone-driven components of the ERG can be functionally separated and characterized in rodents. Our third is to illustrate how the ERG has led to quantitative insights into the regulation by rearing light history of retinal processes. This third goal derives directly from the work of Ted Williams and his colleagues and collaborators, to whom we are deeply indebted and to whom we dedicate this paper.

2. METHODS AND ORIENTATION

The experimental methods we have used for recording and analyzing flash-evoked ERGs have been reported previously, so we will not describe them here except when necessary for clarity: for the human work, see Breton *et al.* (1994); for work with rodents, see Reiser *et al.* (1996) and Lyubarsky and Pugh (1996). In all instances discussed here, ERGs are evoked with homogeneous, full-field, and brief ($\leq$ 1 ms) flashes. As a matter of principle, we prefer to report flash intensities in terms of the currency in which photoreceptors operate, namely, photoisomerizations per cell (symbolized by Φ, the Greek letter "phi"). Our methods of calibrating light incident on the retina and of converting to photoisomerizations are reported in the papers just cited.

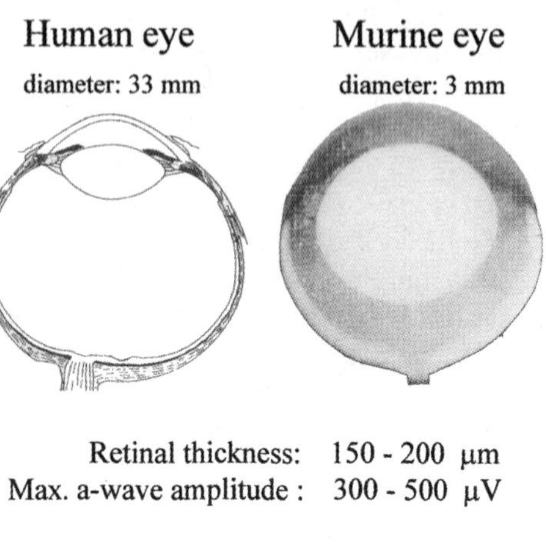

Figure 1. Diagrams of human and murine eyes, scaled to appear about the same size. The human eye schematic is a standard, textbook diagram. The murine eye is a copy of the micrograph of a section of a mouse eye published by Remtulla and Hallett (1985).

Figure 1 presents diagrams of the human and mouse eye. Although the eyes of humans and mice differ in diameter by a factor of ca. 10, the retinas of the two species are of approximately the same thickness and contain the same basic types of cells. In Figure 2, to provide orientation for the remainder of the chapter, we show the schematic of primate retina of Dowling and Boycott (1966); in this schematic we identify the cell types—rods, cones, rod bipolars, cone bipolars—upon which much of the discussion below will focus.

3. THE A-WAVE ORIGINATES IN THE PHOTORECEPTOR CIRCULATING CURRENT

Figure 3 shows a family of corneally recorded murine ERG's, evoked by flashes estimated to produce from 0.095 to 7.5×10^5 Φ per rod. Prominent features include, at low

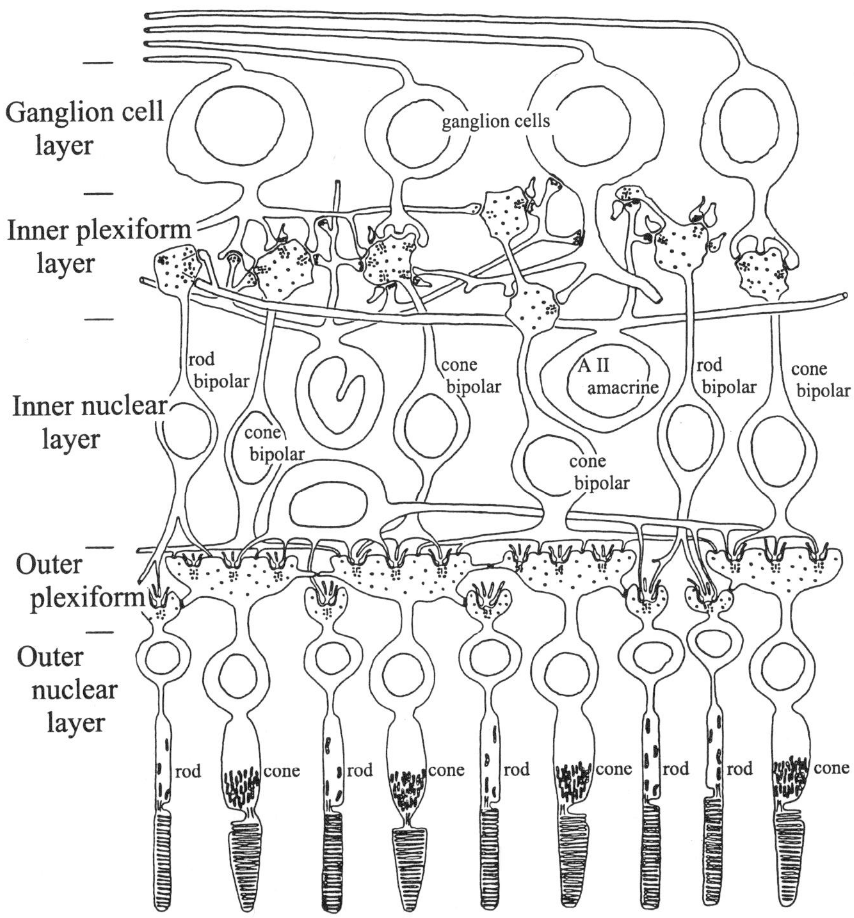

Figure 2. Schematic diagram of longitudinal section of the mammalian retina, illustrating the major cell types, including rods, cones, rod bipolars (RB), cone bipolars (CB), AII amacrine cells, and ganglion cells. Modified from Dowling and Boycott (1966). The retina is shown in the orientation that it would have in the eyes illustrated in Figure 1.

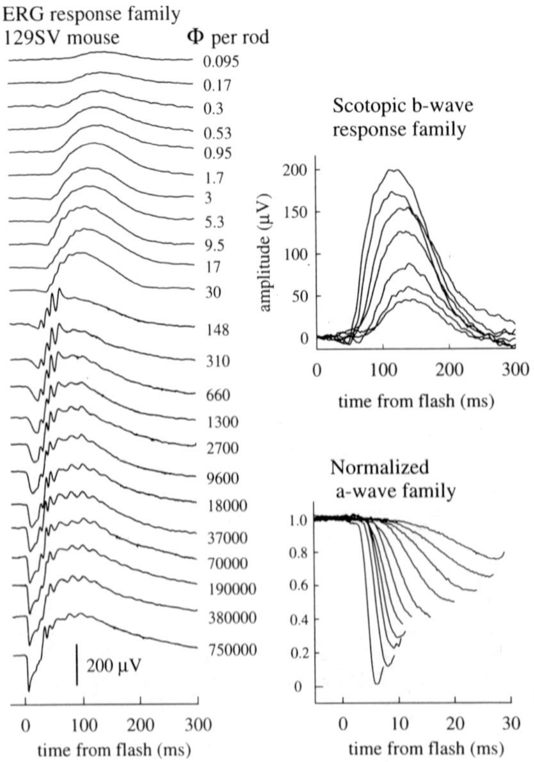

Figure 3. Left panel. Family of murine electroretinograms evoked by full field flashes estimated to produce the number of photoisomerizations per rod given by the number immediately to the right of each curve. ERG's were measured at the cornea, as described by Lyubarsky and Pugh (1996). Data are from Jiang *et al.* (1996, their Fig. 3); the mouse was a 4-month-old of the 129SV strain. Each trace is the average of 2–5 individual responses. Right panels. In the upper panel the responses to the dimmest flashes 0.095 to 9.5 Φ per rod) from the left side have been collected, rescaled, and plotted with a common ordinate to show their saturation behavior (two traces were omitted for clarity). In the lower panel the a-waves from the responses to the 11 most intense flashes have been isolated from the complete ERG traces at the left and plotted on an expanded time base with a common ordinate. The traces were truncated slightly after the point of clear b-wave intrusion and normalized with respect to the amplitude of the response to the most intense flash.

intensities ($\Phi \leq 30$), the scotopic b-wave and at slightly higher intensities the appearance at early times of the negative-going a-wave, followed by the positive-going b-wave and prominent oscillatory potentials. In Section 3 we concentrate on the a-wave. We will summarize four distinct lines of evidence that support the conclusion that the a-wave originates in the suppression of the dark/circulating current of the photoreceptors (primarily of the rods). These lines of evidence are (1) the electrical sign of the a-wave and its saturated magnitude, (2) the magnitude of the saturated a-wave during steady illumination and during the retina's recovery from flashes, (3) the activation kinetics and light sensitivity of the a-wave, and (4) the properties of the pharmacologically isolated a-wave.

3.1. Discovery of the Rod Circulating Current

The understanding of the ionic mechanisms underlying vertebrate rod photoresponses made an enormous leap forward with the investigations of Hagins *et al.* (1970) and Penn and Hagins (1972). These investigations established, with an extracellular current source/sink analysis in the isolated rat retina, that in the dark rods generate a dark or circulating current and that the primary photoresponse of rods was the local suppression of the outer-segment component of this current. In the late 1970s the development of the suction-electrode technique (Baylor *et al.,* 1979a, b) and the application throughout the 1980s and 1990s of suction-electrode and whole-cell patch techniques led to a much deeper appreciation of the mechanisms underlying the circulating current. Nonetheless, the work of Hagins and colleagues provided the essential insights necessary for understanding of how the suppression of the circulating current gives rise to the a-wave.

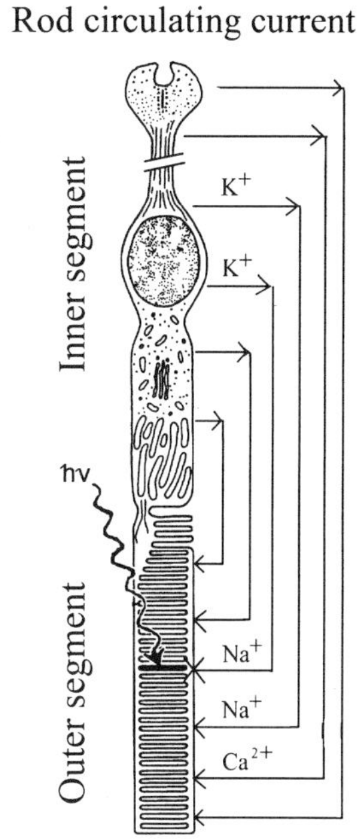

Figure 4. Schematic of a vertebrate rod, illustrating the circulating current. (After Hargrave and McDowell, 1992, their Fig. 2). The rod is shown in the orientation in which it would be found in the eyes illustrated in Figure 1. The captured photon leads to a local suppression of inward outer-segment membrane current as a consequence of the phototransduction process.

Figure 4 provides a schematic of the circulating current of a single rod. In the dark/resting state of the rod, when the membrane potential is ca. −37 mV (Schneeweis and Schnapf, 1995), charges emerge from the inner segment membrane of the rod through conductances primarily selective for K^+; an equal quantity of charge flows into the outer segment through the cGMP-activated conductance, carried primarily by Na^+ (ca. 90%) and Ca^{2+} (ca. 10%). These influxes and effluxes of charge are driven by the electrochemical gradients of the different cations and must be in balance at rest*—otherwise or else the transmembrane potential would not be constant. Moreover, except during the response to intense flashes, when the cGMP-activated conductance can close within a few milliseconds, capacitive currents are negligible (Penn and Hagins, 1972; Cobbs and Pugh, 1987; Breton *et al.,* 1994). Thus, even during the time-varying conductance changes that underlie light responses, a nearly perfect balance of inward and outward currents is maintained, such that the inner segment and outer segment components of the circulating current are mirror images of each other.†

* The complete current balance equation includes currents through all classes of ion channels, as well as currents generated by electrogenic pumps, including the Na^+/Ca^{2+}-K^+ exchanger of the outer segment, and the Na^+/K^+ exchanger of the inner segment. The cGMP-activated current (see Fig. 4) accounts for about 95% of the outer segment membrane current, and the Na^+/Ca^{2+}-K^+ exchange current accounts for the remaining 5%. The sum of all inner-segment currents must balance these two outer-segment currents.

† The approximate equality of the magnitudes of the integrated inner-segment and outer-segment membrane currents during light responses is the biophysical foundation for suction-electrode recordings, in which a photoreceptor inner (or outer) segment is drawn into the suction pipette and the membrane current of the segment in the electrode is measured by means of a salt bridge connected to a current-voltage transducer (Baylor, *et al.,* 1979a, b).

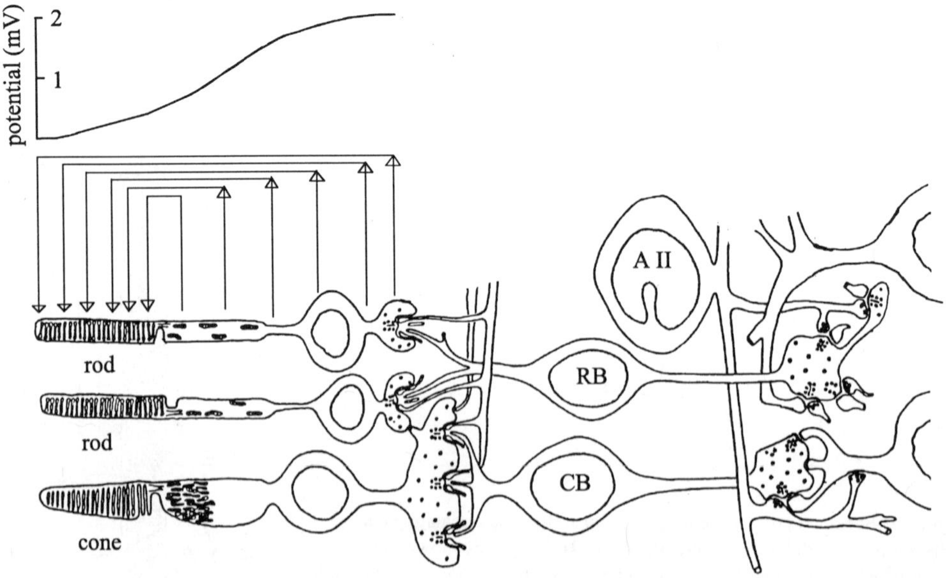

Figure 5. A portion of the retinal diagram of Figure 2 repeated, illustrating how the photoreceptor-layer circulating currents give rise to an extracellular field potential, such that the vitreal side of the retina has a higher, more positive potential, as described in the text and in Appendix I and illustrated here in the inset. When the retina is exposed to a bright flash, the circulating current is reduced to zero, and the field potential collapses, so that the vitreous (or cornea) becomes negative relative to its preflash baseline.

3.2. The Sign and Magnitude of the Field Potential Generated by the Rods' Circulating Currents

Figure 5 shows a small portion of the retinal schematic of Figure 2, illustrating as flow lines the circulating current.* Two features of the balanced inward and outward currents of the rods lead to the generation of an extracellular electric field perpendicular to the retinal surface: first, rods are elongated neurons, and the entire population of rods is oriented approximately perpendicular to the surface of the retina; second, the conductances carrying the inward and outward limbs of the circulating current are located at opposite ends of the cell. To be sure, the electrochemical gradients that drive the inward and outward components of the circulating current are expressed across the cell membrane, but because—from the perspective of the extracellular space—the charges carrying the circulating current have their source at the inner-segment end of the cell and their sink at the outer-segment end, they necessarily give rise to a directional charge flow in the extracellular space. Characterizing this pattern of extracellular current in the terminology of electrical theory, one can properly call the rod circulating current a dipole current. And, because the retina is a layered sheet, one can say that the photoreceptor layer creates a dipole current layer.

* Suction-electrode and whole-cell-electrode recordings have established that cones have circulating currents of magnitude comparable to those of rods. Thus, depending on the relative numbers of rods and cones and their respective circulating current magnitudes, cones' circulating currents contribute to the transreceptor layer field potential. For simplicity we will neglect cone contributions in rodents.

Because of the photoreceptor dipole layer current and the finite resistance of the extracellular space, an electric potential necessarily exists between the rod tips and the rod synaptic terminals. In fact it was the measurement and quantitative characterization of this potential by Hagins *et al.* (1970, their Fig. 7) that led to the inference of the dark or circulating current. For our purposes, however, the bottom line is that, because of the photoreceptor dipole layer current, the vitreal side of the retina is at a higher or more positive electrical potential than the scleral side. Moreover, because the conductance of the intraocular media is relatively high, and there are no major conductive shunts connecting the anterior segment to the posterior pole of the eye,* the cornea itself is positively polarized by the rod dipole layer current. Putting this latter point another way, one can say that a fraction of the rod dipole layer current sinks through the eye media, the anterior segment, and the tears bathing the cornea and returns around the eyeball through various tissues to the scleral side of the retina and in doing so gives the cornea a positive electric potential relative to the scleral side of the neural retina. In Appendix I, we present a simplified calculation that shows that on the basis of (1) the measured rod density of rodents, (2) a circulating-current magnitude of 30 pA for each rod, and (3) the interstitial conductance of the retina, one can expect the fully dark-adapted rod dipole layer current to generate a transretinal field potential of about 1 to 2 mV. This calculation is schematized in the inset in Figure 5.

3.3. The Sign of the a-Wave Response Is That Expected from Suppression of the Rod Dipole Layer Current

Penn and Hagins (1972) and Baylor *et al.* (1979a) demonstrated that a flash that produces a few hundred photoisomerizations per rod completely suppresses the circulating current. Necessarily, then, the dipole layer current in the extracellular space disappears, and the transretinal field potential collapses. This is the reason that the vitreal/corneal a-wave is a negative-going potential, as illustrated in Figure 3.

3.4. The Saturated a-Wave Amplitude a_{max} Is Predicted to Be Proportional to the Rod Circulating Current

In addition to explaining the sign of the a-wave response, the derivation presented in Appendix I of the transretinal potential generated by the rod circulating currents predicts another important feature of the potential. On the assumption that rod circulating currents are uniform across the retina, the derivation (Eq. AI.6) reveals that the transretinal field potential is directly proportional to the rod circulating current. Removing our perspective from the vitreal surface of the retina to the cornea, and considering for the moment only a-wave responses to bright flashes that completely suppress the circulating current and also produce saturated-amplitude a-waves (e.g., the responses to the three most intense flashes in Fig. 3), we have the proportionality relation

$$a_{\text{max}} \propto J_{\text{circ}} \tag{1}$$

where a_{max} is a saturated-amplitude a-wave, and J_{circ} is the (preflash) circulating current of an individual rod. (Hereafter in the text we will omit the subscript "circ" and simply use

---

* Rodieck (1973, Chapter XVIII) reviews the conductivity of the structures of the eye; see Appendix I.

the symbol J to refer to the circulating current unless otherwise stated.) The factors involved in the proportionality relation Eq. 1 are given in Appendix AI.

3.5. Proportionality of a_{max} and Circulating Current during Steady Illumination

Without knowledge of the implicit proportionality constant, Eq. 1 predicts little more than that the a-wave is a negative-going transient when measured at the vitreal side of the retina or at the cornea. We tested the proportionality relation more generally, however, by comparing the data of experimental manipulations that systematically vary the magnitude of the circulating current of isolated rods and vary a_{max} obtained in parallel electroretinographic experiments. For convenience in describing the results of such experiments, it is useful to eliminate the unknown proportionality constant from Eq. 1 by normalizing both a-wave and circulating current to their dark-adapted values, as follows:

$$\frac{a_{max}}{a_{max,dark}} = \frac{J}{J_{dark}} \equiv F \tag{2}$$

where the symbol "≡" means "is defined as"; the symbol F represents the fraction of the maximum circulating current present, i.e., the circulating current at any given time normalized by its value in the fully dark-adapted state. Keep in mind in what follows that in Eq. 2 a_{max} and J refer to the magnitude of the amplitude-saturated a-wave and the magnitude of the residual circulating current, respectively, measured during some manipulation that reduces the magnitudes of both quantities below their dark-adapted levels, $a_{max,dark}$ and J_{dark}.

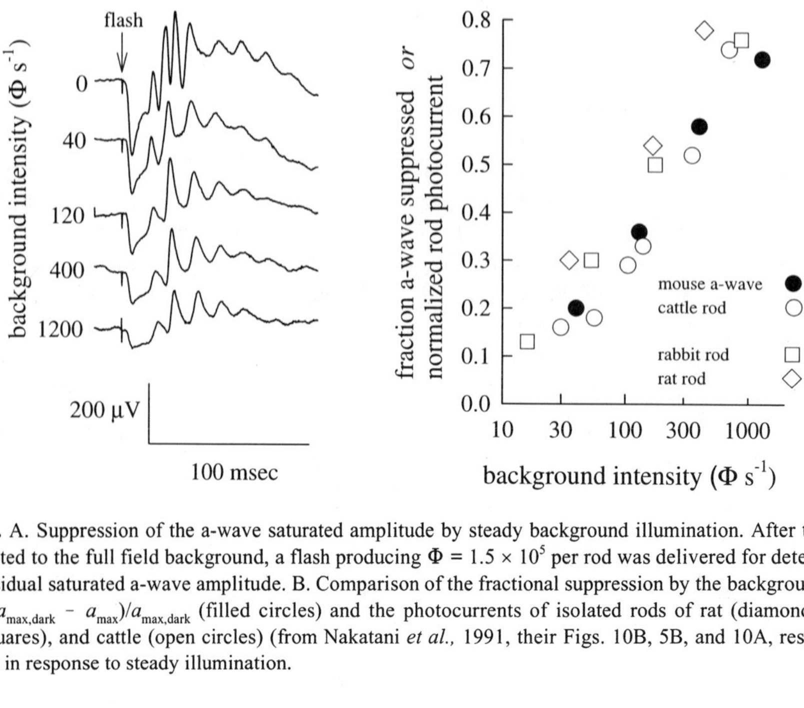

Figure 6. A. Suppression of the a-wave saturated amplitude by steady background illumination. After the mouse was adapted to the full field background, a flash producing $\Phi = 1.5 \times 10^5$ per rod was delivered for determination of the residual saturated a-wave amplitude. B. Comparison of the fractional suppression by the backgrounds of the a-wave, $(a_{max,dark} - a_{max})/a_{max,dark}$ (filled circles) and the photocurrents of isolated rods of rat (diamonds), rabbit (open squares), and cattle (open circles) (from Nakatani *et al.*, 1991, their Figs. 10B, 5B, and 10A, respectively) produced in response to steady illumination.

The results of two experiments in which light was manipulated to vary the circulating-current magnitude are illustrated in Figures 6–7. In the experiment of Figure 6 an initially dark-adapted mouse was presented with *Ganzfeld* illuminations estimated to produce steady rates of photoisomerizations per rod ranging from 0 (dark) to 1200 Φ s^{-1}. After 30 s of exposure to the steady light, a flash of intensity sufficient to saturate the a-wave (Φ = 150,000) was delivered; the resultant ERGs are shown in the left panel. From each ERG response we extract a single number, a_{max}. These values are used to compute $(a_{max,dark} - a_{max})/a_{max,dark}$ to give the fraction of $a_{max,dark}$ suppressed by the steady light and

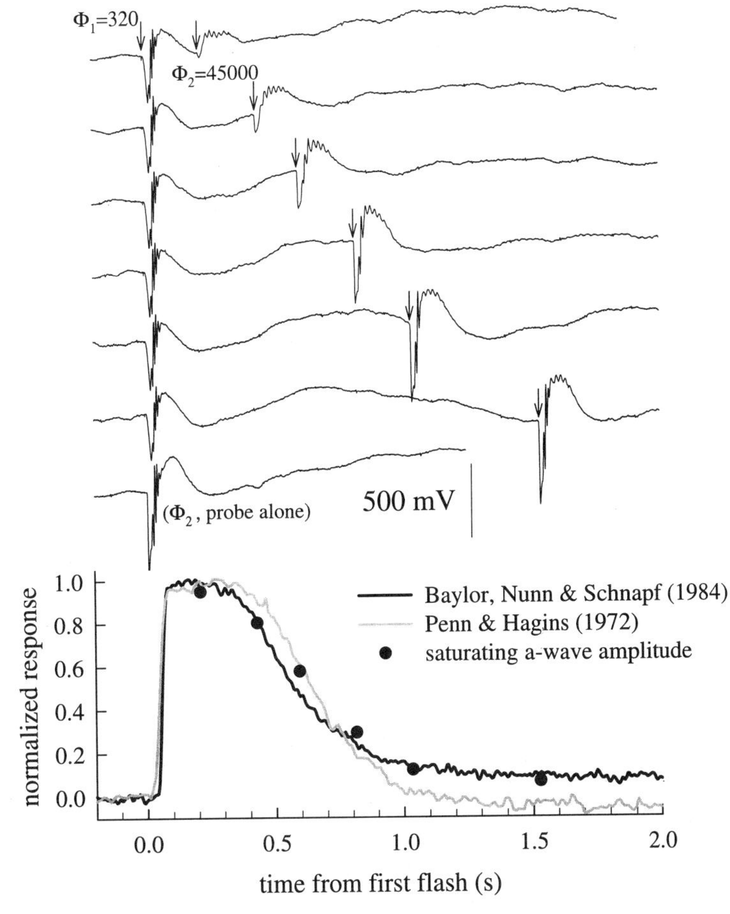

Figure 7. Use of a double-flash ERG protocol to determine the time course of recovery of the rod circulating current in a mouse from the a-wave responses. The upper panel shows a series of electroretinograms: an initial test flash estimated to produce Φ_1 = 320 photoisomerizations per rod is followed at various times by a much more intense probe flash estimated to produce Φ_2 = 45,000 photoisomerizations per rod. The probe flash was chosen so as to produce a very nearly saturated-amplitude a-wave in the fully dark-adapted eye (lowermost trace). The amplitude of the response to the probe flash is seen to recover over a period of a few seconds. In the lower panel, the left-hand side of Eq. 4 is plotted as the filled symbols on exactly the same time base as the upper panel; note that each filled symbol corresponds to one of the probe flash responses, delivered at exactly the same time point as in the upper panel. Plotted along with the symbols are two continuous traces: the first was from the single-primate-rod study of Baylor *et al.* (1984) and was estimated by the investigators to produce Φ = 417; the second is taken from the investigation of Penn and Hagins (1972) of isolated albino rat retina and was estimated by them to produce Φ = 300.

are plotted in panel B as the filled circles. Also plotted in panel B as open symbols are the normalized magnitudes of the photocurrents generated by isolated rods of three mammalian species in response to steady illuminations estimated to produce the photoisomerization rates shown on the abscissa (Nakatani *et al.,* 1991). The term "photocurrent," which is a somewhat misleading holdover from the early days of photoreceptor physiology, needs an explanation: The normalized photocurrent or photoresponse R is defined by

$$R \equiv 1 - F = \frac{J_{\text{dark}} - J}{J_{\text{dark}}} \tag{3}$$

Thus, the ordinate in Figure 6B for the data of the single rods is simply the value of R obtained in response to the steady light intensities specified by the abscissa. According to the linking hypothesis embodied in Eq. 2 and 3, the filled symbols should satisfy

$$\frac{a_{\text{max,dark}} - a_{\text{max}}}{a_{\text{max,dark}}} = R \tag{4}$$

In summary, the agreement in Figure 6 between the filled symbols (left-hand side of Eq. 4) and the open symbols (right-hand side of Eq. 4) supports the generality of the proportionality relation, Eq. 1.

3.6. Proportionality of a_{max} and Circulating Current during Recovery from Flashes

Figure 7 presents a second example of a manipulation that tests the generality of the proportionality relation between saturated a-wave magnitude and circulating current, as embodied in Eq. 4. As seen in Figure 3, the b-wave effectively truncates the a-wave in a time- and intensity-dependent manner. Nonetheless, by presentation of an a-wave-saturating "probe" flash at various times after that of an initial flash, the time course of recovery of the residual circulating current can predictably be followed.* In Figure 7 the comparison is made between "photocurrent" traces measured by Penn and Hagins (1972) in isolated rat retina and by Baylor *et al.* (1984) from an isolated primate rod and, from a mouse, the fraction of the dark-adapted saturated a-wave amplitude that remained suppressed after an initial flash. We estimated this initial flash to produce about the same number of photoisomerizations as were produced in the experiments that yielded the continuous traces. The good agreement between the a-wave data (filled symbols) computed with the left-hand side of Eq. 4 and the continuous traces, which represent $R(t)$ (directly measured right-hand side of Eq. 4), provides further support for the proportionality relation of Eq. 1. Several recent investigations (Birch *et al.,* 1995; Lyubarsky and Pugh, 1996; Pepperberg *et al.,* 1997; Thomas and Lamb, in press) have profitably employed the double-flash technique to examine interesting aspects of circulating current recovery *in situ.*

* B-wave intrusion requires the experimenter to make an important distinction between the flash intensity required to suppress the underlying circulating current and that required to produce a saturated a-wave amplitude. Thus, as illustrated in Figure 7, a test flash producing only $\Phi = 300$ completely suppresses the rod circulating current, but because of b-wave intrusion, a flash producing $\Phi = 45{,}000$ is required to demonstrate unequivocally the complete suppression generated by the $\Phi = 300$ flash.

3.7. Proportionality of a_{max} and *J* for Rods Having Outer Segments of Differing Lengths

A third example of apparently systematic variation in circulating current magnitude occurs as a consequence of the light-rearing dependence of rod outer segment length (Reiser *et al.,* 1996; Section 6, below).

3.8. Light Dependence and Activation Kinetics of the a-Wave Explained by Transduction Cascade Model

Still further evidence for the proportionality relationship between the saturated a-wave and the circulating current comes from kinetic analysis of the rod phototransduction cascade. It is now established that the biochemical mechanism underlying the suppression of the circulating current is a "G-protein cascade": Figure 8 shows a simplified schematic of the phototransduction cascade, focusing on the five primary activation steps. In terms of the G-protein cascade theory, a bright flash completely closes the cGMP-activated channels of the outer-segment plasma membrane, so that the outer-segment limb of the circulating current declines rapidly to zero. The biochemistry and biophysics of the cascade are admittedly complex, but the five activation steps illustrated schematically in Figure 8 can be well approximated analytically and shown to predict quantitatively the time course of suppression of the circulating current of isolated rods and cones by brief flashes (Lamb and Pugh, 1992; Pugh and Lamb, 1993). The kinetic forms of the activation intermediates predicted for the mammalian rod are illustrated in Figure 9. The relatively simple analytical expression approximating the predicted decline in the circulating current deserves note:

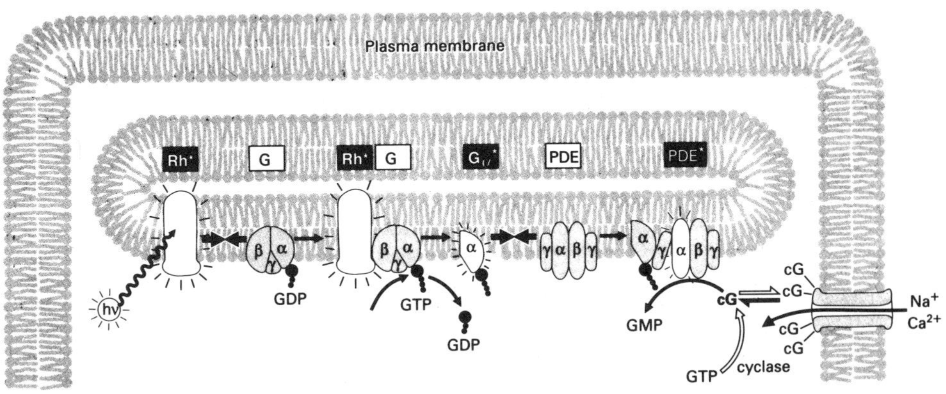

Figure 8. Simplified schematic of the rod phototransduction "G-protein cascade." Only the activation steps are illustrated, and they are as follows: (1) a rhodopsin molecule (Rh) captures a photon and becomes an activated enzyme, Rh*; (2) in its activated form, Rh* catalyzes the exchange of GTP for GDP on the α-subunit of the photoreceptor-specific G-protein transducin; (3) the activated moiety of transducin, G_α-GTP, binds to one of the two γ-subunits of a phosphodiesterase (PDE) molecule, relieving an inhibition and allowing it to enter into its enzymatically active form, PDE*; (4) PDE*s catalyze the hydrolysis of cytoplasmic cyclic GMP, cG, causing the concentration to decline locally; (5) the declining concentration of cGMP causes a net loss of cGMP binding to cGMP-activated cationic channels, which then close, decreasing the inward flow of Na^+ and Ca^{2+}. This decreased cationic influx is the circulating current suppression illustrated in Figure 4, and discussed in the text. (The cascade schematic in this figure shows only the principal activation steps, i.e., those steps leading to the closure of the cGMP-activated channels. For a more detailed analysis of all the activation events, see Lamb and Pugh, 1992; for a review of the biochemistry of all activation and inactivation steps, see Pugh and Lamb, 1993).

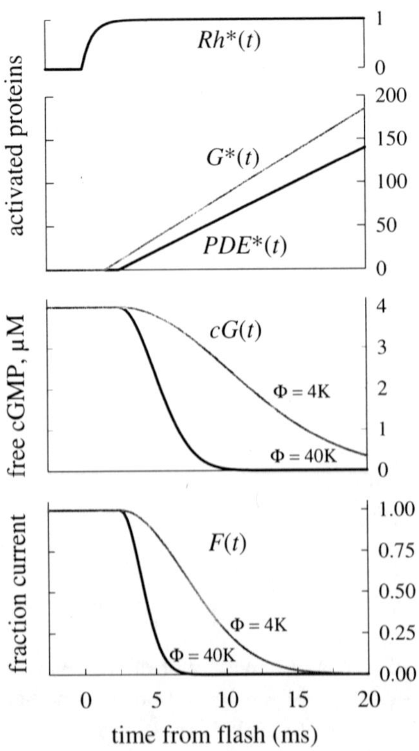

Figure 9. Kinetics of the principal activation steps of the phototransduction cascade, in the absence of inactivation. The top two panels show the *numbers* of activated proteins Rh*, G*, and PDE* in a mammalian rod outer segment as a result of a single captured photon. (1) Photoisomerization causes a unit step *Rh*(t)* of rhodopsin catalytic activity. (2) *Rh** generates a *ramp* of activated G-proteins, *G*(t)*; this ramp is a scaled integral of the *Rh** step, with slope equal to the turnover number of Rh*. (3) the ramp *G*(t)* creates, with some delay and slight loss of gain, a *ramp* of activated PDE catalytic subunits, *PDE*(t)* (the binding reaction of G_α-GTP in Figure 6 has unit stoichiometry: one catalytic subunit PDE* is activated by each bound G_α-GTP.) The two lower panels show the concentration of cGMP, the fraction of cGMP channels open for flashes producing 4000 and 40,000 photoisomerizations (Φ); each individual Φ is assumed to cause a ramp of *G**'s and *PDE**'s, as shown in the upper panels. Thus, (4) *PDE**s catalyze the hydrolysis of cGMP, whose concentration *cG(t)* initially declines as a negative-going *parabola*, the negative integral of the ramp of PDE* activity; (5) rapid and cooperative binding of cGMP to the cGMP-gated channels causes the normalized circulating current, *F(t)* to track the decline of *cG(t)* via a power-law relation, $F(t) = [cG(t)/cG_{\text{dark}}]n_H$, where cG_{dark} is the concentration of cGMP in the outer segment in the dark, and $n_H = 2$–3 is the Hill coefficient of the cGMP-activated channels. Overall, *F(t)* has the form given by Eq. 5 of the text; the amplification constant or coefficient *A* of the argument of the exponential in Eq. 5 can be expressed as the product of the gain factors of the individual cascade steps (Lamb and Pugh, 1992; Pugh and Lamb, 1993).

$$F(t) = e^{-\frac{1}{2}\Phi A(t-t_{\text{eff}})^2} \tag{5}$$

In Eq. 5, *F* is the normalized circulating current (which ranges from 1 in the dark to 0 after a bright flash), Φ is the flash intensity in photoisomerizations per rod, *A* is the "amplification constant" of the rod (in units of $s^{-2}\ \Phi^{-1}$), and t_{eff} is a brief delay (of magnitude a few milliseconds). Eq. 5 can be described as a "delayed Gaussian" function of time.

Lamb and Pugh (1992) showed that Eq. 5 gives a good account of the activation phases of entire families of photocurrent responses of single amphibian rods for flashes producing Φ up to 20,000 or so; above this intensity "rate saturation" sets in, because of

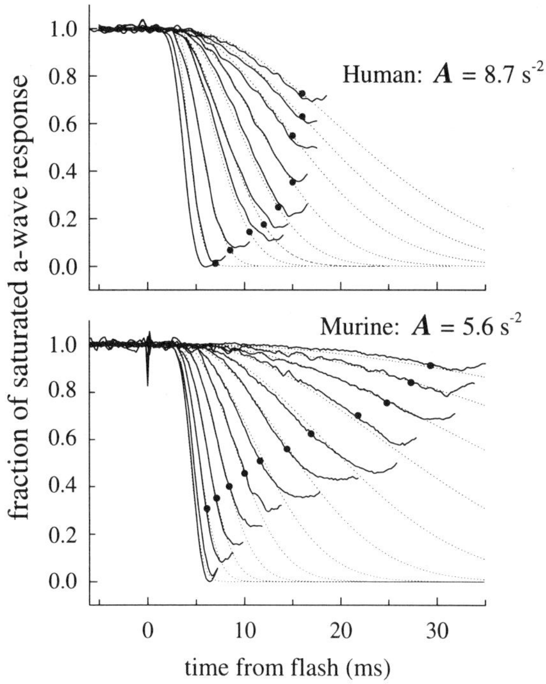

Figure 10. Application of the kinetic analysis of Figure 7 (Eq. 2) to families of human and murine a-waves (data are reprinted from Breton *et al.*, 1994, their Fig. 3, and Lyubarsky and Pugh, 1996, their Fig. 3).

the capacitive current of the membrane (Penn and Hagins, 1972; Cobbs and Pugh, 1987; Breton *et al.*, 1994).* Equation 5 also provides a good account of the responses of isolated macaque rods (Pugh and Lamb, 1993, republished data of Baylor *et al.*, 1984), and isolated human rods (Kraft *et al.*, 1993).

To apply Eq. 5 to a-wave data, we need the relationship between the directly observed a-wave response transient, $a(t)$, and F:

$$\frac{a_{\text{max,dark}} - a(t)}{a_{\text{max,dark}}} = F(t) \tag{6}$$

Equation 6 may at first appear to be in conflict with Eq. 4, but it is not because $a(t)$ and a_{max} are complements of one another: i.e., $a_{\text{max,dark}} = a(t) + a_{\text{max}}(t)$, because a_{max} here always refers to the residual a-wave amplitude that would be measured at any instant were a saturating flash to be delivered.

If Eq. 5 and 6 are combined, the kinetics of the a-wave can be predicted over a wide intensity range, as illustrated in Figure 10, where application to data of a human and a murine subject are shown. Clearly, the G-protein cascade model gives an excellent kinetic account of

* When the photoreceptor capacitive currents are of significant magnitude—as is the case for the response of rods to intense flashes and for most cone responses—one must explicitly convolve Eq. 1 with a first-order filter that represents the effect of the membrane time constant. Thus, a somewhat more complicated formula than Eq. 1 is required (Cideciyan and Jacobson, 1996; Smith and Lamb, in press). For rod responses at relatively low intensities, the effect of the membrane time constant can be lumped into the delay term t_{eff} (Lamb and Pugh, 1992).

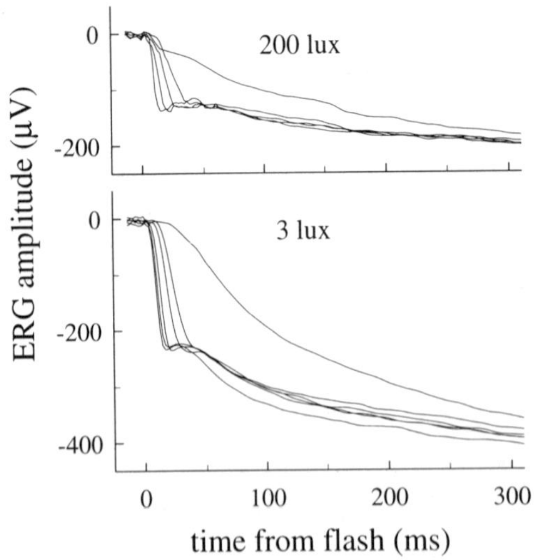

Figure 11. Electroretinograms of isolated retinas of albino rats raised on 12L:12D light cycles of two different illuminations, 200 lux and 3 lux. All recordings were performed at room temperature, 22–23°C. On the basis of the temperature dependence of the circulating current magnitude reported by Penn and Hagins (1972) for the albino rat, the a-wave amplitudes would be expected to be about 8-fold larger at body temperature, 37–38°C, or about 1.9 mV for the 3-lux animals and 1.07 mV for the 200-lux animals. The flashes produced up to approximately 10^6 Φ per rod. (Reprinted from Reiser *et al.*, 1996, their Fig. 1.)

the data, adding further evidence in support of the hypothesis that the a-wave directly reflects the suppression of the rod circulating current. Interestingly, the amplification constant of mammalian rods (A = 5–10 s^{-2} Φ^{-1}) is about 100-fold higher than that of amphibian rods (A = 0.05–0.1 s^{-2} Φ^{-1})! The theoretical analysis underlying the amplification constant gives a quantitative account of this 100-fold effect (Lamb and Pugh, 1992; Pugh and Lamb, 1993).

3.9. Pharmacological Isolation of the a-wave

In many experiments in the literature, pharmacological manipulations have been used to isolate photoreceptor activity from the activity of inner retinal cells. One of the most commonly used manipulations of this kind is the addition of aspartate to the medium bathing the isolated retina (or its injection into the vitreous). Aspartate is an analogue of the universal photoreceptor synaptic transmitter, glutamate, and is widely held to block all synaptic transmission from the photoreceptors to the bipolars and other retinal neurons. Figure 11 presents the data of an experiment in which ERG's were measured from isolated retinas of albino rats exposed to 116 mM aspartate, as reported by Reiser *et al.* (1996). The data are the averaged responses of two groups of albino rats raised under different illumination conditions. (Below, we will specifically consider the effects of rearing illumination on the a-waves.) The ERGs under these conditions have no b-wave and exhibit two distinct vitreal-negative components: a fast component ("fast PIII"), whose initial velocity clearly depends on light intensity at all intensities but whose magnitude clearly saturates, and a slower component ("slow PIII"), whose form appears independent of light intensity at all but the lowest intensity and which saturates at much longer times (not shown). As illustrated in Figure 12, the fast component is clearly the a-wave and is well described by the same kinetic analysis that describes the a-waves of the intact eye.*

* The nature of slow PIII may interest the reader. If, as is likely, the 116 mM aspartate present in the modified Locke's solution completely blocked all synaptic transmission from the photoreceptors by effectively occupying all postsynaptic glutamate receptors, then slow PIII must be attributed to a radially oriented cell type that is "driven" in some way by a glutamate-transmission-independent process triggered by the light response of the rods. One candidate would be Müller cells, which may develop a "circulating current" that is driven by K^+ released from the photoreceptors (Newman and Odette, 1984).

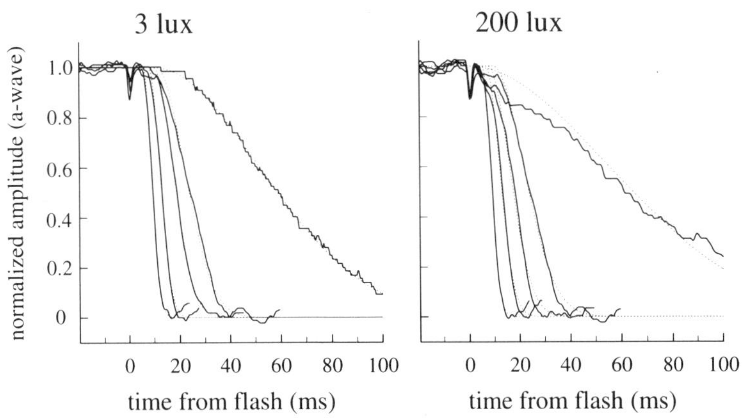

Figure 12. Application of the kinetic analysis of Eq. 2 to the early or "fast PIII" phase of the data of Figure 9. (Reprinted from Reiser *et al.*, 1996, their Fig. 3).

In summary, then, beginning with the pioneering work of Hagins and colleagues in the 1970s, the evidence has mounted steadily, such that the propositions embodied in Eq. 1, 4, and 6 can be put forward as a unified principle, "The First Linking Principle of Mammalian Electroretinography":

(a) The saturated a-wave amplitude of the human and murine ERG is directly proportional to the instantaneous (preflash) circulating current of the rods under all conditions,* and

(b) The complement of the directly observable a-wave transient is directly proportional to the rod circulating current during light responses.

4. THE SCOTOPIC B-WAVE ORIGINATES PRIMARILY IN "DIPOLE CURRENTS" GENERATED BY ROD BIPOLARS

4.1. The Murine Scotopic b-Wave Response Half-Saturates at Less Than 1 Φ Per Rod

Figure 13 illustrates a family of scotopic b-waves collected from a mouse under the same experimental conditions employed to obtain the ERG data of Figures 3 and 7, except here the flashes are only very weak ones, like the uppermost 10 traces of Figure 3. This family of response curves reaches its half-saturation point at an intensity estimated to produce only 1 photoisomerization (Φ) for every 3 rods! There are several reasons for trusting the calibrations and believing that this exquisite sensitivity is a property of the

* An important caveat must be added to this first principle: the circulating currents of cones necessarily contribute to the photoreceptor layer dipole current. Considerable work in humans supports the notion that an a-wave component due to cones can be isolated and characterized (e.g., Hood and Birch, 1995; Cideciyan and Jacobson, 1996; Smith and Lamb, in press). In mice the cone a-wave appears to be largely negligible ($\leq$ 50 μV at cornea) relative to the rod a-wave (typically, 500 μV at cornea) (Lyubarsky and Pugh, 1996).

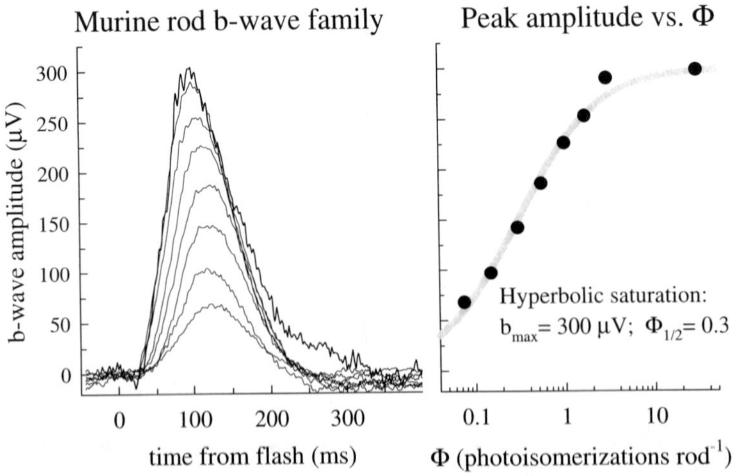

Figure 13. Left panel: Family of b-wave responses of a C57BL/6 mouse to very dim flashes. Right panel: The peak amplitude of the responses in the left panel are plotted with respect to the flash intensity, expressed in photoisomerizations per rod (each filled circle corresponds to the peak amplitude of one of the traces at left, and both panels have the same ordinate scale). The response amplitude vs. intensity data have been fitted with a "Naka-Rushton" or hyperbolic saturation function of the form $b_{peak}/b_{max} = \Phi/(\Phi + \Phi_{½})$. Note that $\Phi_{½} = 0.3$, i.e., 1 photoisomerization for every three rods.

cells of the retina underlying the b-wave. One is given in Figure 10: the recovery of the saturated a-wave amplitude to its fully dark-adapted level is tightly consistent with the recovery time course of the circulating current expected from the classic isolated-rat-retina experiments of Penn and Hagins (1972) and from the single-primate-rod suction-electrode data of Baylor *et al.* (1984). Such consistency implies that the estimation of Φ is accurate. A second reason is given by the kinetic analysis of Figure 10: the amplification coefficients estimated for human- and murine-rod a-waves are very close to those obtained from applying Eq. 5 to the photocurrent responses of isolated mammalian rods (Kraft *et al.,* 1993; Pugh and Lamb, 1993). The calculations involved in estimating A require estimates of Φ, and these estimates made for human and murine ERG experiments and for single-cell electrophysiology are completely independent—although they involve related calculations. A third reason, implicit in the abscissa of Figure 13, is that many other investigators have reported closely comparable scotopic b-wave sensitivity in various mammals. This point is demonstrated in Table 1, in which we provide a summary of published data from a number of investigations in different mammalian species. In summary, we can be confident that the half-saturating intensity in Figure 13 is well estimated in terms of photoisomerizations per rod. Moreover, it is reasonable to identify the b-waves in Figure 13 as "scotopic" or rod-driven, because at such extremely low flash intensities any signals from cones can be expected to be negligible (certainly cones make no contributions to the detection of such dim stimuli in human psychophysical experiments).

What then is the identity of the retinal cell underlying the scotopic b-wave? We will now review four lines of evidence supporting the hypothesis that the b-wave originates in the light-activated currents of the rod bipolars: (1) sensitivity/saturation, (2) electrical sign, (3) kinetics, and (4) molecular biology.

Table 1. Parameters of mammalian scotopic b-wave

Species	ERG method[a]	b_{max} (μV)	$\Phi_{½}$[b]	Reference
Rat	Corneal	520	0.7	Cone (1963)
Rat	Corneal	700	≈ 2	Green and Powers (1982)
Mouse (n = 8)	Corneal	392 (± 110)	2.5 (± 3.2)	Low (1987)
Mouse (n = 4)	Corneal	292 (± 120)[c]	0.9	Jiang *et al.* (1996)
Mouse (n = 4)	Corneal	493 (± 105)[d]	0.7 (± 0.2)	Falsini *et al.* (1997)
Cat	Intravitreal	900	1	Robson and Frishman (1995)
Human	Corneal	—	2.9	Fulton and Rushton (1978)
Human	Corneal	475	3	Peachey *et al.* (1989)
Human (n = 50)	Corneal	245.5	8.4	Hood *et al.* (1994)
Human (n = 57)	Corneal	418 (± 74)	2.2 (± 1.0)	Jacobson *et al.* (1994)
Human (n = 17)	Corneal	355	12	Sandberg *et al.* (1990)

[a]Amplitudes may vary across experiments not only because of the location and type of measuring electrode but also because of the reference electrode location.
[b]Intensity in photoisomerizations per rod ($\Phi_{½}$) that produces a half-maximal scotopic b-wave; in a number of cases in the table, we estimated this value from the published data.
[c]Mice reared under unspecified light conditions having 129SV genetic background and mean age 4 months (range: 3–5 months). The b-wave wave magnitude (790 ± 380 μV) reported by Jiang *et al.* (1996, their Table 1) of these same animals was incorrectly identified as the scotopic b-wave; in fact it was the full rod- + cone-mediated b-wave, and the value in this table gives the correct values for the scotopic b-wave.
[d]Mice with C57BL/6 background, reared at 2.5 lux 12L:12D, having mean age 2.5 months (range: 2–3 months.)

4.2. Sensitivity and Saturation of Scotopic B-Wave May Result from Rod–Rod Bipolar Convergence

The fully dark-adapted vertebrate eye is characterized by a profound general principle: the retina is specialized for the generation and processing of rod responses to single captured photons. As an example of this specialization, consider the behavioral experiments with toads (*Bufo bufo*) done by Aho *et al.* (1993): these investigators demonstrated that the toad achieves 100% accuracy in its tongue-snapping at moving targets under illumination conditions that produce less than 1 Φ per rod per 10 seconds! Mammals have useful vision under "twilight" conditions that produce less than 1 photoisomerization per rod per minute and good vision under conditions that generate less than 5 Φ per rod per second, i.e. less than 1 Φ per rod per integrating time. Savage and Banks (1992), for example, show that human contrast threshold is only 10% at a mean retinal illuminance of 0.05 scotopic trolands; this threshold corresponds to less than 0.5 Φ per rod per second.

The retinal circuitry in cats underlying the processing of rod single-photon responses is thought to be as follows (Freed *et al.*, 1987; Sterling *et al.*, 1988):

1500 rods → 100 RB → 5 AII → 4 CBb_1 → 1 on-beta GC (convergent path)
1 rod → 2 RB → 5 AII → 8 CBb_1 → 2 on-beta GC (divergent path)

where RB means rod bipolar, AII is AII-amacrines, CBb_1 is cone-bipolar subtype 1, and on-beta GC denotes a depolarizing ganglion cell of the beta class.

On the basis of the convergence of signals, the RB appears to be a good candidate for the b-wave generator. If so, then, for the dark-adapted retina, the simultaneous photoisomerization of 1 rhodopsin molecule in each of the 15 or so rods making synapses with 1 RB is sufficient to saturate the RB's response. Put alternatively, it seems plausible that in the fully dark-adapted retina the simultaneous generation of 1 photoisomerization in each rod—suppressing a mere 3% of each rod's circulating current (Penn and Hagins, 1972; Baylor *et al.*, 1979b, Baylor *et al.*, 1984)—could completely activate the signal-transducing channels in the invaginating bipolar dendritic tip and, by dint of the simultaneous activation of the corresponding channels in all the rod-bipolar's dendritic tips, cause its response to saturate.

4.3. The Sign of the Scotopic b-Wave Is Consistent with Its Origin in a Dipole Current of Rod Bipolars, Initiated by an Inward Current Near the Rod–Rod Bipolar Synapse

Rod bipolars are "on-bipolars"; that is, their membrane potentials depolarize in response to light. Substantial evidence supports the conclusion that the depolarization of the rod bipolar in response to light occurs because the diminished release of glutamate at the invaginating ribbon synapse in the rod spherule causes the *opening* of cation-selective channels in the postsynaptic bipolar membrane (Nawy and Copenhagen, 1987; Nawy and Jahr, 1990a, b; Shiells *et al.*, 1981; Shiells and Falk, 1992, 1994). Moreover, it is now also clear that the glutamate-modulated channels of the rod bipolar are in fact cGMP-activated channels, homologous to those in the photoreceptor outer-segment membranes (Nawy and Jahr, 1990a; Nawy and Jahr, 1991; Shiells and Falk, 1992).

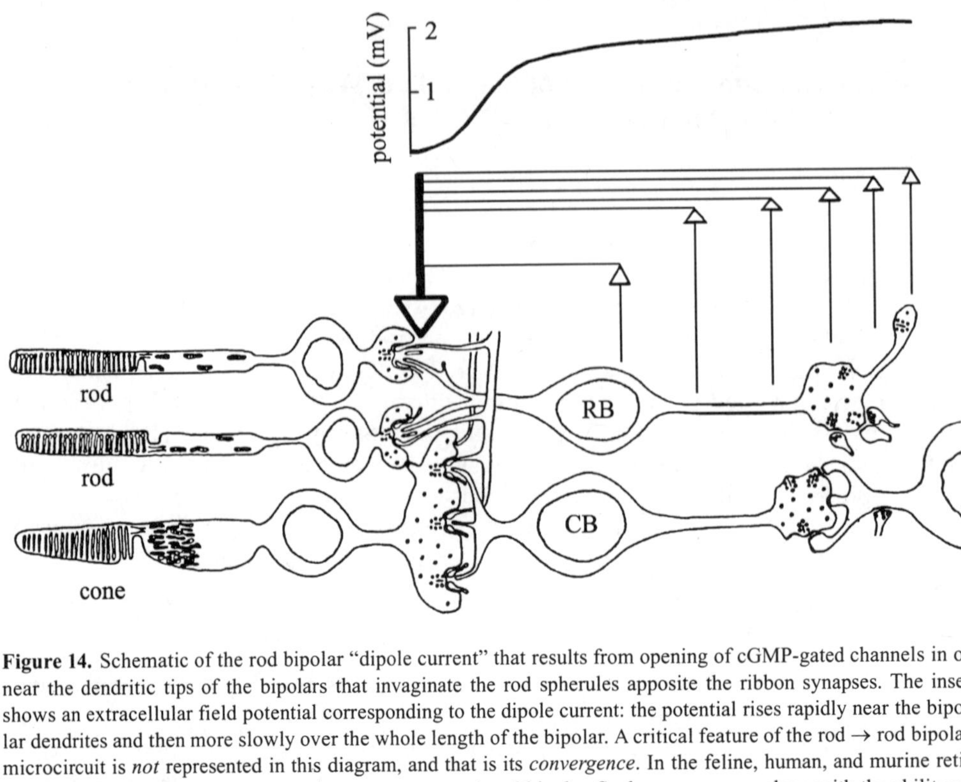

Figure 14. Schematic of the rod bipolar "dipole current" that results from opening of cGMP-gated channels in or near the dendritic tips of the bipolars that invaginate the rod spherules apposite the ribbon synapses. The inset shows an extracellular field potential corresponding to the dipole current: the potential rises rapidly near the bipolar dendrites and then more slowly over the whole length of the bipolar. A critical feature of the rod → rod bipolar microcircuit is *not* represented in this diagram, and that is its *convergence*. In the feline, human, and murine retinas it is expected that 15 or more rods converge on each rod bipolar. Such convergence, along with the ability of the circuit to transmit faithfully the single photon responses of the rods, is thought to underlie the ultrasensitive b-wave saturation function in Figure 12.

The rod bipolar is well situated in the retina to be the generator of the scotopic b-wave, as has been argued by Xu and Karwoski (1994) on the basis of a source/sink analysis. To illustrate this point, Figure 14 sketches the form of the rod dipole current expected on the assumption that the cGMP-activated channels activated by decreased glutamate release reside in or near the invaginating dendritic tips of the rod bipolars. The light-triggered opening of cGMP-activated channels in the rod bipolar dendrites creates a sink for inward charge flow, in the same way that the open cGMP-activated channels of the rod outer segment constitute a sink in the rod outer segment in the dark. Moreover, the same electrophysiological principles that produce an equality of the inner- and outer-segment limbs of the photoreceptor dipole current are in force: thus, the rod bipolar's photoresponse generates a radial dipole current. The direction of flow of the rod-bipolar dipole current in light is the same as that of the rod dipole current in the dark: thus, the rod bipolar dipole layer current generates an extracellular field that causes the vitreal side of the retina to be at a higher or more positive electrical potential than the scleral side. In other words, the rod bipolar's response to light should create a vitreal or corneal ERG component that is positive-going, like the scotopic b-wave.*

In closing, it is interesting to derive an estimate of the saturated magnitude of the dipole current of an individual rod bipolar, in effect "inverting" the calculation in Appendix I used to derive the magnitude of the saturated a-wave. (1) From Table 1, we find that b_{max}, the saturated scotopic b-wave amplitude, ranges roughly from 300 to 700 μV at the cornea. For humans and mice the typical values are about 400–500 μV; let us take the value 450 μV. Interestingly, a_{max}, taken as the saturated amplitude of the corneal a-wave in the dark, is of about the same magnitude in two representative mammalian populations: human, 463 ± 81 μV (Breton *et al.,* 1994); C47BL/6 mouse, 465 ± 135 μV (Lyubarsky and Pugh, 1996). (2) As noted above, Sterling *et al.* (1988) found on average 1 rod bipolar for every 15 rods in cats, and we can expect this ratio to be similar in other mammalian retinas. (3) The overall thickness of the bipolar layer is roughly the same as that of the photoreceptor layer, and the interstitial conductivity should be approximately the same as that of the inner-segment region of the photoreceptor layer (see Appendix I). It follows from points (1) to (3) that the saturated b-wave response corresponds to a saturated rod-bipolar cGMP-activated current roughly 15-fold greater than that of the circulating current of an individual rod. If the mammalian rod circulating current is taken to be 30 pA, then the saturated rod bipolar current *in situ* would be 450 pA.

4.4. Knockout of the mGluR6 Glutamate Receptor Eliminates Murine b-Wave Responses

Because all vertebrate photoreceptors hyperpolarize in response to light, the fact that rod bipolars depolarize in response to light leads to an insight: the synapse connecting the rods to the rod bipolars is a sign-inverting synapse. The most common glutamate receptors involved in chemical synaptic transmission are ionotropic receptors, which are nonspecific cationic channels, opened by the binding of glutamate (Seeburg, 1993; Peng *et al.,* 1995;

* Other hypotheses about the generator of the b-wave should not be dismissed without consideration. A long-standing candidate for the dipole generator underlying the b-wave is the Müller cell (see, e.g., Newman and Odette, 1984). The Müller cell has a radial orientation and retinal positioning similar to the bipolars'. Increased inward current near the photoreceptor terminals, presumably carried by K^+, would create a source/sink distribution somewhat like that of an on-bipolar cell. We think, however, that the total body of relevant evidence about the scotopic b-wave is most consistent with a rod bipolar generator current.

Lerma *et al.,* 1997). Such ionotropic receptors necessarily make sign-conserving synapses with presynaptic cells, such as the photoreceptor → horizontal cell and photoreceptor → off-bipolar cells. A substantial body of evidence has now accumulated that the mechanism underlying the electrical sign inversion of the photoreceptor → on-bipolar cells is a second-messenger cascade, somewhat similar to that of the rod cascade (Fig. 6).

There is histological, pharmacological, and molecular-biological evidence that the first component of the rod bipolar response is a metabotropic glutamate receptor (mGluR), which is a seven-transmembrane helix receptor homologous to rhodopsin and other visual pigments (see http://receptor.mgh.harvard.edu/GCRDBHOME.html). Recent histological investigations have revealed a number of distinct mGluRs in the retina (Hartveit *et al.,* 1995, Peng *et al.,* 1995), and a unique subtype, mGluR6, has been localized to rod bipolars (Nakajima *et al.,* 1993; Masu *et al.,* 1995).

Pharmacological evidence supporting the involvement of an mGluR receptor in on-bipolar signalling began to accumulate years ago in experiments in which certain glutamate analogues, in particular APB (2-amino-4-phosphonobutyrate), were found to block responses of on-cells in the retina selectively (Shiells *et al.,* 1981; Slaughter and Miller, 1981; Stockton and Slaughter, 1989). In time it became clear that these analogues were specific ligands for metabotropic glutamate receptors (mGluRs) (cf. Tian and Slaughter, 1995). Combined physiological and pharmacological analysis with APB also provided substantial evidence for linkage of the b-wave and on-bipolar responses in amphibians (e.g., Stockton and Slaughter, 1989; Gurevich and Slaughter, 1993; Tian and Slaughter, 1995) and in mammals (Euler *et al.,* 1996).

Compelling molecular-biological evidence for the hypothesis that a specific metabotropic receptor is the first link in on-bipolar responses comes from the work of Nakajima *et al.* (1993), who cloned a novel glutamate receptor, mGluR6, and localized it in on-bipolar cells, and from Masu *et al.* (1995). The latter investigators developed a line of transgenic mice with mGluR6 "knocked out"; absence of mGluR6 expression was established by immunocytochemistry. Histologically, mGluR6-KO animals appear to have normal retinas and, in particular, a normal bipolar cell layer. Strikingly, the ERG's of the animals exhibit no b-waves; moreover, no "on-cell" responses were recordable from the optic tectum.

The complete mGluR6-activated cascade of the bipolars has not yet been determined. Histological work by Vardi *et al.* (1993) has established the presence of a G-protein of the "G_0" class in the invaginating tips of the rod bipolars, and in the dendrites of cone on-bipolars, and this G-protein seems likely to be the second component of the rod bipolar cascade. Moreover, it seems nearly certain on the basis of the work of Nawy and Jahr (1990a, 1991) and Shiells and Falk (1992, 1994) that the final (electric signal) effector of the rod bipolar cascade is a cGMP-activated channel. It follows logically that both a guanylate cyclase and a cGMP-phosphodiesterase are present. There are candidates for identity of the cyclase (reviewed by Pugh *et al.,* in press) but none for the phosphodiesterase.

4.5. The Kinetics of the Activation of the b-wave Are Consistent with a Second-Messenger Cascade

Enzymes constitute "integrating steps" or low-pass filters, converting an impulsive input into a step (t^0 function), a step input into a ramp (t^1 function), and a ramp into a parabola (t^2 function). For example, as illustrated in Figure 9, an impulsive input (a photon captured by a rhodopsin molecule) leads to a step of R* enzymatic activity, the step of R* activity leads to a ramp of G* and PDE*, and the ramp of PDE* leads (initially) to a declining parabola of cGMP concentration in the outer segment* (the circulating current tracks the cGMP concentration through a power law—compare the two lowermost panels of Fig. 9).

In an elegant pair of papers, Robson and Frishman (1995, 1996) have pharmacologically isolated the on-response pathways of the cat retina, measured scotopic b-waves, and applied to their activation phases a kinetic analysis of the rod bipolar response based on a second-messenger cascade. They take as the "input" to the rod bipolar the activation phase of the rod response, which has the form of a parabola, $A\Phi t^2$; it bears emphasis that parabolic activation kinetics should hold especially well at the single-photon and dim-flash rod responses that provide the characteristic input to the rod bipolar (Table 1; see above; see Pugh and Lamb, 1993, their Fig. 10). Robson and Frishman (1995, their Fig. 12) show that t^5 activation kinetics very nicely describes both their pharmacologically isolated cat-rod b-wave responses and the rod-bipolar responses of the dogfish (Ashmore and Falk, 1980). The metabotropic cascade can be expected to contribute three additional integrating steps to the three contributed by the rod: (1) mGluR6 activation/inactivation by glutamate binding/unbinding, (2) G_o* activation by mGluR6* and terminal phosphate hydrolysis of G_o*, (3) cGMP hydrolysis by PDE* and synthesis by guanylate cyclase. The total of six integrating steps predictably produces t^5 activation kinetics.

4.6. Summary

In summary, then, substantial evidence now supports the following hypothesis, which we proposed should be called "The Second Linking Principle of Mammalian Electroretinography":

> The scotopic b-wave of the fully dark-adapted mammalian ERG is directly proportional to the light-induced, glutamate-regulated dipole current of the rod bipolars.

5. A MURINE B-WAVE COMPONENT CAN BE ISOLATED THAT PROBABLY ORIGINATES IN CONE ON-BIPOLAR CELLS

Rod bipolars are strictly "on-bipolars," but cone bipolars cells can be either "on" or "off" in their response to light. Cone on-bipolars and rod bipolars have a number of similarities: (1) when the respective photoreceptors hyperpolarize that make ribbon synapses with rod bipolars and cone on-bipolars, both types of bipolars depolarize; (2) responses of both are blocked by the metabotropic receptor antagonist APB; (3) both express a G-protein of the "G_o" class in their dendritic tips (Vardi *et al.,* 1993); and (4) responses of both are apparently lost in the mGluR6 knockout mouse, which exhibits no positive-going b-wave responses in its ERG's and which has no "on-responses" in the optic tectum (Masu *et al.,* 1995).

Point 4, needs some exposition, which is best done in the context of experimental results. Figure 15 illustrates a series of whole-field corneal ERG's collected from a mouse. In each trial (represented by a single trace) an intense conditioning flash (CF) was presented; this flash is estimated to have photoisomerized about 1% of the rhodopsin in the

* Note that at early times the delayed Gaussian formula, Eq. 5, describes a declining parabola: thus for small values of the argument, $1 - \exp[-\frac{1}{2}\Phi A(t - t_{eff})^2] \approx -\frac{1}{2}\Phi A(t - t_{eff})^2$.

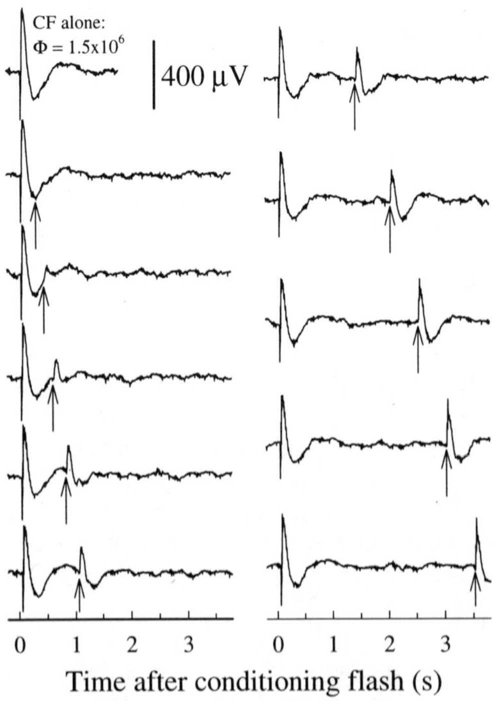

Figure 15. Isolation of cone-driven b-waves in the murine retina by a double-flash protocol. Each individual trace—beginning at top left and going down, then to top right and going down—presents the corneal ERG obtained in response to a pair of flashes. The first flash produce about 1.5×10^6 Φ, approximately a 1% bleach of the rhodopsin; under these conditions, the rod circulating current does not begin to show recovery before 4 s (Penn and Hagins, 1972, their Figs. 8, 9). The second flash was 1/10 as intense, generating 1.5×10^6 Φ in rods, and produces a response driven solely by cones. The cone-driven response reaches its saturated amplitude in 3 s. By analogy with the analysis that leads to identification of the rod-bipolar dipole current as the origin of the scotopic b-wave, it is proposed that the cone-driven b-waves in this figure arise from the cone on-bipolars.

rods upon each presentation, and completely suppresses the rod circulating current for at least 4 s (see, Penn and Hagins, 1972, their Fig. 8). At various times after the presentation of the CF, a second, probe flash was presented, as indicated by the upward-pointing arrows. As can be seen, a corneal-positive ERG response to the probe flash appears by about 0.5 s, and this response component has reached its saturating amplitude by about 2.5 s after the CF; the half-time of its recovery is about 1 s. Because the rod circulating current is completely suppressed, this response component can safely be attributed to cone-driven cells; by conventional nomenclature it is a "b-wave." By analogy with the analysis applied to the scotopic b-wave, it is natural to attribute this cone b-wave to a dipole current originating in cone on-bipolar cells. Does this hypothesis make sense?

Yes, it does. Cones have been estimated to constitute about 3% of the total number of photoreceptors in the mouse retina (Carter-Dawson and LaVail, 1979; Szel *et al.*, 1992). Although no anatomical survey of the murine retinal bipolar cell types has been published, it seems likely, by analogy with cat and primate, that cone on-bipolars will be nearly as numerous as rod bipolars (Freed *et al.*, 1987; Sterling *et al.*, 1988). Because the rod bipolars and cone bipolars occupy the same retinal layer (and are thus exposed to comparable extracellular resistance profiles), it is plausible that cone on-bipolars generate the typically 200–300 μV corneal cone-driven potentials seen in experiments such as that in Figure 15, as compared to the 400–500 μV scotopic b_{max} values attributed to rod bipolars (Table 1).

Using the double-flash isolation technique of Figure 15, we have recently measured the action spectrum of the murine cone-driven b-wave (Lyubarsky *et al.*, 1997) and established that it reflects two distinct cone types, a UV-sensitive cone ($\lambda_{max} \approx 360$ nm) and a green-sensitive cone ($\lambda_{max} \approx 510$ nm), in C57BL/6 mice. The b-wave responses driven by UV-sensitive cones appear to be more sensitive than those driven by the green-sensitive cones, consistent with previous flicker photometric measurements (Jacobs *et al.*, 1991; Deegan and Jacobs,

1993). Anatomical investigations of the numbers and synaptic connectivity of mouse cone-bipolars should prove very valuable as the mouse eye is developed for the investigation of retinal disease and signal processing. Certainly, the notion that the rodent eye has little or no cone function should be extirpated from the literature. Indeed, save for the absence of a cone-dominated fovea, the murine eye appears to have considerable functional similarity to that of humans. One of our research goals has been to help lay the foundation for the use of mice in the study of eye disease and in therapeutic intervention in genetic disease. The demonstrable ability to characterize cone function with the murine ERG considerably extends the utility of mouse models of genetic eye disease and opens the possibility of assessing therapeutic interventions aimed at preserving cone function under disease conditions that cause rods to deteriorate. We have recently used the ERG methods presented here to characterize rod and cone function in mice with a rhodopsin transgene that causes disease symptoms mimicking human retinitis pigmentosa and have demonstrated that cone function survives much longer than that of rods (Falsini *et al.,* 1997).

6. EFFECTS OF LIGHT-REARING HISTORY ON RETINAL CELL RESPONSES AS MANIFEST IN RODENT ELECTRORETINOGRAPHY

The investigations of Williams and others have established that a number of properties of the rods of albino rats are regulated by light-rearing history. These properties include the length of the outer segment (Penn and Williams, 1986; Schremser and Williams, 1995a; Reiser *et al.,* 1996), the rhodopsin content of the outer segment (Schremser and Williams, 1995a, b), the rate of regeneration of rhodopsin after a substantial bleaching exposure (Penn and Williams, 1986), and the amounts of the transduction cascade proteins transducin and arrestin (Organisciak *et al.,* 1991). These "photostatic" regulations are not obligatorily associated with light damage to the retina: they can occur with no loss of photoreceptor nuclei in the outer nuclear layer.

An example of an effect of light-rearing history on retinal-cell electrical properties was reported by Reiser *et al.* (1996). They measured the transretinal ERG's of isolated retinas of albino rats reared under 3-lux and 200-lux *Ganzfeld* illumination conditions (12L:12D). As illustrated in Figure 11, the saturated amplitude of the a-wave was lower in the animals reared at 200 lux than in those reared at 3 lux. The reduced a-wave amplitudes of the 200-lux animals was well accounted for by the combination of an observed fact and a hypothesis: the fact is that the outer segment lengths of the animals reared at 200 lux were reduced on average to 68% the length of those reared at 3 lux; the hypothesis is that the circulating currents of the rods of both groups of animals are proportional to the outer segment lengths and have the same proportionality constant. This hypothesis can be considered an extension of Eq. 1 and can be reasonably justified.* Reiser *et al.* found a_{max} in

* The justification rests on three points. (a) The cGMP-activated channels can be expected to be expressed at a constant surface density, and the number per unit length of outer segment membrane should therefore be the same in the long rods of the 3-lux animals and in the shorter rods of the 200-lux animals. (b) The voltage dependence of the rod cGMP-activated channel is very weak in the voltage operating range of the channels (Baylor and Nunn, 1986), and thus the total inward current through the cGMP-channels should be nearly proportional to the outer segment length. (c) The resistance of the interstitial space in the outer-segment region is much lower than that in the inner-segment region, and the two groups should not differ in the latter, because the outer nuclear region's thickness was not different, as shown by histology.

200-lux animals was reduced to 56% that of the animals reared at 3 lux, in good agreement with the expectation.

In part, the findings of Reiser *et al.* (1996) are significant in that they provide a foundation for electrophysiological investigation of one of the principal features of "photostasis"—the regulation of the length of rod outer segments. Measurement of a_{max} in a standard manner can be used to estimate the relative length of the rods quantitatively and noninvasively. This method should, for example, make possible measurements on a single cohort of animals reared under different nondamaging illumination conditions over their life history. Somewhat more speculative applications can be made with animals having certain types of eye disease, particularly early in their history.

The significance of carefully controlling the light-rearing history of all rodents—not just albino rats—used in vision research cannot be overstated at this time when many lines of transgenic rodents are being generated to test hypotheses about retinal mechanisms, including those that lead to hereditary eye disease. Recently, for example, Naash *et al.* (1996) have demonstrated increased susceptibility to light damage in mice with rhodopsin point mutations, and Wang *et al.* (1997) have shown that these same transgenic mice exhibit accelerated photoreceptor degeneration relative to controls when reared on a 12L:12D cycle with illumination of about 75 lux.

Inspired by the investigations of Williams and colleagues, we have developed customized rearing environments for mice, with precisely regulated light, temperature and air filtration. Each environment is a light-tight box 2 × 2 × 6 feet, the interior of which is painted flat white and which holds 16 individual mouse retainers and typically 50–70 mice. The interior lighting in the environment is diffuse, effected by low-wattage incandescent bulbs placed behind diffusing screens in the ceilings of the boxes. Measurements with a light meter show that the lighting is direction-independent. The illumination in the box can be varied from 0 to 30 photopic lux. From calculations presented in Appendix II, we estimate that 1 photopic lux illumination produces 650 photoisomerizations per rod per second for a mouse with a fully dilated pupil and about one-third this amount for a natural pupil in a steady illumination of 2–3 lux.

In Figure 16 we provide summary data that may prove useful for characterizing the effects of aging and light rearing on properties of the murine ERG. All the results were collected from mice of the C57BL/6 strain. In the upper panel of Figure 16 is given the amplitude of the scotopic b-wave for a flash estimated to produce 1.5 Φ per rod on average. This flash produces a saturated or very nearly saturated scotopic b-wave in fully dark-adapted mice of this strain (Fig. 13; Table 1). In the middle panel of Figure 16 we give the saturated amplitude of the a-waves of the same animals when they are fully dark adapted, obtained in response to a flash that produces 2×10^6 Φ per rod. Finally, in the lowest panel we provide the saturated amplitude of the isolated cone b-wave of the same animals, measured with the double-flash protocol illustrated in Figure 15.*

The data in Figure 16 come from two groups of animals: the ERG's of one group, represented by the open symbols, were measured shortly after their arrival from the commercial supplier, Charles River Laboratories, Inc. The other animals, of exactly the same origin and strain, were born and reared exclusively in our light-controlled environments

* The flash employed to collect the saturating cone b-wave response was the same as that used to generate the saturated a-wave responses reported in the middle panel, but we are unsure of the absolute numbers of photoisomerizations per cone, because mouse cone collecting areas are not yet known. Rough calculations show that the flash should isomerize approximately 1% of the pigments in the two cone classes.

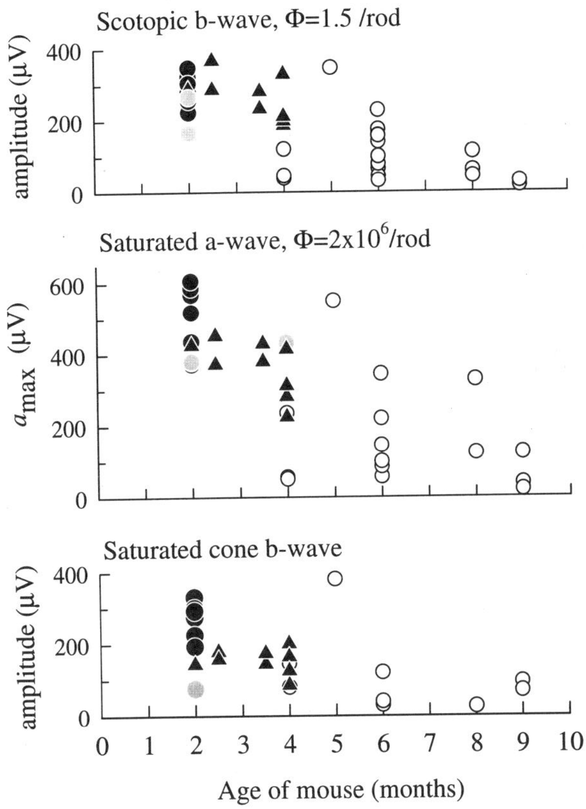

Figure 16. Evidence of light-rearing history effects in pigmented (C57BL/6) mice. The common abscissa for all three panels is the age of the mouse at the time of ERG testing. The uppermost panel plots the b-wave peak amplitude in response to a flash estimated to produce on average 1.5 Φ per rod (a nearly saturated scotopic b-wave; see Fig. 13). The middle panel shows the saturated amplitude of the a-wave, obtained in response to a flash producing 2×10^6 Φ per rod The lowermost panel shows the saturated amplitude of the cone driven b-wave, determined as in Figure 14, 3 sec after the conditioning flash. The cone b-wave saturating test flash was the same as that used to generate the saturated a-waves in the middle panel. Data are shown for four populations of mice. *Open circles* represent the data of animals tested within a few days of receipt from the commercial supplier, Charles River Laboratories. *Filled symbols* represent data of animals born and reared in 12L:12D controlled-light environments as follows: filled circles, darkness from birth then 0.5 lux for 4 weeks prior to testing; filled triangles, 2.5 lux from birth. Gray circles represent a few animals from the commercial supplier that were maintained at 0.5 lux for three weeks prior to testing.

on 12L:12D cycles with the light cycle illuminance specified in the figure caption. All animals were fully dark adapted overnight in the same manner prior to the experiments.

Perhaps the most striking feature of the data is the apparent effect of aging per se. All three saturation parameters decline rapidly after 2 months of age. At 4 months—when there are data both from animals whose ERG's were measured a few days after their arrival from the supplier and from animals raised in the controlled-light environments—it can be seen that the new arrivals have lower average amplitudes for all three ERG parameters, scotopic b-wave amplitude, saturated a-wave amplitude, and saturated cone-driven b-wave.

Another, stronger trend is also clear: animals that are reared under precisely controlled lighting conditions have noticeably less variability in any given parameter. This point

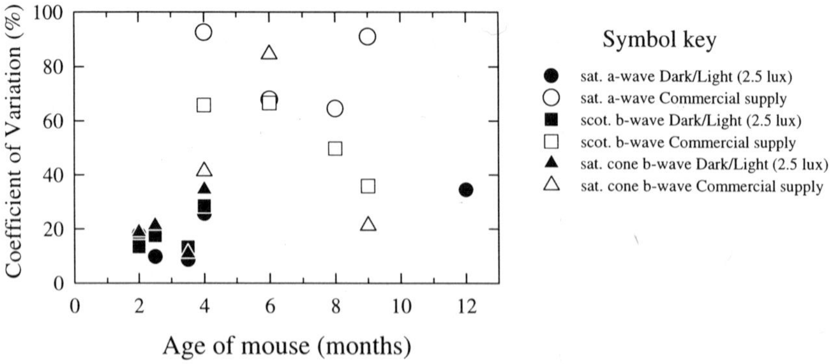

Figure 17. Coefficients of variation for the murine ERG parameters of Figure 16 as a function of age. The coefficient of variation is the mean response divided by the standard deviation, expressed as a percentage. The points are all derived from the data of Figure 16, except for the filled circle at 12 months, which represents results obtained recently from 5 mice who were born and have been maintained for the entire period in the controlled rearing conditions.

is made more clearly in Figure 17, where we plot the coefficient of variation as a function of age. Note that on this graph we have added a point from five in-house-reared animals of age 12 months and that the coefficient of variation is small for these animals, despite their reduced saturated a-wave amplitudes.

We emphasize in closing that we do not consider Figures 16–17 to represent an experiment in controlled light rearing. Rather, we hope that by publishing these data we can add to the weight of testimony of the other participants in the symposium: light-rearing conditions exert important regulation on many of the properties of the retina, and it behooves everyone investigating rodent retinas to follow the lead of Williams and colleagues and rear animals under precisely controlled illumination conditions.

ACKNOWLEDGMENTS

Our work was supported by NIH EY-02660 and the Penn Initiative for Retinitis Pigmentosa. B. Falsini was supported in part by funds from the Fulbright Foundation.

LITERATURE CITED

Aho, A. C., Donner, K., Helenius, S., Larsen, L., and Reuter, T. (1993). Visual performance of the toad (*Bufo bufo*) at low light levels: retinal ganglion cell responses and prey-catching accuracy. *Journal of Comprehensive Physiology A* 172:671–682.

Ashmore, J. F., and Falk, G. (1980). Responses of rod bipolar cells in the dark-adapted retina of the dogfish, *Scyliorinus canicula. Journal of Physiology (London)* 300:115–150.

Baylor, D. A., and Nunn, B. J. (1986). Electrical properties of the light-sensitive conductance of rods of the salamander *Ambystoma tigrinum. Journal of Physiology (London)* 371:115–145.

Baylor, D. A., Lamb, T. D., and Yau, K.-W. (1979a). The membrane current of single rod outer segments. *Journal of Physiology (London)* 288:589–611.

Baylor, D. A., Lamb, T. D., and Yau, K.-W. (1979b). Responses of retinal rods to single photons. *Journal of Physiology (London)* 288:613–634.

Baylor, D. A., Nunn, B. J., and Schnapf, J. L. (1984). The photocurrent, noise and spectral sensitivity of rods of the monkey *Macaca fascicularis. Journal of Physiology (London)* 357:575–607.

Berson, E. L. (1992). Electrical phenomena in the retina. In: Hart, W. M., Jr. (ed.). *Adler's Physiology of the Eye.* Mosby-Year Book, St. Louis, Missouri. pp. 506–567.

Birch, D. G., Hood, D. C., Nusinowitz, S., and Pepperberg, D. R. (1995). Abnormal activation and inactivation mechanisms of rod transduction in patients with autosomal dominant retinitis pigmentosa and the Pro-23-His mutation. *Investigative Ophthalmology and Visual Science* 36:1603–1614.

Breton, M. E., Schueller, A. W., Lamb, T. D., and Pugh, E. N., Jr. (1994). Analysis of ERG a-wave amplification and kinetics in terms of the g-protein cascade of phototransduction. *Investigative Ophthalmology and Visual Science* 35:295–309.

Carter-Dawson, L. D., and LaVail, M. M. (1979). Rods and cones in the mouse retina. I. Structural analysis using light and electron microscopy. *Journal of Comparative Neurology* 188:245–262.

Cideciyan, A. V., and Jacobson, S. G. (1996). An alternative phototransduction model for human rod and cone ERG a-waves: normal parameters and variation with age. *Vision Research* 36:2609–2621.

Cobbs, W. H., and Pugh, E. N., Jr. (1987). Kinetics and components of the flash photocurrent of isolated retinal rods of the larval salamander *Ambystoma tigrinum. Journal of Physiology (London)* 394:529–572.

Cone, R. A. (1963). Quantum relations of the rat electroretinogram. *Journal of General Physiology* 46:1267–1286.

Daiger, S. P., Sullivan, L. S., and Rodriguez, J. A. (1995). Correlation of phenotype with genotype in inherited retinal degeneration. *Behavioral and Brain Sciences* 18:452–467.

Deegan, J. F., II and Jacobs, G. H. (1993). On the identity of the cone types of the rat retina. *Experimental Eye Research* 56:375–377.

Dowling, J. E., and Boycott, B. B. (1966). Organization of the primate retina: electron microscopy. *Proceedings of the Royal Society London B* 166:80–111.

Euler, T., Schneider, H., and Waessle, H. (1996). Glutamate responses of bipolar cells in a slice preparation of the rat retina. *Journal of Neuroscience* 16:2934–2944.

Falsini, B., Lyubarsky, A. L., Pugh, E. N., Jr., and Wong, F. (1997). Rod photoreceptor dysfunction in transgenic mice expressing P347S rhodopsin gene mutation: ERG evidence. *Investigative Ophthalmology and Visual Science* 38:886.

Freed, M. A., Smith, R. G., and Sterling, P. (1987). Rod bipolar array in the cat retina: pattern of input from rods and GABA-accumulating amacrine cells. *Journal of Comparative Neurology* 266:445–455.

Fulton, A. B., and Rushton, W. A. H. (1978). The human rod ERG: correlation with psychophysical responses in light and dark adaptation. *Vision Research* 18:793–800.

Green, D. G., and Powers, M. K. (1982). Mechanisms of light adaptation in rat retina. *Vision Research* 22:209–216.

Gurevich, L., and Slaughter, M. M. (1993). Comparison of the waveforms of the on bipolar neuron and the b-wave of the electroretinogram. *Vision Research* 33:2431–2435.

Hagins, W. A., Penn, R. D., and Yoshikami, S. (1970). Dark current and photocurrent in retinal rods. *Biophysical Journal* 10:380–412.

Hargrave, P., and McDowell, J. H. (1992). Rhodopsin and phototransduction. *International Review of Cytology* 137B:49–97.

Hartveit, E., Brandstatter, J. H., Enz, R., and Waessle, H. (1995). Expression of the mRNA of seven metabotropic glutamate receptors (mGluR1 to 7) in the rat retina. An *in situ* hybridization study on tissue sections and isolated cells. *European Journal of Neuroscience* 7:1472–1483.

Hood, D. C., and Birch, D. G. (1995). Phototransduction in human cones measured using the a-wave of the ERG. *Vision Research* 35:2801–2810.

Hood, D. C., Shady, S., and Birch, D. (1994). Understanding changes in the b-wave of the ERG caused by heterogeneous receptor damage. *Investigative Ophthalmology and Visual Science* 35:2477–2488.

Jacobs, G. H., Neitz, J., and Deegan, J. F. (1991). Retinal receptors in rodents maximally sensitive to ultraviolet light. *Nature* 353:655–656.

Jacobson, S. G., Kemp, C. M., Cidecyan, A. V., Macke, J. P., Sung, C.-H., and Nathans, J. (1994). Phenotype of stop codon and splice site rhodopsin mutations causing retinitis pigmentosa. *Investigative Ophthalmology and Visual Science* 35:2521–2534.

Jiang, H., Lyubarsky, A., Dodd, R., Vardi, N., Pugh, E. N., Baylor, D., Simon, M. I., and Wu, D. (1996). Phospholipase C β4 is involved in modulating the visual response in mice. *Proceedings of the National Academy of Sciences of the USA* 93:14598–14601.

Koutalos, Y., and Yau, K. W. (1993). A rich complexity emerges in phototransduction. *Current Opinion in Neurobiology* 3:513–519.

Kraft, T. W., Schneeweis, D. M., and Schnapf, J. L. (1993). Visual transduction in human rod photoreceptors. *Journal of Physiology (London)* 464:747–765.

Lamb, T. D., and Pugh, E. N., Jr. (1992). A quantitative account of the activation steps involved in phototransduction in amphibian photoreceptors. *Journal of Physiology (London)* 449:719–757.

Lem, J., and Makino, C. L. (1996). Phototransduction in transgenic mice. *Current Opinion in Neurobiology* 6:453–458.

Lerma, J., Morales, M., Vicente, M. A., and Herreras, O. (1997). Glutamate receptors of the kainate type and synaptic transmission. *Trends In Neuroscience* 20:9–12.

Low, J. C. (1987). The corneal ERG of the heterozygous retinal degeneration mouse. *Graefe's Archives of Clinical and Experimental Ophthalmology* 225:413–417.

Lyubarsky, A. L., and Pugh, E. N., Jr. (1996). Recovery phase of the murine rod photoresponse reconstructed from electroretinographic recordings. *Journal of Neuroscience* 16:563–571.

Lyubarsky, A. L., Falsini, B., and Pugh, E. N., Jr. (1997). Different functional properties of the components of the mouse ERG driven by UV and green cones. *Investigative Ophthalmology and Visual Science* 38:885.

Masu, M., Iwakabe, H., Tagawa, Y., Miyoshi, T., Yamashita, M., Fukuda, Y., Sasaki, H., Hiroi, K., Nakamura, Y., Shigemoto, R., Takada, M., Nakamura, K., Nakao, K., Katsuki, M., and Nakanishi, S. (1995). Specific deficit of the ON response in visual transmission by targeted disruption of the mGluR6 gene. *Cell* 80:757–765.

Naash, M. I., Peachey, N. S., Li, Z.-Y., Gryczan, C. C., Goto, Y., Blanks, J., Milam, A. H., and Ripps, H. (1996). Light-induced acceleration of photoreceptor degeneration in transgenic mice expressing mutant rhodopsin. *Investigative Ophthalmology and Visual Science* 37:775–782.

Nakajima, Y., Iwakabe, H., Akazawa, C., Nawa, H., Shigemoto, R., Mizuno, N., and Nakanishi, S. (1993). Molecular characterization of a novel retinal metabotropic glutamate receptor mGluR6 with a high agonist selectivity for L-2-amino-4-phosphonobutyrate. *Journal of Biological Chemistry* 268:11868–11873.

Nakatani, K., Tamura, T., and Yau, K.-W. (1991). Light adaptation in retinal rods of the rabbit and two other non-primate mammals. *Journal of General Physiology* 97:413–435.

Nawy, S., and Copenhagen, D. R. (1987). Multiple classes of glutamate receptor on depolarizing bipolar cells in retina. *Nature* 325:56–58.

Nawy, S., and Jahr, C. E. (1990a). Suppression by glutamate of cGMP-activated conductance in retinal bipolar cells. *Nature* 346:269–271.

Nawy, S., and Jahr, C. E. (1990b). Time-dependent reduction of glutamate current in retinal bipolar cells. *Neuroscience Letters* 108:279–283.

Nawy, S., and Jahr, C. E. (1991). cGMP-gated conductance inretinal bipolar cells is suppressed by the photoreceptor transmitter. *Neuron* 7:677–683.

Newman, E. A., and Odette, L. L. (1984). Model of electroretinogram *b*-wave generation: A test of the K^+ hypothesis. *Journal of Neurophysiology* 51:164–182.

Organisciak, D. T., Xie, A., Wang, H.-M., Jiang, Y.-L., Darrow, R. M., and Donoso, L. A. (1991). Adaptive changes in visual cell transduction protein levels: effect of light. *Experimental Eye Research* 53:773–779.

Peachey, N. S., Alexander, K. R., and Fishman, G. (1989). The luminance-response function of the dark-adapted human electroretinogram. *Vision Research* 29:263–270.

Peng, Y.-W., Blackstone, C. D., Huganir, R. L., and Yau, K.-W. (1995). Distribution of glutamate receptor subtypes in the vertebrate retina. *Neuroscience* 66:483–497.

Penn, R. D., and Hagins, W. A. (1972). Kinetics of the photocurrent of retinal rods. *Biophysical Journal* 12:1073–1094.

Penn, J. S., and Williams, T. P. (1986). Photostasis: regulation of daily photon-catch by rat retinas in response to various cyclic illuminances. *Experimental Eye Research* 43:915–928.

Pepperberg, D. R., Birch, D. G., and Hood, D. C. (1997). Photoresponses of human rods *in vivo* derived from paired-flash electroretinograms. *Visual Neuroscience* 14:73–82.

Pugh, E. N., Jr. (1986). Vision: physics and retinal physiology. In: Atkinson, R. C., Herrnstein, R. J., Lindzey, G., and Luce, R. D. (eds.). *Stevens' Handbook of Experimental Psychology.* Wiley, New York. pp. 75–163.

Pugh, E. N., Jr. and Lamb, T. D. (1993). Amplification and kinetics of the activation steps in phototransduction. *Biochimica et Biophysica Acta* 1141:11–149.

Pugh, E. N., Jr., Duda, T., Sitaramayya, A., and Sharma, R. K. (1997). Photoreceptor guanylate cyclases: a review. *Bioscience Reports* 17:429–473.

Reiser, M. A., Williams, T. P., and Pugh, E. N., Jr. (1996). The effect of light history on the aspartate-isolated fast-PIII responses of the albino rat retina. *Investigative Ophthalmology and Visual Science* 37:221–229.

Remtulla, S., and Hallett, P. E. (1985). A schematic eye for the mouse, and comparisons with the rat. *Vision Research* 25:21–31.

Robson, J. G., and Frishman, L. J. (1995). Response linearity and kinetics of the cat retina: the bipolar cell component of the dark-adapted electroretinogram. *Visual Neuroscience* 12:837–850.

Robson, J. G., and Frishman, L. J. (1996). Photoreceptor and bipolar-cell contributions to the cat electroretinogram: a kinetic model for the early part of the flash response. *Journal of the Optical Society of America* 13:613–622.

Rodieck, R. W. (1973). *The Vertebrate Retina—Principles of Structure and Function.* Freeman, San Francisco.

Sandberg, M. A., Miller, S., and Berson, E. L. (1990). Rod electroretinograms in an elevated cyclic guanosine monophosophate-type human retinal degeneration. *Investigative Ophthalmology and Visual Science* 31:2283–2287.

Savage, G. L., and Banks, M. S. (1992). Scotopic visual efficiency: constraints by optics, receptor properties, and rod pooling. *Vision Research* 32:645–656.

Schneeweis, D. M., and Schnapf, J. L. (1995). Photovoltage of rods and cones in the macaque retina. *Science* 268:1053–1056.

Schremser, J.-L., and Williams, T. P. (1995a). Rod outer segment (ROS) renewal as a mechanism for adaptation to a new intensity environment. I. Rhodopsin levels and ROS length. *Experimental Eye Research* 61:17–24.

Schremser, J.-L., and Williams, T. P. (1995b). Rod outer segment (ROS) renewal as a mechanism for adaptation to a new intensity environment. II. Rhodopsin synthesis and packing density. *Experimental Eye Research* 61:25–32.

Seeburg, P. H. (1993). The TINS/Tips lecture. The molecular biology of mammalian glutamate receptor channels. *Trends in Neuroscience* 16:359–364.

Shastry, B. S. (1997). Signal-transduction in the retina and inherited retinopathies. *Cellular and Molecular Life Sciences* 53:419–429.

Shiells, R. A., Falk, G., and Naghshineh, S. (1981). Action of glutamate and aspartate analogues on rod horizontal and bipolar cells of the dogfish retina. *Nature* 294:592–594.

Shiells, R. A., and Falk, G. (1992). The glutamate receptor-linked cyclic GMP cascade of retinal on-bipolar cells is pertussis and cholera toxin-sensitive. *Proceedings of the Royal Society London* B247:17–20.

Shiells, R. A., and Falk, G. (1994). Responses of rod bipolar cells isolated from dogfish retinal slices to concentration-jumps of glutamate. *Visual Neuroscience* 11:1175–1183.

Slaughter, M. M., and Miller, R. F. (1981). 2-amino-4-phosphonobutyric acid: a new pharmacological tool for retina research. *Science* 211:182–185.

Smith, N. P., and Lamb, T. D. (1997). The A-wave of the human electroretinogram recorded with a minimally invasive technique. *Vision Reseach* 37:2943–2952.

Sterling, P., Freed, M. A., Smith, R. G. (1988). Architecture of rod and cone circuits to the on-beta ganglion cell. *Journal of Neuroscience* 8:623–642.

Stockton, R. A., and Slaughter, M. M. (1989). B-Wave of the electroretinogram. *Journal of General Physiology* 93:101–122.

Szel, A., Rohlich, P., Caffe, A. R., Juliusson, B., Aguire, G., and van Veen, T. (1992). Unique separation of two spectral classes of cones in the mouse retina. *Journal of Comparative Neurology* 325:327–342.

Thomas, M. M., and Lamb, T. D. (1997). Kinetics of post-bleach recovery of maximal response and amplification constant in human rod photoreceptors, determined from the a-wave of the electroretinogram. *Journal of Physiology (London)* 504P:P30.

Tian, N., and Slaughter, M. M. (1995). Correlation of dynamic responses in the ON bipolar neuron and the b-wave of the electroretinogram. *Vision Research* 35:1359–1364.

Vardi, N., Matesic, D. F., Manning, D. R., Liebman, P. A., and Sterling, P. (1993). Identification of a G-protein in depolarizing rod bipolar cells. *Visual Neuroscience* 10:473–478.

Wang, M., Lam, T. T. M., Tso, M. O. M., and Naash, M. I. (1997). Expression of a mutant opsin gene increases the susceptibility of the retina to light damage. *Visual Neuroscience* 14:551–562.

Wong, F. (1997). Investigating retinitis pigmentosa—a laboratory scientist perspective. *Progress in Retinal and Eye Research* 16:353–373.

Wyszecki, G., and Stiles, W. S. (1982). *Color Science, 2nd ed.* Wiley, New York.

Xu, Z., and Karwoski, C. J. (1994). Current source density analysis of retinal field potentials II. Pharmacological analysis of the b-wave and M-wave. *Journal of Neurophysiology* 72:96–105.

APPENDIX I: DERIVATION OF TRANSRECEPTOR-LAYER POTENTIAL FROM THE ROD CIRCULATING CURRENT AND ROD DENSITY

This appendix presents a derivation of the transreceptor-layer potential generated by the photoreceptor layer circulating currents. We have employed a faithful geometrical layout, but the biophysics (the membrane current density and the extracellular resistance profile) has been simplified. The reader interested in more details is encouraged to examine the classic paper of Hagins *et al.* (1970).

Figure A1 serves to explain the derivation graphically, and Table A1 identifies the relevant variables and parameters. First, consider the bottom of the figure, where a drawing of a representative rod shows the geometrical layout of the retina. We take the x coordinate to be along the longitudinal axis of the rods, which is the radial axis of the globe. We set $x = 0$ at the tips of the outer segments; the total length of the rod is $L = 100$ μm. The outer segment is 25 μm long, so the inner segment → outer segment (i/o) junction is at $x_{i/o} = 25$. The graph presents four panels, which we will now describe from the bottom upwards. Initially, consider only the unbroken curves in the panels.

Panel D illustrates the membrane current *density* of a individual rod assumed to have a circulating current $J_{circ} = 30$ pA. The membrane current density is assumed uniform in each segment, inner and outer. Uniformity of the membrane current density as a function of longitudinal position along the outer segment was found by Hagins *et al.* (1970) to be consistent with their extracellular potentiometric data and was directly confirmed in measurements of the membrane currents of single toad rods with the suction-electrode method by Baylor *et al.* (1979a). Uniformity of the inner-segment membrane current density is a

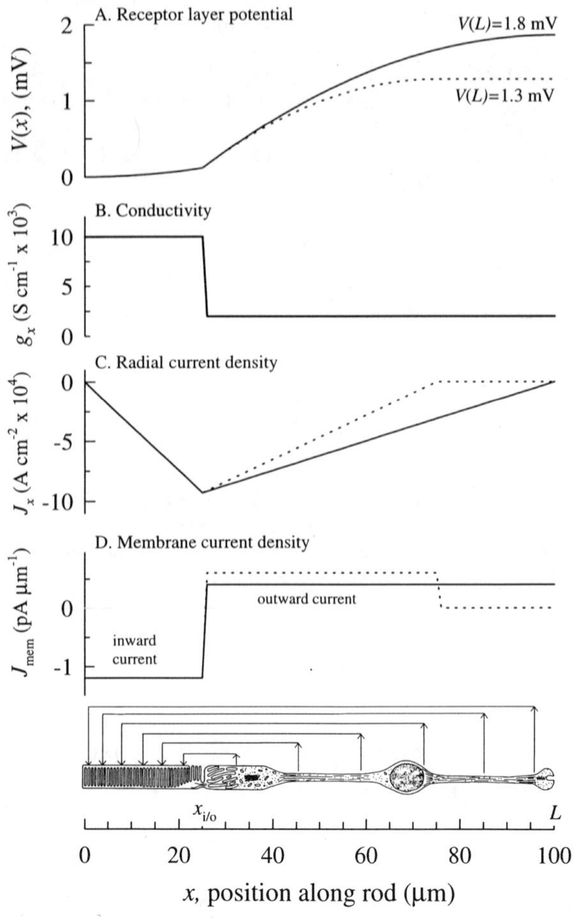

Figure A1.

Table A1. Variables and parameters determining the transreceptor dipole layer potential

Symbol	Magnitude/unit	Description	References[a]
x	100 μm	Radial coordinate of retina starting at outer-segment tips	
L	100 μm	Total rod length	1, 2
$x_{i/o}$	25 μm	Position of inner segment[b]—outer segment junction	1
J_{circ}	30 pA	Rod circulating current	1, 3
ρ	3×10^7 rods cm^{-2}	Rod density	1, 2, 4
$g_{x,os}$	0.01 S cm^{-1}	Extracellular conductivity in outer-segment layer	1
$g_{x,is}$	0.002 S cm^{-1}	Extracellular conductivity in inner-segment layer	1
j_x	A cm^{-2}	Extracellular radial current density	
$J_{x,max}$	ρJ_{circ} A cm^{-2}	Maximum radial current density (at outer segment—inner segment junction)	

[a]1, Hagins *et al.* (1970); 2, Penn and Williams (1986); 3, Baylor *et al.* (1984); 4, Reiser *et al.* (1996).
[b]The term "inner segment" is used here to refer to the entire length of the rod from the base of the outer segment to the synaptic terminal, in other words, to all but the outer segment.

rough approximation (which will be relaxed when we consider the dashed curves). The outer-segment membrane current density is −30 pA per 25 μm = −1.2 pA μm^{-1}. The outward inner-segment current density, 0.4 pA μm^{-1}, is less than that of the outer segment because the inner segment length is 3 times longer than the outer segment. The total inward and outward membrane currents necessarily sum to zero, as explained in the discussion in section 3.2 of the circulating current.

Panel C presents the extracellular radial current density. At the inner segment → outer segment junction, radial current j_x is maximal and necessarily satisfies

$$j_{x,max} = -\rho J_{circ} \tag{AI.1}$$

Hagins *et al.* (1970) empirically discovered the radial current to be maximal at the inner segment → outer segment junction, but this fact now follows from the well-established nature of the circulating current. In short, the relation expressed in Eq. AI.1 holds because, as illustrated in the drawing of the rod, all current sources are in the inner segment and all sinks in the outer segment. For murine retinas $\rho = 3 \times 10^7$ rods cm^{-2} is the spatial density of rods (see Table A1), and thus $j_{x,max} = -9 \times 10^{-4}$ A cm^{-2}. (The negative sign of j_x comes from the coordinate system, because the current is flowing to the left, i.e. in the negative direction on the *x*-axis.) The linear radial current profiles shown in panel C follow from the assumptions that the membrane current is uniform in each segment, from the boundary condition that the radial current must decline approximately to zero at the rod tips and at the synaptic terminals. In sum, the radial current distribution follows directly from the membrane current distribution.

To derive the transretinal potential due to the radial current, we need a relation from the electrical theory of volume conductors:

$$\begin{aligned} \vec{j} &= g\vec{E} \\ &= -g\vec{\nabla}V \end{aligned} \tag{AI.2}$$

In Eq. AI.2 $\vec{j}$ is the current vector, *g* the volume conductivity (which in the general case is a 3 × 3 matrix or tensor), $\vec{E}$ the electric field, $\vec{\nabla}$ the gradient operator and, *V* the electric potential. For a one-dimensional flow, as is the case for the extracellular current of the photoreceptor layer of the retina, Eq. AI.2 becomes simply

$$j_x = -g_x \frac{dV}{dx} \tag{AI.3}$$

where g_x is the conductivity in the x direction. Given a known radial current profile, j_x, and a known distribution of conductivity, g_x, Eq. AI.2 can readily be integrated to yield $V(x)$ and, specifically, the transreceptor layer potential, $V(L) - V(0)$, as follows.

We take the conductivity profile to be that of Panel B: in the outer-segment region the conductivity is relatively high, 0.01 S cm^{-1}, because of the substantial interstitial space between the outer segments; in the inner-segment region where the ellipsoids and the nuclei of the ONL are highly packed the conductivity drops to 0.002 S cm^{-1}, one-fifth the value in the outer-segment layer (Table 1). (Uniform conductivity is an oversimplication; see Hagins *et al.*, 1970.)

To obtain the potential profile $V(x)$, we first write piecewise linear formulas for the radial current distribution, using $x_{i/o}$ to represent the x position of the inner- → outer-segment junction:

$$\begin{aligned} j_x &= j_{x,\max} \frac{x}{x_{i/o}}, & 0 \le x \le x_{i/o} \\ &= j_{x,\max} \frac{L-x}{L-x_{i/o}}, & x_{i/o} \le x \le L \end{aligned} \tag{AI.4}$$

By intregrating Eq. AI.3, we arrive at

$$\begin{aligned} V(x) &= -\int_0^x \frac{j_x(x')}{g_x(x')}\, dx' \\ &= \frac{j_{x,\max} x^2}{2 g_{x,os} x_{i/o}}, \qquad 0 \le x \le x_{i/o} \\ &= j_{x,\max} \left[\frac{x_{i/o}}{2 g_{x,os}} + \frac{\left[(L - x_{i/o})^2 - (L-x)^2 \right]}{2 g_{x,is} (L - x_{i/o})} \right], \qquad x_{i/o} \le x \le L \end{aligned} \tag{AI.5}$$

Panel A plots Eq. AI.5. Finally, then, we arrive at the following expression for the potential difference across the photoreceptor layer:

$$V(L) - V(0) = \rho J_{\text{circ}} \left[\frac{x_{i/o}}{2 g_{x,os}} + \frac{L - x_{i/o}}{2 g_{x,is}} \right] \tag{AI.6}$$

where we have substituted $j_{x,\max} = \rho J_{\text{circ}}$ from Eq. AI.1. For the parameter values in Table A1, we find $V(L)$ = 1.8 mV. (Note that in the integration care must be taken to express all quantities in comparable units; thus, for the CGS units we adopted, $x_{i/o} = 25 \times 10^{-4}$ cm, etc.).

The dashed curves in panels A–D show the distributions used for a second set of calculations. The form of the radial current distribution in panel C in this case is a good approximation to that deduced by Hagins *et al.* (1970, their Fig. 13); the transreceptor layer potential in this case is 1.3 mV.

An important generality that follows from the analysis that yields Eq. AI.6 is that the transreceptor potential is directly proportional to the circulating current. This proportionality holds regardless of the exact distribution of the membrane current and can be seen from the steps of the derivation to depend only on the fact that Eq. AI.1 holds at the position of the inner segment $\rightarrow$ outer segment junction, which in turn depends only on the principle that the outer-segment membrane contains exclusively the sinks of the rod circulating current.

It bears emphasis that the potential derived applies across the receptor layer. As noted in the body of this paper (section 3.2), it is not expected that there will be much loss in this potential across the remainder of the retina or through the ocular media. The essential problem in comparing the magnitude predictions of Eq. AI.6 with corneally measured a-waves, however, is that the reference electrode references the potential not at $x = 0$ but rather at some intermediate level in the return path through the anterior segment, extraocular fluid, and typically volume conductance of the animal's head. Crudely, we estimate the effect of the position of the reference electrode (in the forehead) in our experiments to diminish the measured corneal potential to one-third to one-fifth of its transretinal magnitude. Hagins *et al.* (1970, their Fig. 6b) report a transretinal saturated a-wave in the albino rat of about 2.5 mV.

It is useful to conclude with an explicit formula for the proportionality relation, Eq. 1, now generalized to any dipole current generator in the retina and expressed for the potential recorded at the cornea:

$$V_{\text{cornea}} - V_{\text{ref}} = c_{\text{electrodes}} \rho_{\text{cell type}} \bar{r}_{\text{layer}} J_{\text{dipole}} \tag{AI.7}$$

where V is the potential in volts, $c_{\text{electrodes}}$ is a dimensionless positive factor less than unity, $\rho_{\text{cell type}}$ is the spatial density in the retina of a specific cell type, and J_{dipole} is the dipole or circulating current of the specific type of cell. In the case of the rods, by deriving Eq. AI.6 we have explained all the terms except $c_{\text{electrodes}}$ and $\bar{r}_{\text{layer}}$. The "lumped transretinal-layer resistance," $\bar{r}_{\text{layer}}$, is the rightmost term, in rectangular brackets in Eq. AI.6, and appears to be about 2 ohms cm^2 in magnitude for the rod layer; it seems likely that this value will be roughly the same for the bipolar cell layer, given the overall thickness of the layer and cell packing density. From comparison of published transretinal and corneal a-wave magnitudes (for conventional reference electrode positions), it appears that $c_{\text{electrodes}}$ is likely to be 0.2 to 0.3 in most instances, though the lumping of the conductances of the distributed return paths for the currents in corneal recordings must be viewed with caution. Because cell density counts are now available, Eq. AI.7 may prove useful for estimating the dipole currents of different cell types, as we have argued for the rod bipolar in Section 4. For the specific case of the corneally recorded saturated a-wave amplitude, then, Eq. AI.6 gives, in reasonable agreement with observations,

$$\begin{aligned} V_{\text{cornea}} - V_{\text{ref}} &= 0.3 \times 3 \times 10^7 \left(\text{rods cm}^{-2}\right) \times 2 \left(\Omega \text{ cm}^2\right) \times 30 \times 10^{-12} \left(A\ rod^{-1}\right) \\ &= 0.54 \text{ mV.} \end{aligned}$$

APPENDIX II: HOMOGENENOUS LIGHT-REARING ENVIRONMENTS, QUANTIFYING THEIR ILLUMINATION, AND CONVERTING ILLUMINANCE TO RATES OF PHOTOISOMERIZATION IN RODS

Because various functional properties of the retina are under the regulation of light-rearing history, animals must be reared in enclosed environments that allow precision control of illumination. Unfortunately, many animal-rearing quarters present enormous variations in illumination intensity, especially when cages are stacked on top of one another. To achieve good control of illumination, a condition approximating that in a *Ganzfeld** should be created. In such environments, all light sources should be hidden behind diffusing screens, and the interior of the environment should be painted flat white. A lux meter should be used to determine the directional variation of the intensity of light impinging on the floor or cage area in which the animals live. We now discuss the calculations needed for conversion of illumination levels into rates of photoisomerizations in the rods of mice. We emphasize that these calculations properly apply only to *Ganzfeld* or homogeneous illumination environments. (For a complete discussion of light measurement relevant to vision, see Wyszecki and Stiles, 1982; a shorter treatment can be found in Pugh, 1986.)

First, some background. Because most light sources are broad band, comprising photons of many energies (or equivalently, photons of many frequencies or wavelengths), one ideally would like to know the spectral distribution of any light source to be able determine its visual effectiveness. Photometric units were invented in an effort to circumvent the need for measuring the spectral distributions of light sources and to yield instead a single number characterizing the visual effectiveness or "luminous efficiency" of any particular spectral distribution for the human visual system. Thus, an International Commission on Illumination (CIE) was established in the early part of this century, and that commission defined and maintains international photometric units and standards.

There are two types of luminous efficiency measures, scotopic and photopic. It is now conclusively established that luminous efficiency under scotopic conditions is completely determined by the efficacy with which rhodopsin in the rods captures light of different wavelengths that impinges on the cornea; the CIE function describing the wavelength dependence of this efficacy is called the scotopic luminous efficiency function and is symbolized by V'_{λ}. The physiological mechanism underlying photopic spectral sensitivity is more complex than that underlying the scotopic: the photopic luminous efficiency function for humans is probably determined by a combination of L- and M-cone signals to the so-called "parasol" ganglion cells. Nonetheless for its physiological complexity, the CIE has *defined* the human photopic luminosity function, V_{λ} (which peaks at 555 nm), and this function is now incorporated into all illumination meters or lux meters. The following are the international photometric conversion standards (Wyszecki and Stiles, 1982, p. 258):†

$$
\begin{aligned}
1 \text{ photopic lumen} &\equiv 1/683 \text{ watts } (\lambda = 555 \text{ nm}) \\
&= 4.1 \times 10^{15} \text{ photons s}^{-1} \ (\lambda = 555 \text{ nm})
\end{aligned}
$$

---

* A *Ganzfeld* ("total field" in German) illumination condition is one in which light impinges on the eye in equal quantities from all directions; as an example, a *Ganzfeld* results if one puts half a ping-pong ball over the eye.

† The conversion from watts (joules s^{-1}, an energy unit) to photons (a quantal unit) is based on Planck's Law: $\epsilon = h\nu = hc/\lambda$, where ϵ is the energy of an individual photon (in joules), h is Planck's constant (6.625×10^{-34} joules s), c is the speed of light (3×10^8 m s^{-1}), and λ is the wavelength of light (in meters).

$$\begin{aligned}1 \text{ scotopic lumen} &\equiv 1/683 \text{ watts } (\lambda = 555 \text{ nm})\\ &= 1/1700 \text{ watts } (\lambda = 507 \text{ nm})\\ &= 1.5 \times 10^{15} \text{ photons s}^{-1} (\lambda = 507 \text{ nm})\end{aligned}$$

where "≡" means "defined as."

Irradiance is a physical quantity that refers to the flux density of photons falling on a surface: irradiance can be expressed in quantal units such as photons cm^{-2} or in energy units such as watts m^{-2}. *Illuminance* refers to the visually effective flux density of light falling on a surface. The international standard unit of illuminance is the photopic (scotopic) *lux*, which is simply 1 lumen m^{-2} (for conversion to lux from other units of illumination, the reader should consult Wyszecki and Stiles, 1982, their Table 1 (2.4.4)).

Two cautionary comments are called for before we proceed to illustrate conversion from lux to photoisomerizations per rod. The first is that illumination used in animal cages is usually broad band, not monochromatic, and the exact conversion from lumens to photons s^{-1} depends on the spectral distribution of the source. Thus, 1 photopic lumen can be equivalent to from 1 to 3 scotopic lumens, depending on the spectral distribution of the broad-band source. The second is that the factor needed for the photopic → scotopic conversion is specified for the *human eye,* with its particular ocular-medium transmissivity. Because the rodent eye media are more transmissive in the blue and UV than human eye media, some light sources (such as fluorescent lights) may produce considerably more light captured in rodent rods than would be expected from application of the human photopic → scotopic conversion. For computational purposes here, we will assume that the factor 1.5 applies for the conversion. Thus, we assume that 1 photopic lumen is equivalent to 2.25×10^{15} photons s^{-1} (at $\lambda = 507$ nm).

We next compute the *retinal irradiance* in a rodent eye produced by a homogenous or *Ganzfeld* illumination environment in which we have measured the illuminance I in *photopic lux.* That is, we wish to estimate Q, the effective number of photons s^{-1} ($\lambda = 507$ nm) impinging on each unit area of retinal surface. To do so, we need four numbers: (1) the illuminance I (in photopic lux or lumens m^{-2}); (2) A_p, the area of the pupil (in mm^{-2}); (3) the surface area of the retina (in μm^2). Thus,

$$Q\left(\text{photons s}^{-1}\mu\text{m}^{-2} \text{ of retina}\right) = \frac{I\left(\text{photons s}^{-1}\text{m}^{-2} \text{ at cornea}\right) \times 10^{6}\left(\text{mm}^2\text{m}^{-2}\right) \times A_{\text{pupil}}\left(\text{mm}^2\right)}{A_{\text{retina}}\left(\mu\text{m}^2\right)} \tag{AII.1}$$

Several assumptions are hidden in expression AII.1. The one that most deserves emphasis is that the illumination is implicitly *Ganzfeld* in character: this assumption allows one to conclude that all the light entering into the pupil is distributed uniformly over the retinal surface. Another assumption, which was discussed above, is that in applying Eq. AII.1 one must have correctly converted from photopic lux to scotopic lux and thence to rhodopsin-equivalent photons at 507 nm.

Application of Eq. AII.1 is also complicated by the dependence of pupil size on retinal illuminance. We have found the dark adapted pupil of 4- to 8-week-old C57BL/6 mice to have a diameter of about 2 mm and thus an area of 3.1 mm^2. The retinal surface area of the mouse (see Fig. 1) can be assumed to be that of a hemisphere of radius 1.5 mm, or about $2\pi(1.5)^2 = 14$ mm^2 (Remtulla and Hallett, 1985).

With the various assumptions and concerns now laid out, we turn to the final conversion, that from retinal illuminance to photoisomerizations per rod per second or Φ s^{-1}.

Only a single additional factor is needed, and that is a_C, the end-on collecting area of the individual rod at the λ_{max} of rhodopsin. Because of the general stereotypy of mammalian rods, a_C can be expected to be about the same value for rods of different species. We have previously estimated it to be 1.3 μm^2 for murine rods (Lyubarsky and Pugh, 1996).

By combining all the factors discussed, we arrive at the following relations:

$$
\begin{aligned}
1 \text{ photopic lux} &\equiv 1 \text{ photopic lumen m}^{-2} \\
&\sim 2.25 \times 10^{15} \text{ photons s}^{-1}\text{ m}^{-2}\ (\lambda = 507 \text{ nm}) \\
&\sim 7.0 \times 10^{9} \text{ photons s}^{-1} \text{ entering mouse pupil } (3.1 \text{ mm}^2) \\
&\sim 500 \text{ photons s}^{-1}\ \mu\text{m}^{-2} \text{ of retinal surface area} \\
&= 650 \text{ photoisomerizations rod}^{-1}\text{ s}^{-1}
\end{aligned}
$$

where the symbol "~" means "is equivalent to." In a previous publication (Lyubarsky and Pugh, 1996) we estimated the fully dilated mouse pupil to be 2 mm in diameter and thus to have an area of 3.1 mm^2, but in laboratory rearing environments the size of the pupil can change in response to the illumination. In an ongoing series of experiments, we are determining the dependence of the mouse pupil size on illumination. Our preliminary data show that at an illuminance of 2–3 lux the area of the adult C57BL/6 mouse pupil is about 1 mm^2. Thus, under 2–3 lux illumination, with pupil free to contract, for C57BL/6 mice 1 photopic lux ~ 210 photoisomerizations per rod per second.

8

EFFECTS OF PHOTORESPONSE PROLONGATION ON RETINAL RODS OF TRANSGENIC MICE

Clint L. Makino,[1] John G. Flannery,[2] Jeannie Chen,[3] and Robert L. Dodd[4]

[1]Department of Ophthalmology
Harvard Medical School and the Massachusetts Eye and Ear Infirmary
243 Charles Street
Boston, Massachusetts 02114
[2]Department of Visual Science and Neuroscience
School of Optometry
569 Minor Hall
University of California, Berkeley
Berkeley, California 94720-2020
[3]Mary D. Allen Laboratory for Vision Research
Doheny Eye Institute
Departments of Ophthalmology and Cell and Neurobiology
University of Southern California
School of Medicine, BMT 401
1333 San Pablo Street
Los Angeles, California 90033
[4]Department of Neurobiology
Fairchild Building
Stanford University School of Medicine
Stanford, California 94305

1. INTRODUCTION

At the edge of darkness, retinal rod photoreceptors report the presence of single photons. Within seconds after a stepped increase in the level of illumination, however, rods lose their exquisite sensitivity. By adapting in this way, rods preserve their ability to signal changes in lighting over a range of background intensities that spans several log units. On a slower time scale, rods undergo other remarkable adaptations to the level of illumination. These have been studied extensively in albino rats (e.g. by Williams and Baker, 1980; Penn and Anderson, 1991), which are particularly sensitive to changes in environmental lighting because their irises lack pigmentation (reviewed by Lanum, 1978). In response to an increase in lighting, rods of albino rats altered the length and structure of their outer segments, changed their rates of rhodopsin regeneration and opsin synthesis, and even reduced their numbers through cell death when the light was made bright enough or when the exposure was prolonged. Because the incident light normally passes through the long axis of the outer segment, shorter outer segments catch fewer photons. Similarly, the fewer the rhodopsins per unit length of outer segment, the lower the photon catching ability of the rod. Thus albino rats proved capable of "photostasis," keeping their quantal catch constant at approximately 10^{16} photons per eye per day when the lighting varied by three log units of intensity. Furthermore, rats initially reared under bright lighting but then transferred into dimmer lighting compensated appropriately to maintain the same daily quantum catch (Penn and Williams, 1986; Penn *et al.,* 1989; Penn and Anderson, 1991; Schremser and Williams, 1995a, b). The signal for photostasis is likely to be carried in part by the retina's neural response to light, because severing one optic nerve of a rat afforded some protection to that eye against light-dependent retinal damage but had no effect on the intact eye (Bush and Williams, 1991). This result further suggests that brain centers external to the eye are involved. It should therefore be possible to evoke photostasis by delivery of the appropriate neural signal to these centers, in the absence of a change in the illumination, for example by tampering with the phototransduction machinery such that the rods falsely signal a higher (or lower) lighting exposure.

The phototransduction cascade is summarized in Figure 1 (reviewed by Koutalos and Yau, 1993; Pugh and Lamb, 1993; Hurley, 1994; Yarfitz and Hurley, 1994; Bownds and Arshavsky, 1995; Helmreich and Hofmann, 1996). The visual pigment rhodopsin is a member of a large class of G-protein-coupled, seven-transmembrane-helix receptors. Photoexcitation causes rhodopsin to become catalytically active, enabling it to facilitate the exchange of a GTP for a bound GDP on the G-protein transducin. With GTP bound, transducin splits into Tα-GTP and Tβγ. Tα binds to the γ subunit of a phosphodiesterase (PDE). The PDE is released from a standing inhibition and rapidly hydrolyzes cGMP. The plasma membrane of the rod outer segment contains a large number of cation-specific channels that are held open by cGMP. The influx of Na^+ and, to a lesser extent, Ca^{++} through these channels in darkness keeps the rod partially depolarized. By decreasing the cGMP concentration, a single quantum of light causes hundreds of these channels to close and prevents a million Na^+ ions from entering the rod. The resultant hyperpolarization spreads passively through the cell to the synaptic terminal, where it modulates the release of the neurotransmitter glutamate onto second-order neurons.

Recovery of the photoresponse requires shutoff of the activated proteins and restoration of the cGMP level to that present in fully dark-adapted rods. Rhodopsin's activity is shut off after phosphorylation by rhodopsin kinase, followed by the binding of arrestin. Transducin shuts itself off through an intrinsic GTPase activity that is accelerated by its interaction with PDEγ and an as yet unidentified factor. Guanylate cyclase raises the levels of cGMP, and the membrane channels reopen.

In separate experiments, we interfered genetically with the photoresponse shutoff in mouse rods in three different ways. Rods of some transgenic mice were induced to express a mutant, truncated rhodopsin (S334ter) lacking the phosphorylation sites on its carboxyl terminus. Rods from a second type of mouse had intact rhodopsin but lacked arrestin. In a third type of mouse, rods were made to express a mutant, GTPase-deficient transducin. Recovery of the flash responses in rods from each type of mouse was much slower than normal, as though photic stimulation persisted for an extended period after the flash. Evidence that photoresponse prolongation evoked a photostasis type of adaptation is presented below. A summary of some previous work on these mice is included (Raport *et al.,*

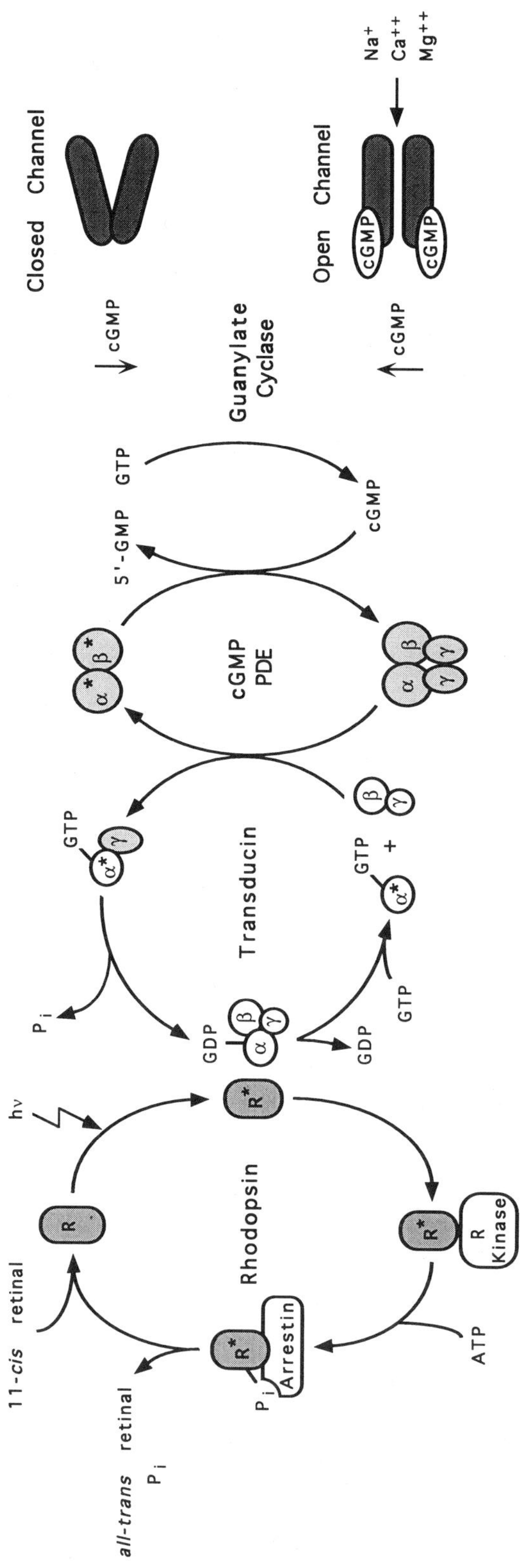

Figure 1. Molecular mechanism of phototransduction.

1994; Chen *et al.*, 1995; Xu *et al.*, in press). Interestingly, the mice reacted quite differently to one of the genetic manipulations.

2. METHODS

2.1. Transgenic Mice

Generation of the three types of transgenic mice has been described elsewhere (Raport *et al.*, 1994; Chen *et al.*, 1995; Xu *et al.*, in press). Briefly, an 11-kb BAM HI fragment containing the murine gene for rhodopsin was subjected to site-specific mutagenesis to change codon 334 to a stop codon. The construct was injected into the nucleus of a fertilized mouse oocyte. Reverse transcriptase polymerase chain reaction (RT-PCR) measurements indicated that the ratio of transgenic to endogenous transcript was about 0.12 in our S334ter mice. Western blot analysis of rod-outer-segment-enriched preparations from S334ter mice confirmed the presence of truncated opsin protein. S334ter was not detectable in control retinas.

In order to knock out arrestin, we replaced the native arrestin gene in embryonic stem (ES) cells with a modified gene, which lacked some promoter regions as well as the first two exons, by homologous recombination. Positive ES-cell clones were injected into blastocysts to produce chimeric animals. Progeny were screened by PCR for germ-line transmission. Densitometric analysis of Western blots from retinal preparations indicated a reduced arrestin content in hemizygous knockout (A+/−) mice and an absence of arrestin in homozygous knockout (A−/−) mice.

A fusion gene was constructed from 4.4 kb of the mouse rod opsin promoter, to ensure photoreceptor specific expression, and the cDNA for human cone transducin. A point mutation was introduced that changed glutamine 204 to a leucine (Q204L). The gene construct was added to the mouse genome by pronuclear injection of the transgene into fertilized, single-celled embryos. Q204L mRNA transcripts were detected with RT-PCR, and protein expression was quantified with Western analysis using antibody A1.1 (Raport *et al.*, 1994). Localization of the Q204L transducin to outer segments was demonstrated immunohistochemically with an antibody that recognized human cone Tα but not rod Tα (Lerea *et al.*, 1989).

Mice were maintained on a 12 hour light:12 hour dark regimen at the California Institute of Technology, Stanford University, and the University of Florida. One female arrestin knockout mouse was placed in darkness during her pregnancy, and her offspring were kept in darkness for their entire lives.

2.2. Physiological Recording

Photoresponses were analyzed by suction-electrode recording as described previously (Raport *et al.*, 1994). TΔ47 transducin transgenic mice were dark adapted for a minimum of 1.5 hours, the TΔ2 transducin transgenic mouse was dark adapted for 4 hours, and all other mice were dark adapted overnight. Under dim red or infrared light, the mouse was anesthetized with CO_2 and decapitated, and its eyes were removed. All operations thereafter were carried out under infrared illumination. Retinas were isolated into chilled, oxygenated Leibovitz's L-15 medium (GIBCO, Grand Island, NY). Retinas were mounted photoreceptor-side-up on a glass capillary array (Gallileo Electro-optics, Sturbridge, MA), stabilized with gentle suction, and separated from the vitreous humor by a razor slid between the retina and the capillary array. Small samples of retina were chopped finely and the pieces loaded into an experimental chamber. In some experiments, the floor of the chamber was coated with poly-l-lysine to cause the tissue to adhere (Mazia *et al.*, 1975). The chamber was installed on the stage of an inverted microscope outfitted with an infrared-sensitive, closed-circuit television system. The tissue was perfused with heated, enriched, bicarbonate-buffered medium equilibrated with 95% O_2/5% CO_2, pH 7.4, at 35–38 °C. An outer segment of an isolated rod or a rod attached to a small piece of tissue was pulled into a silanized, glass pipette filled with a HEPES-buffered medium. Cells were stimulated with flashes or with steps of light from a shuttered, tungsten-halogen source. The intensity of the light was controlled with neutral-density filters and the spectral composition regulated with narrow-band interference filters. Photoresponses were recorded with a current-to-voltage converter. Records were low-pass filtered at 30 Hz (−3 dB, eight-pole-Bessel filter) and digitized at 100, 200, or 400 Hz. Some records were also digitally low-pass filtered at 8 Hz by convolution with a Gaussian function.

The HEPES-buffered medium contained (mM): Na^+, 141.5; K^+, 3.6; Mg^{++}, 2.4; Ca^{++}, 1.2; Cl^-, 140.8; succinate, 3; l-glutamate, 0.5; HEPES buffer, 10; EDTA, 0.02; glucose, 10. In some experiments, the HEPES buffer concentration was 3 mM, the l-glutamate concentration was 3 mM, and BME amino-acid supplement (Sigma, St. Louis, MO) and BME vitamin supplement (Sigma or GIBCO) were added. The bicarbonate-buffered medium contained (mM): Na^+, 144; K^+, 3.6; Mg^{++}, 2.4; Ca^{++}, 1.2; Cl^-, 123.3; HCO_3^-, 20; succinate, 3; l-glutamate, 0.5; HEPES buffer, 10; EDTA, 0.02; glucose, 10; plus the vitamin and amino-acid supplements.

2.3. Histology

Mice were anesthetized and euthanized. The corneas were marked for orientation and the eyes enucleated. Eyes were opened at the ora serrata and fixed overnight at 4 °C in 0.1 M phosphate buffer, pH 7.4, containing 2% formaldehyde and 2.5% glutaraldehyde, or in 0.1 M sodium cacodylate buffer, pH 7.4, containing 4% paraformaldehyde and 2.5% glutaraldehyde. Eyes were bisected through the optic nerve head along the superior-inferior axis, postfixed in 2% osmium tetroxide, dehydrated in a graded ethanol series and embedded either in epon or in epon/araldite epoxy resin. One-μm-thick sections passing through the optic disc and including the entire hemisphere were counterstained with methylene blue for light microscopy. For electron microscopy, 60- to 70-nm-thick sections were placed on uncoated copper or butvar-coated mesh grids and stained with either 2% or 7% aqueous uranyl acetate and lead citrate. Grids were examined and photographed with a Philips CM10 or a Hitachi electron microscope.

3. RESULTS

3.1. Phosphorylation-Deficient Rhodopsin

Transgenic mouse rods contained a mixture of normal and mutant S334ter rhodopsins. S334ter lacked the phosphorylation sites of the carboxyl terminus that normally mediate shutoff of photoexcited rhodopsin's catalytic activity (Fig. 2). Averaged photoresponses of a wild type and a transgenic mouse rod to flashes of increasing strength are shown in Figure 3. The rising phase of the S334ter rod responses appeared to be normal, as was the initial recovery, but then a slowly declining "tail" prolonged the final recovery

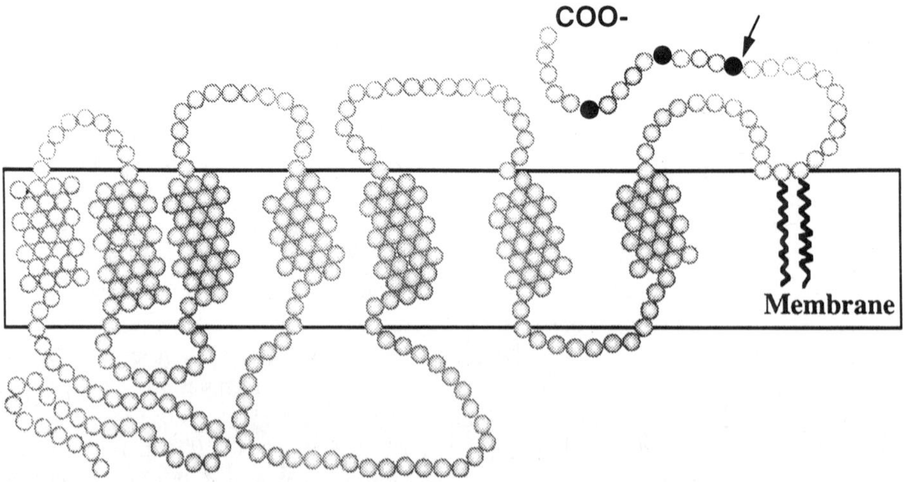

Figure 2. Schematic diagram of rhodopsin. The carboxyl terminus is located intracellularly, and the amino terminus faces the intradiscal space. Filled circles identify the sites where photoexcited rhodopsin becomes phosphorylated (Ohguro *et al.*, 1994). The arrow marks the truncation site in S334ter. Reprinted with permission from Chen *et al.* (1995). Copyright 1997, American Association for the Advancement of Science.

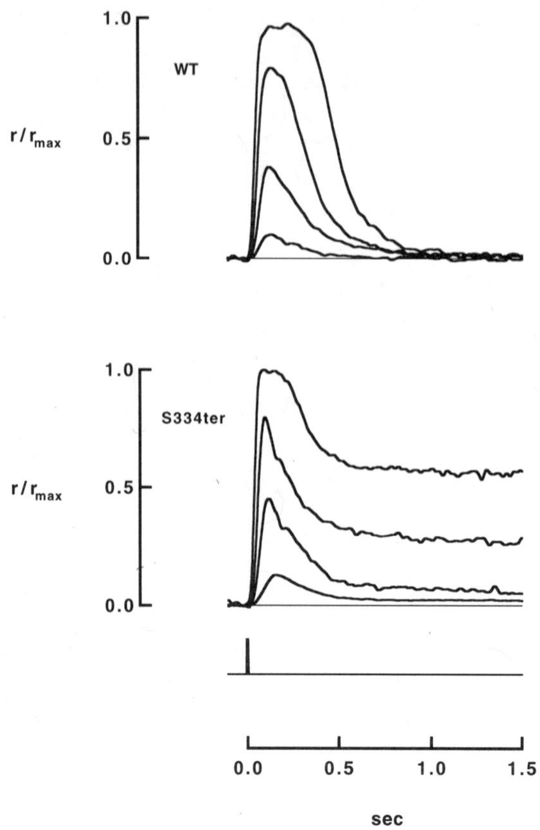

Figure 3. Averaged, normalized photocurrent responses of a wild-type rod (upper panel) and a rod expressing phosphorylation-deficient S334ter rhodopsin (lower panel). The lowermost trace shows the stimulus monitor. Flash strengths at 500 nm were 10, 36.2, 117, and 746.6 photons μm^{-2} for the wild-type rod and 4.4, 30.2, 110, and 355 photons μm^{-2} for the S334ter rod.

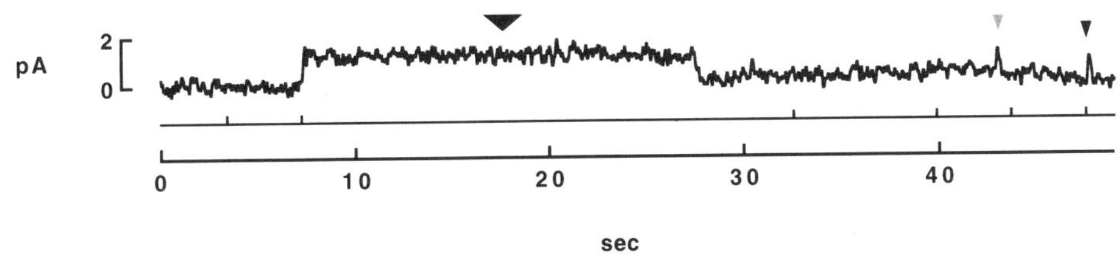

Figure 4. Responses of an S334ter rod to dim flashes at 500 nm. Prolonged (large arrowhead) and normal responses (small black arrowhead) occurred in random order. Spontaneous, quantal-response-like events sometimes resulted from the thermal activation of rhodopsin (small gray arrowhead). Each flash delivered an average of 4.4 photons μm^{-2}.

at all flash strengths. The nature of the tail was explored further with very dim flashes that, on average, elicited less than one photoisomerization per trial. Normal single-photon responses and failures were observed to interleave randomly with odd responses that were about 1.5 times larger than normal and variable in duration (Fig. 4). Some responses persisted for over 30 s.

It is not known how the prolonged response was shut off. The exponential distribution of prolonged response durations suggested a stochastic, first-order mechanism with a mean delay of 4.4 s (Chen *et al.,* 1995). The occasional reappearance of late prolonged-response-like events after the apparently complete recovery of a prolonged response indicated that shutoff was slightly reversible. Possible mechanisms for shutoff include binding of a splice variant of arrestin (Palczewski *et al.,* 1994; Smith *et al.,* 1994b; Palczewski and Smith, 1996) to photoactivated, unphosphorylated S334ter; phosphorylation of S334ter at a site distant from its carboxyl terminus (Wilden and Kuhn, 1982; McDowell *et al.,* 1985; Newton and Williams, 1991); and a rapid thermal transition of photoexcited S334ter to a catalytically inactive state.

S334ter rods were slightly more sensitive to flashes than were wild-type rods. A flash strength ($i_{1/2}$) of 37 ± 7 photons μm^{-2} (mean ± SD, n = 7) elicited a half-maximal response in S334ter rods, compared to 49 ± 27 photons μm^{-2} for wild type rods (n = 77). The ratio of prolonged to normal single-photon responses was 0.15 (Chen *et al.,* 1995), so if S334ter rods contained the normal amount of normal rhodopsin but also contained an additional 15% S334ter, then transgenic mouse rods should be about 22% more sensitive, because photoisomerization of an S334ter molecule leads to a photoresponse that is almost 1.5 times larger than that due to photoisomerization of a normal rhodopsin. The predicted $i_{1/2}$ for a S334ter rod would be 38 photons μm^{-2}, similar to that observed.

The average prolonged single-photon response could be simulated by the sum of a train of 32 normal single-photon responses equally spaced in time (Fig. 5). Because ~13% of the visual-pigment photoisomerizations cause prolonged responses, S334ter rods could be considered to respond to a virtual or apparent "equivalent" light containing five times more photons than are actually present. A similar conclusion would obtain if the integration time of the averaged dim flash response from an S334ter rod were divided by the integration time of the normal single-photon response, where integration time is defined as the integral of the photoresponse divided by its amplitude. The latter estimate would be slightly higher because determination of the averaged dim flash response considers the reversibility of the prolonged response shutoff. The virtual light would even affect the rod in complete darkness. In primate rods, spontaneous thermal activations of rhodopsin give rise to photoresponse-like events about once every 2.6 min (Baylor *et al.,* 1984). Mouse

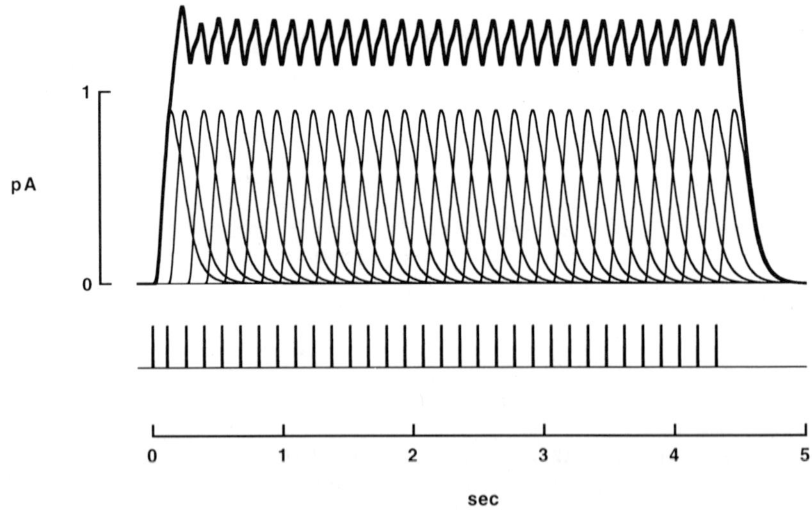

Figure 5. Summed train of normal single-photon responses. Thirty-two normal single-photon responses from an S334ter rod, offset in time and added together, produce a waveform having the same amplitude and duration as the mean prolonged single-photon response from this rod.

rods have a smaller volume and fewer rhodopsins than primate rods so the frequency of thermal events would probably be somewhat lower. If S334ter has the same thermal stability as normal rhodopsin, the "dark light" would be about five times more intense in an S334ter rod than in a wild type rod.

In wild-type rods, the response to a step of dim- to moderate-intensity light rises to a peak and then declines in less than 5 s to a lower plateau that is maintained for the duration of the light exposure (Fig. 6). The decline is one manifestation of light adaptation (Baylor and Hodgkin, 1974; Nakatani *et al.,* 1991). The resemblance of the averaged *flash* responses of a transgenic mouse rod to the responses of a wild-type rod to *steps* of light that gradually dim

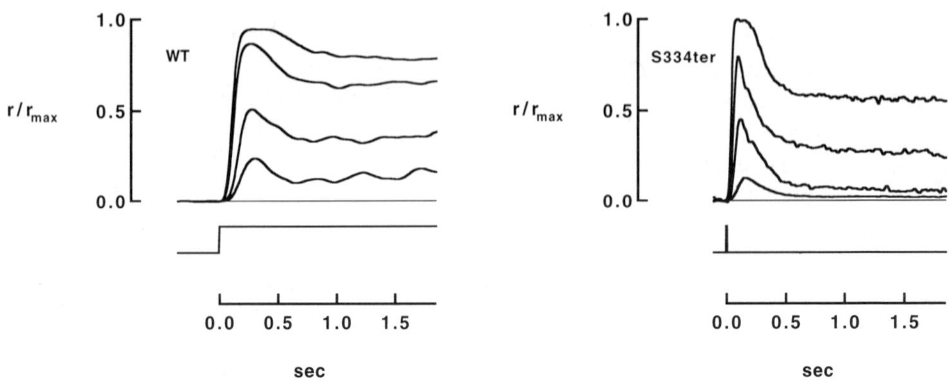

Figure 6. Comparison of averaged step responses of a wild-type rod (left) to averaged flash responses from an S334ter rod (right). The step intensities were 81.7, 297, 989, and 1730 photons μm^{-2} s^{-1} at 500 nm. Flashes delivered 4.4, 30.2, 110, and 355 photons μm^{-2} at 500 nm.

over many seconds suggested a second method for estimating the equivalent light. Responses of a wild-type rod to declining ramped steps of light were approximated with responses to 4.4-s steps of steady light. A half-maximal plateau response was elicited in a wild-type rod by a photon dose of 4154 photons μm^{-2}, delivered over 4.4 s. In comparison, a flash strength of 241 photons μm^{-2} at 500 nm presented to an S334ter rod gave rise to a half-maximal tail amplitude measured at 1 s after the flash. The equivalent light in an S334ter rod, found from the ratio of the two photon dosages, was 17 times greater than the actual light. The equivalent light estimated from this second method was threefold higher than that from the first method because the second method made provisions for light adaptation, but both methods really apply only to S334ter rods in darkness or to rods absorbing single photons sporadically. A third method was needed for the higher levels of illumination typical of the rearing conditions for most of the mice in this study.

The intensity of steady light producing a half-maximal response, once the photoresponse stabilized, was compared for wild-type and transgenic mouse rods. A light intensity of 944 ± 348 photons μm^{-2} s^{-1} (n = 4) produced a half-maximal, steady-state response. The S334ter rod response to a step of light did not decline to a lower plateau level. Instead, an initially rapid rate of rise was followed by a slower increase that finally stabilized at 8–9 s after light onset (Fig. 7). Measured at 10 s after light onset, S334ter rods gave a half-maximal, steady-state response to an intensity of 33 ± 9 photons μm^{-2} s^{-1} (n = 3). If the early differences between the step responses of the two types of rods are ignored, the equivalent light in S334ter rods was 29 times more intense than the light actually received. This third estimate for the equivalent light probably best described the mice of this study, but it should be noted that values obtained from the three methods differed by less than an order of magnitude.

The rods of mice expressing S334ter decreased in number with age. Three months after birth, the outer nuclear layer, which contains the rod nuclei, was up to 20% thinner than that of a wild-type mouse, and after 1 year, it was up to 40% thinner (Fig. 8). The rod outer segments were also slightly shorter than those of wild-type rods.

3.2. Knockout of Arrestin

Hemizygous deletion of arrestin (A+/−) reduced the level of arrestin to about 30% of normal but did not affect the rod photoresponse kinetics (Fig. 9) (Xu *et al.*, in press).

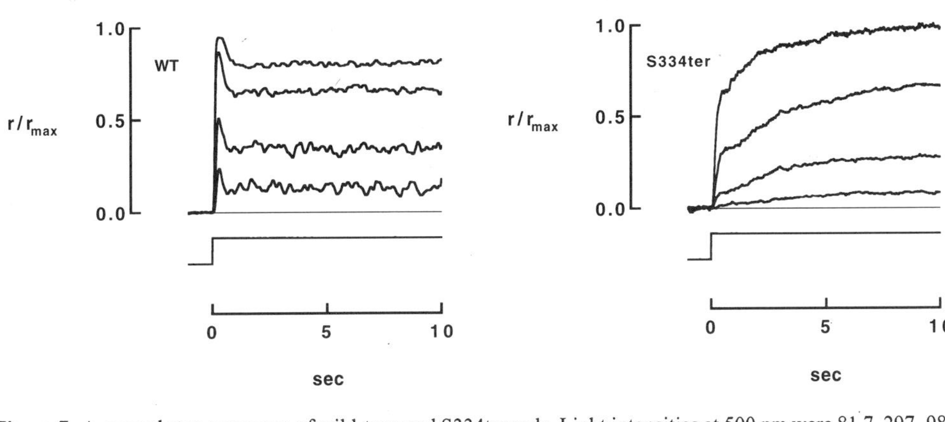

Figure 7. Averaged step responses of wild-type and S334ter rods. Light intensities at 500 nm were 81.7, 297, 989, and 1730 photons μm^{-2} s^{-1} for the wild type rod and 6.5, 24, 78.5, and 285 photons μm^{-2} s^{-1} for the S334ter rod.

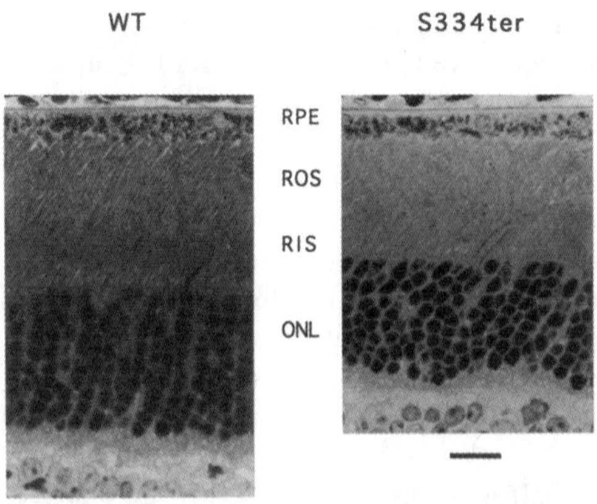

Figure 8. Retinas of wild-type (left) and S334ter (right) mice at 1 year of age. The rod loss from this S334ter retina was somewhat more gradual than in other mice. Scale bar denotes 15 μm. Abbreviations: RPE, retinal pigment epithelium; ROS, rod outer segments; RIS, rod inner segments; ONL, outer nuclear layer.

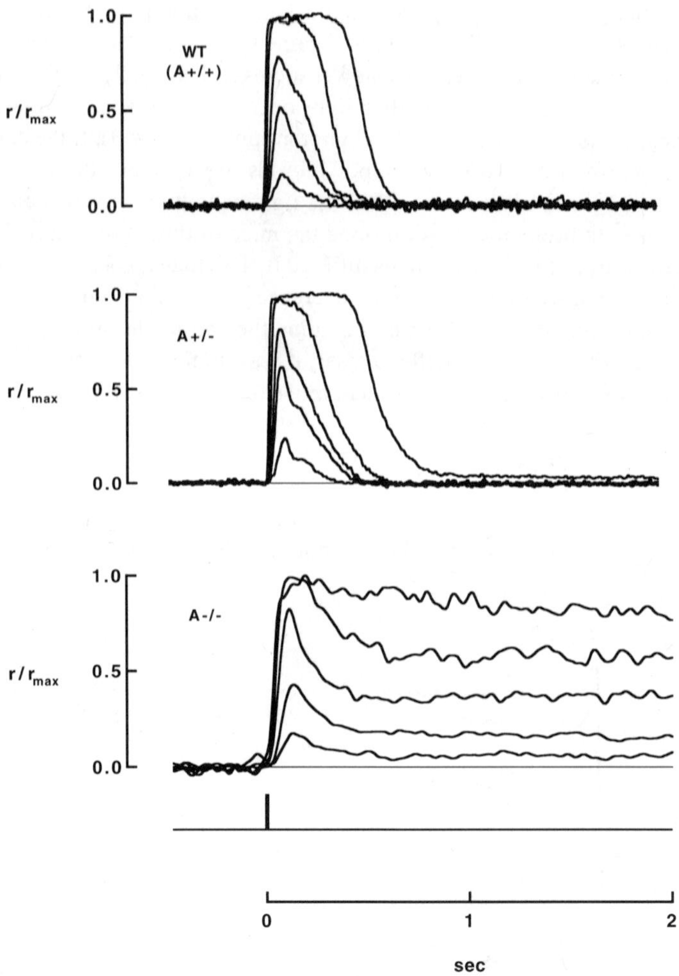

Figure 9. Averaged responses of wild type (upper panel), A+/− (middle panel), and A−/− (lower panel) rods to flashes at 500 nm. Flash strengths were 8.3, 53.7, 99.4, 361, and 1210 photons μm^{-2} for the wild type rod; 20.9, 76, 133, 431, and 3010 photons μm^{-2} for the A+/− rod; and 14.1, 51.1, 166, 603, and 2030 photons μm^{-2} for the A−/− rod.

The time to peak and the integration time of the dim flash response were normal, as was the rod's sensitivity to flashes. The morphology of retinas from A+/– mice was also normal. Thus partial deletion of arrestin did not significantly alter phototransduction because arrestin binding was not normally a rate-limiting step in the photoresponse recovery. Equivalent light was not present in A–/– rods, and photostasis was undisturbed.

Full deletion of arrestin (A–/–) produced averaged photoresponses that were qualitatively similar to those of S334ter (Fig. 9), but the nature of the quantal responses was quite different. S334ter rods showed two types of single-photon responses, normal and prolonged, but A–/– rods showed only one. Each A–/– response rose normally, recovered about 30 to 60%, then exhibited a tail that declined exponentially with a time constant of 42 ± 58 s (n = 15) (Xu *et al.*, in press). A likely explanation is that partial quench of photoexcited rhodopsin's activity by phosphorylation initiated recovery, but full recovery with a normal time course required complete quench of rhodopsin's activity by arrestin. In the absence of arrestin, the photoresponse had a tail whose recovery may have been governed by the thermal transition of the catalytically active metarhodopsin II state of rhodopsin to the inactive metarhodopsin III state, a process that occurs in 60 to 120 s at 37°C (Ebrey, 1968; Cone and Cobbs, 1969; Baumann and Bender, 1973; Bennett, 1980). Recovery of the tail may have been accelerated by multiple phosphorylation of rhodopsin (Wilden and Kuhn, 1982), where each added phosphate partially reduced rhodopsin's catalytic activity (Miller *et al.*, 1986; Arshavsky *et al.*, 1987; Philippov *et al.*, 1990; Wilden, 1995).

The integration time of the A–/– dim flash response was 17 ± 3 s (n = 9), equivalent to the summation of 66 normal single photon responses. Comparison of A–/– rod flash responses to responses of wild-type rods to 42-s steps of light gave an equivalent light in an A–/– rod that was 105 times greater than actual. A proper comparison of A–/– rod step responses to those of wild-type rods has not yet been made.

The sensitivity of A–/– rods was slightly subnormal; $i_{1/2}$ was 71 ± 23 photons μm^{-2} (n = 11). Preliminary results suggested that rod outer segments from the central retina of one A–/– mouse were 15 μm in length, 35% shorter than the 23 μm long rod outer segments from one wild-type mouse (Fig. 10). Shorter outer segment length would be expected to raise $i_{1/2}$ proportionately because of a reduction in collecting area. In addition to their shorter length, A–/– rod outer segments developed structural abnormalities (Figs. 10, 11).

Interestingly, rhodopsin levels in 4-week-old A–/– mouse retinas were up to 20-fold lower than normal. This difference could not be explained by shortening of outer segments alone. The outer nuclear layer thickness was normal, so the decreased rhodopsin was not

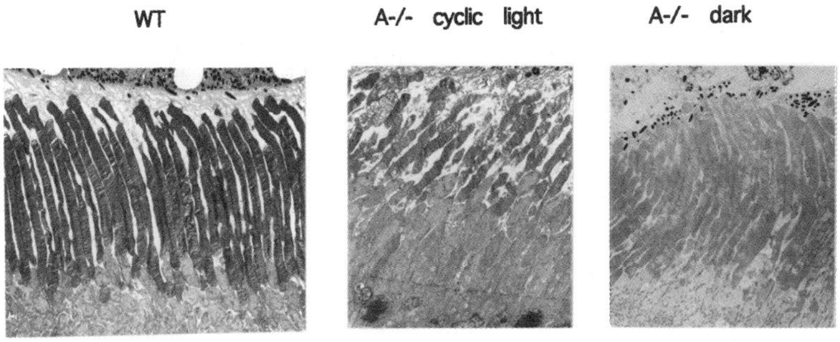

Figure 10. Longitudinal sections of rods from wild-type (left), A–/– cyclic-light-reared (middle), and A–/– dark-reared (right) mice. Rods of wild type and A–/– cyclic-light-reared mice were from the central retina. ×1650.

Figure 11. Cross sections of rod outer segments from the central retina of wild-type (left panel) and A−/− (right panel) mice. Connecting cilia of A−/− rods in the upper left corner indicate proximity to the inner segment. ×6600.

due to a loss of rods. To check whether some rods lacked outer segments, we evaluated their areal density in sections taken tangential to the central retina at the level of the outer segments (Fig. 11). In one wild-type mouse, there were 4.9×10^5 rod outer segments mm^{-2}. In A−/− retinas, we restricted the analysis to sections taken near the junction of the inner and outer segments to minimize a spuriously low cell count due to shortened outer segments. Isolated connecting cilia or the extensions of the cilia into the rod outer segments served as indicators of the appropriate retinal depth. Nevertheless, the density of 3.4×10^5 rod outer segments mm^{-2} determined for one A−/− mouse was a lower limit because tiering of the rods caused the rod inner segment–outer segment juncture to vary in depth by as much as 10 μm. The accuracy of the method could be improved if individual rods were followed from inner to outer segment in serial sections. The combination of low outer segment count and short outer segments still cannot account for the observed reduction in retinal rhodopsin. Additional factors have yet to be identified.

One litter of mice, born and reared in constant darkness, had retinas with the proper number of rods, outer segments that were normal in length, and normal rhodopsin content. At the electron-microscopic level, rod outer segments still contained a few abnormalities. The mother of this dark-reared litter, who was born and reared in cyclic light but placed in constant darkness for 6 weeks surrounding her pregnancy, also had lengthened rod outer segments. It should be noted, however, that in contrast to the changes in retinal histology and rhodopsin content, the A−/− rod photoresponse properties appeared to be independent of the mouse's light exposure history.

3.3. GTPase-Deficient Transducin

Several lines of transgenic mouse rods expressing different amounts of the mutant, Q204L human cone transducin α subunit were produced (Raport *et al.,* 1994). In each line, levels of the endogenous Tα, Tβ, and Tγ were unchanged. In the TΔ2 line, expression of

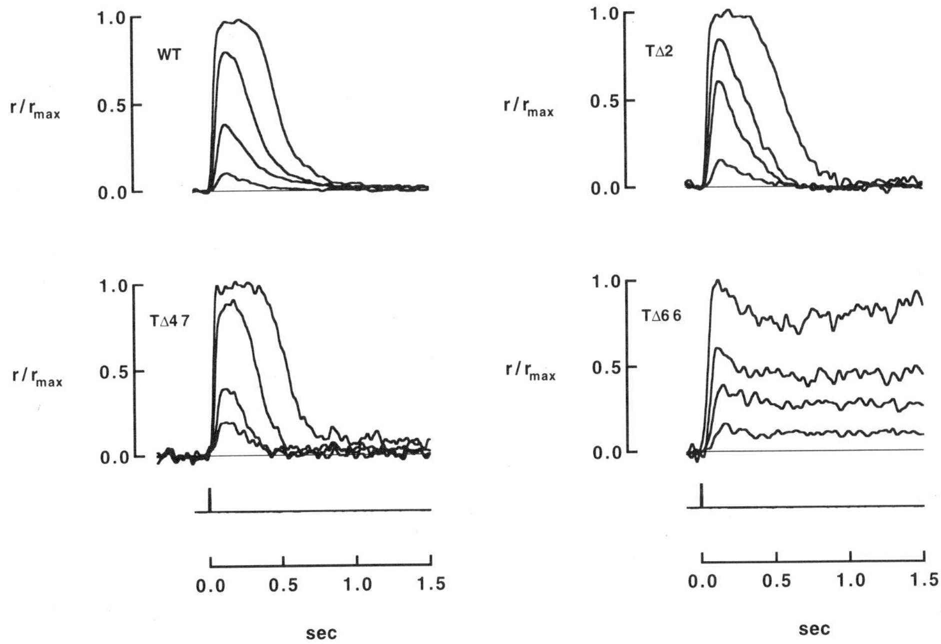

Figure 12. Flash families of wild-type and transgenic mouse rods expressing GTPase-deficient transducin. TΔ2 retinas contained similar amounts of native and mutant transducin, although transgene expression across the retina was nonuniform. All rods in TΔ47 and TΔ66 retinas appeared to express the transgene. There was twice to three times as much mutant as native transducin in TΔ47 retinas and four to six times as much mutant as native transducin in TΔ66 retinas. Flash strengths at 500 nm were 10, 36.2, 117, and 746.6 photons μm^{-2} for the wild type rod; 3.6, 22.5, 40.5, and 144 photons μm^{-2} for the TΔ2 rod; 56.5, 130, 371, and 1290 photons μm^{-2} for the TΔ47 rod; and 510, 1700, 5810, and 22500 photons μm^{-2} for the TΔ66 rod.

Q204L approached that of the wild type Tα, yet no detectable effects on the rod photoresponse were observed (Fig. 12). The time to peak of the dim-flash response was 130 ± 20 ms (n = 5) and the integration time was 230 ± 10 ms (n = 4). Possibly, the mutant human cone Tα interacted with mouse rhodopsin and rod Tβγ with a much lower efficiency than did the native rod Tα. It was also possible that, because of the small sample size and nonuniform transgene expression across the retina, none of the recorded rods expressed Q204L. Interestingly, rod sensitivity appeared to be slightly high; $i_{1/2}$ was 26 ± 17 photons μm^{-2} (n = 7).

Even in the TΔ47 line, where the level of Q204L was two to three times higher than that of the wild-type Tα, the photoresponse kinetics were essentially normal (Fig. 12) (Raport *et al.,* 1994). Sensitivity of the rod to flashes, however, decreased nearly twofold; $i_{1/2}$ was 87 ± 39 photons μm^{-2} (n = 8). This $i_{1/2}$ value is likely to be a lower limit because some rods did not respond to the brightest flash available ($i > 8 \times 10^4$ photons μm^{-2} at 500 nm). It is not yet clear whether the high $i_{1/2}$ was a reaction to the transgenic increase in Tα content directly or whether the change was photostasis related. This question could be approached by rearing of TΔ47 mice in darkness.

Sensitivity decreased further in TΔ66 mice, whose retinal levels of Q204L exceeded that of the endogenous Tα by four to six times (Raport *et al.,* 1994). Many rods were unresponsive to unattenuated flashes. In rods that did respond, $i_{1/2}$ was 1783 ± 1627 photons μm^{-2} (n = 7), 30-fold greater than normal. Flash responses were small and markedly prolonged (Fig. 12). The time to peak of the dim-flash response was slightly extended, 220 ± 50 ms (n = 3), and the integration time was 2400 ± 910 ms (n = 2), 10 times greater than normal. In some rods, the initial, "rapid" recovery before the onset of the tail was negligible. The slow response recovery may have been due to the eventual hydrolysis of GTP by Q204L or to a GTP hydrolysis-independent mechanism (e.g. Erickson *et al.,* 1992).

The equivalent light in a TΔ66 rod due to its longer dim-flash response-integration time appeared to have been completely compensated for by the increase in $i_{1/2}$. The reduction in maximal response amplitude may also be important because it is also a feature of light adaptation in normal rods. Quantification of the dark current is technically difficult, however; measurement depends in part on the efficiency of current collection by the electrode during recording. In addition, the relation between dark current and neurotransmitter release at the rod synapse is not known for transgenic or for wild-type rods.

The morphologies of both the TΔ47 and TΔ66 mouse retinas were remarkably intact (Fig. 13). The outer nuclear layer was unchanged, and rod outer segment length was nearly normal. Inclusions were observed in the inner segments of some rods. At the electron-microscopic level, the inclusions appeared to consist of autophagic vesicles (Remé, 1977) containing membranous whorls (Fig. 14), which are sometimes seen after a light stress (Moriya *et al.,* 1986). The appearance of similar whorls in rods after treatment with cytochalasin D (Williams *et al.,* 1988; Vaughan and Fisher, 1989) and in rods of rds mice expressing a transgene for mutant rds/peripherin (Kedzierski *et al.,* 1997) suggests that disc morphogenesis was disturbed in TΔ47 and TΔ66 rods. Rod outer segments also showed occasional disruptions in the organization of the discs in the outer segment, but the two types of disturbances were not necessarily concurrent within a rod.

What made the transducin transgenic mouse special? How did it cope with photoresponse prolongation differently than the other two types of transgenic mice? Biochemical analyses of TΔ47 and TΔ66 retinas by Raport *et al.* (1994) revealed substantial but highly specific alterations. The amount of catalytic α subunit of cGMP PDE was reduced by about 50% in both TΔ47 and TΔ66 retinas. The catalytic β subunit of PDE was low in TΔ47 and undetectable in TΔ66. As a result, light-induced PDE activity was reduced by 63% in TΔ47 and virtually abolished in TΔ66 retina (Fig. 15). The presence of normal levels of PDE mRNA suggested that its level was regulated at translation or by increased protein degradation. Interestingly, cGMP levels in both TΔ47 and TΔ66 were low in the

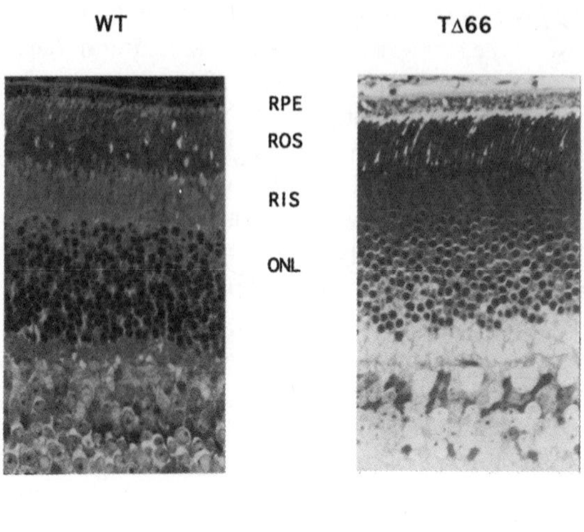

Figure 13. Retinas of a wild-type mouse at 8 weeks (left) and a TΔ66 mouse at 13 weeks (right). Scale bar denotes 20 μm.

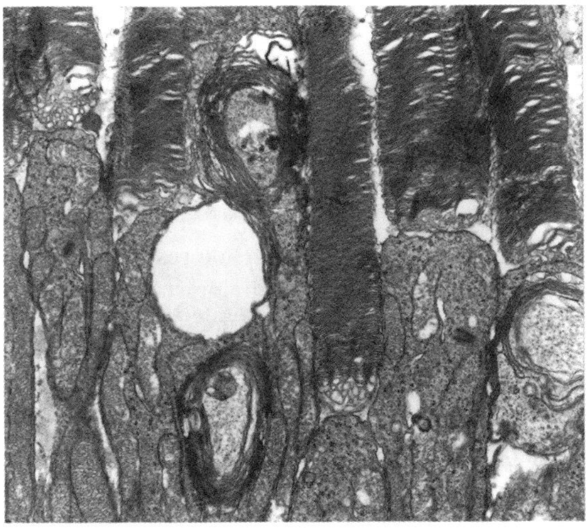

Figure 14. Membranous whorls in the inner segments of TΔ66 rods. ×15,680.

face of decreased PDE activity, rather than high. This pattern was accounted for in part by a decrease in Ca^{++}-sensitive guanylate cyclase activity in TΔ66, but a full explanation is not yet available. The low cGMP concentration in TΔ66 retinas provides the basis for the reduction in response amplitude observed in single rods electrophysiologically. The levels of rhodopsin, PDEγ, phosducin, arrestin, recoverin, and native Tα and β were normal.

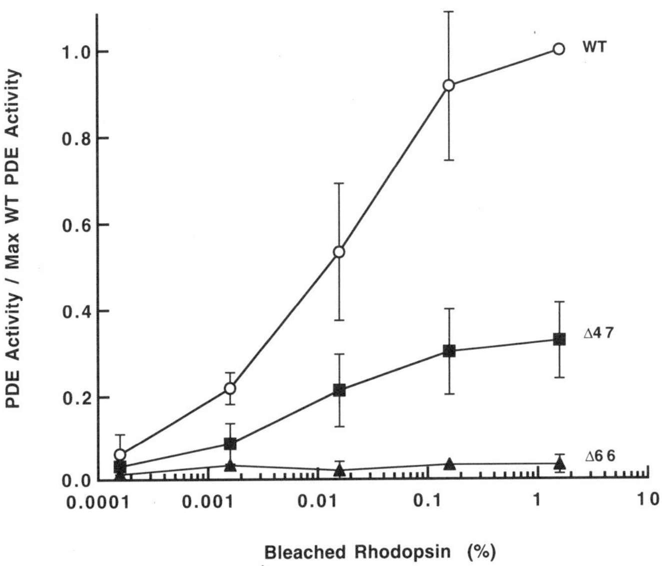

Figure 15. Phosphodiesterase activity of transgenic mouse retinas. ^{3}H-cGMP and ^{14}C-GMP were added to retinal homogenates of dark-adapted mice and exposed to light for up to 10 s. Three min later, the reaction mixture was quenched, and phosphodiesterase hydrolysis products were subjected to quantitative analysis (Hurley and Stryer, 1982; Raport *et al.,* 1994). Reprinted with permission from Raport *et al.* (1994). Copyright 1997, Association for Research in Vision and Ophthalmology.

4. DISCUSSION

4.1. Prolonged Photoresponses and Photostasis

Three perturbations of the phototransduction shutoff pathways were genetically engineered into rods of transgenic mice: expression of the phosphorylation-deficient S334ter rhodopsin, deletion of arrestin, and expression of the GTPase-deficient human cone transducin. All three perturbations prolonged the photoresponses to flashes, mimicking the effect of continued photoexcitation. The morphological changes in the retinas of the S334ter and A-/- mice appeared to be related to photostasis. There were fewer rod outer segments and those present were disrupted and shorter than normal. Part of the degeneration in S334ter mice and its progressive nature were likely to have been caused by a mechanism independent of photostasis, however. It is now known that transgenic mouse rods degenerate upon expression of Q344ter, a different truncated rod opsin lacking the last five carboxyl-terminal amino acids of rhodopsin (Sung *et al.,* 1994). All of the phosphorylation sites are intact in Q344ter. The deleted segment was not known to participate directly in phototransduction, and expression of Q344ter had only minor effects on the cell's flash response. Q344ter's defect was its aberrant localization throughout the rod. The ectopic accumulation of these rhodopsins within the rod inner segment probably impaired cellular function and led to cell death. Because the amino acids missing from Q344ter were also missing from S334ter, the retinal degeneration in the S334ter mouse may have had two etiologies, one stemming from the inappropriate localization of rhodopsin and one due to an increased equivalent light. In future studies, the photostasis component could be studied further in transgenic mice expressing a mutant, full-length, phosphorylation-deficient opsin or in S334ter mice reared in darkness.

Evidence for photostasis was stronger in A-/- mouse retinas. A large increase in equivalent light in A-/- rods due to photoresponse prolongation was accompanied by a decrease in the number of outer segments, a decrease in the outer segment length, structural disturbances in the discs, and a decreased retinal rhodopsin content. For the most part, these changes were absent in A-/- mice reared in darkness. Furthermore, the rod outer segments of a mouse born and reared in cyclic light lengthened when the mouse was later placed in darkness. A+/- had normal photoresponses, normal retinal morphology, and normal retinal rhodopsin content.

Additional support for equivalent-light-induced photostasis is emerging from studies on other transgenic mice. Electroretinographic measurements in transgenic mice expressing the mutant V20G, P23H, P27L opsin indicated that the responses to bright flashes recovered more slowly than normal (Goto *et al.,* 1995, 1996). Functional and histological loss of rods was observed in these transgenic mice as well as in mice expressing the mutant P23H opsin (Olsson *et al.,* 1992; Naash *et al.,* 1993, 1996). Rod outer segments were shortened and disrupted. Retinal degeneration was milder in animals reared in darkness (Naash *et al.,* 1996).

Dark rearing produced no obvious changes in the course of retinal degeneration in rds mice (Sanyal and Hawkins, 1986), vitiligo mice (Smith *et al.,* 1994a), "nervous" mice (LaVail *et al.,* 1993), or mice expressing the mutant K296E opsin (Li *et al.,* 1995). Although dark rearing was originally reported to slow the degeneration in Royal College of Surgeons (RCS) rat retinas (Dowling and Sidman, 1962), this result was later attributed to an interaction with eye pigmentation and not the RCS condition (LaVail and Battelle, 1975). Thus light exposure history was ruled out as a nonspecific factor contributing to all types of retinal degeneration. The studies on mice expressing K296E opsin are intriguing.

In vitro, K296 opsin mutants constitutively activated transducin (Cohen *et al.,* 1992; Robinson *et al.,* 1992). The presence of K296E in rods should have produced a significant equivalent background light, yet the electroretinogram was characteristic of the dark-adapted state (Li *et al.,* 1995). Further investigation revealed that K296E was constitutively phosphorylated and bound to arrestin, i.e. permanently quenched. In spite of this quench, the rods still degenerated. The underlying cause is under investigation.

Observations consistent with photostasis have been made in *Drosophila.* Substantial reduction of arrestin in mutant flies prolonged the recovery of their electrical response to flashes and caused a light-dependent retinal degeneration (Dolph *et al.,* 1993). The degeneration was prevented by the norpA mutation, which interrupted the phototransduction cascade and eliminated the electrical response. A similar light- and phototransduction-dependent retinal degeneration resulted from the rdgB mutation (Harris and Stark, 1977; Stark *et al.,* 1983).

High expression of GTPase-deficient transducin gave rise to prolonged photoresponses resembling those of S334ter and A−/− rods, but retinal morphology was little affected. Instead, there was a significant down-regulation in both the production and hydrolysis of cGMP. These compensatory changes eliminated the equivalent light. It is tempting to attribute the changes in TΔ66 to photostasis, however, because TΔ47 rods were also less sensitive even though they did not have prolonged photoresponses. Furthermore, little is known about links between overexpression of transducin and the regulation of cGMP metabolism because our initial attempts to cause overexpression of wild-type rod transducin in rods of transgenic mice were inconclusive (Raport *et al.,* 1994). Nonetheless our results from TΔ66 mice indicate that rod degenerative changes need not occur as an inevitable consequence of photostasis. Rods are capable of adjusting their phototransductive power by down-regulating cGMP metabolism. For reasons unknown, the retina does not do so in true photostasis and irreversibly destroys rod photoreceptors instead.

Light-exposure history can cause the outer segment concentration of rhodopsin to change in a direction counter to that of the light intensity (Organisciak and Noell, 1977; Battelle and LaVail, 1978; Penn and Williams, 1986; Penn and Anderson, 1987; Penn *et al.,* 1989; Schremser and Williams, 1995b), although substantial changes in lighting are required and seem to produce no more than a twofold difference in rhodopsin concentration. There was no evidence that rhodopsin concentration within individual rods changed in the transgenic mice of this study. Perhaps the pupillary regulation of retinal illuminance in these pigmented mice obviated their need to change rod outer segment rhodopsin concentration. A similar argument might explain why levels of endogenous transducin were normal both in transducin-transgenic mice and in A−/− mice and why arrestin levels were normal in transducin transgenic mice, as light-exposure history is known to affect the levels of these two proteins, too (Farber *et al.,* 1991; Mirshahi *et al.,* 1991; Organisciak *et al.,* 1991). Alternatively, apparent discrepancies between real and equivalent light-induced photostasis may have been due to complications associated with the specific transgenic manipulations.

4.2. Retinal Disease

It has been proposed that some retinal diseases arise from a photostasis response to the equivalent light stemming from a mutation in one of the proteins involved in phototransduction (Fain and Lisman, 1993). Certain types of autosomal dominant retinitis pigmentosa (ADRP) are caused by mutations in rhodopsin that delete some or all of its phosphorylation sites (Horn *et al.,* 1992; Apfelstedt-Sylla *et al.,* 1993; Kim *et al.,* 1993;

Restagno *et al.,* 1993). Night blindness can result from mutations in rhodopsin kinase thought to produce a nonfunctional protein (Yamamoto *et al.,* 1997). Patients with these types of mutations would be expected to suffer from visual disturbances like those described in S334ter mice. The P23H opsin mutant was introduced into transgenic mouse rods because it was known to cause ADRP in humans (Dryja *et al.,* 1990). The G90D and the A292E opsin mutants are constitutively active *in vitro* and cause congenital stationary night blindness (Sieving *et al.,* 1995; Dryja *et al.,* 1993; Rao *et al.,* 1994). Oguchi disease, a different type of stationary night blindness, can be caused by null mutations in the arrestin gene (Fuchs *et al.,* 1995). Persons heterozygous for the arrestin mutation were asymptomatic. Thus the arrestin mutation in humans produces a condition similar to that described in A–/– and A+/– mice. A missense mutation, G38D, in Tα has been linked to the Nougaret form of stationary night blindness (Dryja *et al.,* 1996). Mutation of the homologous glycine in the ras oncogene protein significantly reduces its GTPase activity (Gibbs *et al.,* 1984; McGrath *et al.,* 1984; Sweet *et al.,* 1984), suggesting that rod function in Nougaret patients may be similar to that in the GTPase-deficient transducin transgenic mouse rods.

If retinal disease were caused by photostasis, then it should be treatable by diminution of the daily light exposure. A therapy consisting of monocular occlusion for 6 to 8 hours per day was explored in two RP patients, but after a 5-year period progression of the disease in the two eyes of each patient was equal (Berson, 1980). One patient is now known to have had point mutations in two rod-outer-segment structural proteins, rds and ROM1 (Dryja *et al.,* in press), so the retinal disease in that patient was probably unrelated to photostasis. It is also possible that the reduction in daily light exposure was insufficient to affect the course of the disease or that beneficial effects were too small to detect because of the advanced state of disease prior to the beginning of the therapy. A reduction in daily light exposure might still improve the prognosis for RP patients with other genotypes.

The severity of the retinal degeneration in the transgenic mouse seemed to differ from that in the corresponding human disease. For example, Oguchi disease is described as a stable condition, whereas A–/– mice experienced a slow but progressive retinal degeneration. Age (O'Steen *et al.,* 1974; Penn *et al.,* 1985) and genetic background (LaVail *et al.,* 1987) affect the extent to which the retina reacts to an overexposure to photons. Perhaps these factors were more significant in mice than in their human counterparts.

4.3. The Neural Signal for Photostasis

The rat retina strives to maintain a daily quantal catch of 10^{16} photons per eye (Penn and Williams, 1986). If the quantal catch is distributed uniformly over the ~1.7×10^7 rods in the retina (Lashley, 1932; Hagins *et al.,* 1970; Reiser *et al.,* 1996), each rod might be expected to absorb 5.9×10^8 quanta per day. Over a 12-hour period, photoisomerizations would average about 13,600 $\text{rod}^{-1}\ \text{s}^{-1}$. Bleaching rates of this magnitude keep rat rods in saturation (Nakatani *et al.,* 1991), so this set point seems sensible in that greater photon flux exposures would not produce any increase in the rod response. It raises a question, however: How would the rod "know" when its set point has been exceeded? Increasing the intensity over that producing saturation lengthens the period required for full recovery after the light is extinguished, so the eye may take its cue from the length of the dark period or try to maintain a constant dark-adaptation period.

Perhaps the assumption that light is absorbed uniformly across the retina is not valid. The physiological optics focus diffuse light onto the centrally located rods, so lighting that saturates central rods might fall within the dynamic range of peripheral rods.

Overexposure to light has been reported to cause a preferential loss of rods in the superior central region of the retina (Noell *et al.,* 1966; Grignolo *et al.,* 1969; LaVail, 1976; Rapp and Williams, 1980). Perhaps the fate of these central rods depended upon signals generated by the peripheral ones.

A third possibility is that photostasis is mediated in part by cones. Vision in rats is photopic when the rate of bleaching exceeds 10^3 photoisomerizations rod^{-1} s^{-1} (Green, 1971). Although the action spectrum for photostasis is similar to the absorption spectrum of rhodopsin (Noell *et al.,* 1966; Williams and Howell, 1983), the rat cone spectral sensitivity is not very different (Dodt and Echte, 1961; Green, 1971); its maximum is near 510 nm (Neitz and Jacobs, 1986). Indeed the rod and cone signaling pathways converge in the mammalian retina at the level of the cone bipolar cell, if not before (reviewed by Kolb, 1994; Vaney, 1997). Overstimulation by light might be mediated by cones, while insufficient illumination might be signaled by rods.

ACKNOWLEDGMENTS

We thank H. Covington for excellent technical support and E. Nekrasov for assistance in preparing the figures. This work was supported by Career Development Awards from Research to Prevent Blindness (JC, JGF, CLM), Lion's of Massachusetts (CLM), E. Mathilda Ziegler Foundation for the Blind (CLM), the National Eye Institute, EY-08980 (JGF), the Steinbach Fund (JC), and the National Institute of General Medical Sciences (RLD).

LITERATURE CITED

Apfelstedt-Sylla, E., Kunisch, M., Horn, M., Ruther, K., Gerding, H., Gal, A., and Zrenner, E. (1993). Ocular findings in a family with autosomal dominant retinitis pigmentosa and a frameshift mutation altering the carboxyl terminal sequence of rhodopsin. *British Journal of Ophthalmology* 77:495–501.

Arshavsky, V. Y., Antoch, M. P., Dizhoor, A. M., Rakhilin, S. V., and Philippov, P. P. (1987). Interaction of visual transduction enzymes in a detergent solution. In: Ovchinnikov, Y. A., ed. *Retinal Proteins.* VNU Science Press, Utrecht. pp. 497–503.

Battelle, B.-A., and LaVail, M. M. (1978). Rhodopsin content and rod outer segment length in albino rat eyes: modification by dark adaptation. *Experimental Eye Research* 26:487–497.

Baumann, C., and Bender, S. (1973). Kinetics of rhodopsin bleaching in the isolated human retina. *Journal of Physiology (London)* 235:761–773.

Baylor, D. A., and Hodgkin, A. L. (1974). Changes in time scale and sensitivity in turtle photoreceptors. *Journal of Physiology (London)* 242:729–758.

Baylor, D. A., Nunn, B. J., and Schnapf, J. L. (1984). The photocurrent, noise and spectral sensitivity of rods of the monkey *Macaca fascicularis. Journal of Physiology (London)* 357:575–607.

Bennett, N. (1980). The decay of metarhodopsin II in cattle rod outer segment membranes: protonation and spectral changes. *Biochemical and Biophysical Research Communications* 96:1695–1701.

Berson, E. L. (1980). Light deprivation and retinitis pigmentosa. *Vision Research* 20:1179–1184.

Bownds, M. D., and Arshavsky, V. Y. (1995). What are the mechanisms of photoreceptor adaptation? *Behavioral and Brain Sciences* 18:415–424.

Bush, R. A., and Williams, T. P. (1991). The effect of unilateral optic nerve section on retinal light damage in rats. *Experimental Eye Research* 52:139–153.

Chen, J., Makino, C. L., Peachey, N. S., Baylor, D. A., and Simon, M. I. (1995). Mechanisms of rhodopsin inactivation *in vivo* as revealed by a COOH-terminal truncation mutant. *Science* 267:374–377.

Cohen, G. B., Oprian, D. D., and Robinson, P. R. (1992). Mechanism of activation and inactivation of opsin: role of Glu113 and Lys296. *Biochemistry* 31:12592–12601.

Cone, R. A., and Cobbs, W. H., III (1969). Rhodopsin cycle in the living eye of the rat. *Nature* 221:820–822.

Dodt, E., and Echte, K. (1961). Dark and light adaptation in pigmented and white rats as measured by electroretinogram threshold. *Journal of Neurophysiology* 24:427–445.

Dolph, P. J., Ranganathan, R., Colley, N. J., Hardy, R. W., Socolich, M., and Zuker, C. S. (1993). Arrestin function in inactivation of G protein-coupled receptor rhodopsin *in vivo. Science* 260:1910–1916.

Dowling, J. E., and Sidman, R. L. (1962). Inherited retinal dystrophy in the rat. *Journal of Cell Biology* 14:73–109.

Dryja, T. P., McGee, T. L., Reichel, E., Hahn, L. B., Cowley, G. S., Yandell, D. W., Sandberg, M. A., and Berson, E. L. (1990). A point mutation of the rhodopsin gene in one form of retinitis pigmentosa. *Nature* 343:364–366.

Dryja, T. P., Berson, E. L., Rao, V. R., and Oprian, D. D. (1993). Heterozygous missense mutation in the rhodopsin gene as a cause of congenital stationary night blindness. *Nature Genetics* 4:280–283.

Dryja, T. P., Hahn, L. B., Reboul, T., and Arnaud, B. (1996). Missense mutation in the gene encoding the α subunit of rod transducin in the Nougaret form of congenital stationary night blindness. *Nature Genetics* 13:358–360.

Dryja, T. P., Hahn, L. B., Kajiwara, K., and Berson, E. L. (in press). Dominant and digenic mutations in the *peripherin/RDS* and *ROM1* genes in retinitis pigmentosa. *Investigative Ophthalmology and Visual Science.*

Ebrey, T. G. (1968). The thermal decay of the intermediates of rhodopsin *in situ. Vision Research* 8:965–982.

Erickson, M. A., Robinson, P., and Lisman, J. (1992). Deactivation of visual transduction without guanosine triphosphate hydrolysis by G protein. *Science* 257:1255–1258.

Fain, G. L., and Lisman, J. E. (1993). Photoreceptor degeneration in vitamin A deprivation and retinitis pigmentosa: the equivalent light hypothesis. *Experimental Eye Research* 57:335–340.

Farber, D. B., Seager-Danciger, J. S., and Organisciak, D. T. (1991). Levels of mRNA encoding proteins of the cGMP cascade as a function of light environment. *Experimental Eye Research* 53:781–786.

Fuchs, S., Nakazawa, M., Maw, M., Tamai, M., Oguchi, Y., and Gal, A. (1995). A homozygous 1-base pair deletion in the arrestin gene is a frequent cause of Oguchi disease in Japanese. *Nature Genetics* 10:360–362.

Gibbs, J. B., Sigal, I. S., Poe, M., and Scolnick, E. M. (1984). Intrinsic GTPase activity distinguishes normal and oncogenic *ras* p21 molecules. *Proceedings of the National Academy of Sciences of the USA* 81:5704–5708.

Goto, Y., Peachey, N. S., Ripps, H., and Naash, M. I. (1995). Functional abnormalities in transgenic mice expressing a mutant rhodopsin gene. *Investigative Ophthalmology and Visual Science* 36:62–71.

Goto, Y., Peachey, N. S., Ziroli, N. E., Seiple, W. H., Gryczan, C., Pepperberg, D. R., and Naash, M. I. (1996). Rod phototransduction in transgenic mice expressing a mutant opsin gene. *Journal of the Optical Society of America A* 13:577–585.

Green, D. G. (1971). Light adaptation in the rat retina: evidence for two receptor mechanisms. *Science* 174:598–600.

Grignolo, A., Orzalesi, N., Castellazzo, R., and Vittone, P. (1969). Retinal damage by visible light in albino rats: an electron microscope study. *Ophthalmologica* 157:43–59.

Hagins, W. A., Penn, R. D., and Yoshikami, S. (1970). Dark current and photocurrent in retinal rods. *Biophysical Journal* 10:380–412.

Harris, W. A., and Stark, W. S. (1977). Hereditary retinal degeneration in *Drosophila melanogaster:* a mutant defect associated with the phototransduction process. *Journal of General Physiology* 69:261–291.

Helmreich, E. J. M., and Hofmann, K.-P. (1996). Structure and function of proteins in G-protein-coupled signal transfer. *Biochimica et Biophysica Acta* 1286:285–322.

Horn, M., Humphries, P., Kunisch, M., Marchese, C., Apfelstedt-Sylla, E., Fugi, L., Zrenner, E., Kenna, P., Gal, A., and Farrar, J. (1992). Deletions in exon 5 of the human rhodopsin gene causing a shift in the reading frame and autosomal dominant retinitis pigmentosa. *Human Genetics* 90:255–257.

Hurley, J. B. (1994). Termination of photoreceptor responses. *Current Opinion in Neurobiology* 4:481–487.

Hurley, J. B., and Stryer, L. (1982). Purification and characterization of the γ regulatory subunit of the cyclic GMP phosphodiesterase from retinal rod outer segments. *Journal of Biological Chemistry* 257:11094–11099.

Kedzierski, W., Lloyd, M., Birch, D. G., Bok, D., and Travis, G. H. (1997). Generation and analysis of transgenic mice expressing P216L-substituted rds/peripherin in rod photoreceptors. *Investigative Ophthalmology and Visual Science* 38:498–509.

Kim, R. Y., Al-Maghtheh, M., Fitzke, F. W., Arden, G. B., Jay, M., Bhattacharya, S. S., and Bird, A. C. (1993). Dominant retinitis pigmentosa associated with two rhodopsin gene mutations: leu-40-arg and an insertion disrupting the 5'-splice junction of exon 5. *Archives of Ophthalmology* 111:1518–1524.

Kolb, H. (1994). The architecture of functional neural circuits in the vertebrate retina. The Proctor lecture. *Investigative Ophthalmology and Visual Science* 35:2385–2404.

Koutalos, Y., and Yau, K.-W. (1993). A rich complexity emerges in phototransduction. *Current Opinion in Neurobiology* 3:513–519.

Lanum, J. (1978). The damaging effects of light on the retina. Empirical findings, theoretical and practical implications. *Survey of Ophthalmology* 22:221–249.

Lashley, K. S. (1932). The mechanism of vision. V. The structure and image-forming power of the rat's eye. *Journal of Comparative Psychology* 13:173–200.

LaVail, M. M. (1976). Survival of some photoreceptor cells in albino rats following long-term exposure to continuous light. *Investigative Ophthalmology* 15:64–70.

LaVail, M. M., and Battelle, B.-A. (1975). Influence of eye pigmentation and light deprivation on inherited retinal dystrophy in the rat. *Experimental Eye Research* 21:167–192.

LaVail, M. M., Gorrin, G. M., Repaci, M. A., Thomas, L. A., and Ginsberg, H. M. (1987). Genetic regulation of light damage to photoreceptors. *Investigative Ophthalmology and Visual Science* 28:1043–1048.

LaVail, M. M., White, M. P., Gorrin, G. M., Yasumura, D., Porrello, K. V., and Mullen, R. J. (1993). Retinal degeneration in the nervous mutant mouse. I. Light microscopic cytopathology and changes in the interphotoreceptor matrix. *Journal of Comparative Neurology* 333:168–181.

Lerea, C. L., Bunt-Milam, A. H., and Hurley, J. B. (1989). α transducin is present in blue-, green-, and red-sensitive cone photoreceptors in the human retina. *Neuron* 3:367–376.

Li, T., Franson, W. K., Gordon, J. W., Berson, E. L., and Dryja, T. P. (1995). Constitutive activation of phototransduction by K296E opsin is not a cause of photoreceptor degeneration. *Proceedings of the National Academy of Sciences of the USA* 92:3551–3555.

McDowell, J. H., Curtis, D. R., Bakar, U. A., and Hargrave, P. A. (1985). Phosphorylation of rhodopsin: localization of phosphorylated residues in the helix V-helix VI connecting loop. *Investigative Ophthalmology and Visual Science* 26:291.

McGrath, J. P., Capon, D. J., Goeddel, D. V., and Levinson, A. D. (1984). Comparative biochemical properties of normal and activated human ras p21 protein. *Nature* 310:644–649.

Mazia, D., Schatten, G., and Sale, W. (1975). Adhesion of cells to surfaces coated with polylysine. *Journal of Cell Biology* 66:198–200.

Miller, J. L., Fox, D. A., and Litman, B. J. (1986). Amplification of phosphodiesterase activation is greatly reduced by rhodopsin phosphorylation. *Biochemistry* 25:4983–4988.

Mirshahi, M., de Kozak, Y., Tarraf, M., Razaghi, A., Thillaye, B., and Faure, J.-P. (1991). Early disappearance of alpha-transducin in light-induced photoreceptor degeneration in albino rats. *Current Eye Research* 10:993–1000.

Moriya, M., Baker, B. N., and Williams, T. P. (1986). Progression and reversibility of early light-induced alterations in rat retinal rods. *Cell and Tissue Research* 246:607–621.

Naash, M. I., Hollyfield, J. G., Al-Ubaidi, M. R., and Baehr, W. (1993). Simulation of human autosomal dominant retinitis pigmentosa in transgenic mice expressing a mutated murine opsin gene. *Proceedings of the National Academy of Sciences of the USA* 90:5499–5503.

Naash, M. L., Peachey, N. S., Li, Z.-Y., Gryczan, C. C., Goto, Y., Blanks, J., Milam, A. H., and Ripps, H. (1996). Light-induced acceleration of photoreceptor degeneration in transgenic mice expressing mutant rhodopsin. *Investigative Ophthalmology and Visual Science* 37:775–782.

Nakatani, K., Tamura, T., and Yau, K.-W. (1991). Light adaptation in retinal rods of the rabbit and two other nonprimate mammals. *Journal of General Physiology* 97:413–435.

Neitz, J., and Jacobs, G. H. (1986). Reexamination of spectral mechanisms in the rat (*Rattus norvegicus*). *Journal of Comparative Psychology* 100:21–29.

Newton, A. C., and Williams, D. S. (1991). Involvement of protein kinase C in the phosphorylation of rhodopsin. *Journal of Biological Chemistry* 266:17725–17728.

Noell, W. K., Walker, V. S., Kang, B. S., and Berman, S. (1966). Retinal damage by light in rats. *Investigative Ophthalmology* 5:450–473.

Ohguro, H., Johnson, R. S., Ericsson, L. H., Walsh, K. A., and Palczewski, K. (1994). Control of rhodopsin multiple phosphorylation. *Biochemistry* 33:1023–1028.

Olsson, J. E., Gordon, J. W., Pawlyk, B. S., Roof, D., Hayes, A., Molday, R. S., Mukai, S., Cowley, G. S., Berson, E. L., and Dryja, T. P. (1992). Transgenic mice with a rhodopsin mutation (Pro23His): a mouse model of autosomal dominant retinitis pigmentosa. *Neuron* 9:815–830.

Organisciak, D. T., and Noell, W. K. (1977). The rod outer segment phospholipid/opsin ratio of rats maintained in darkness or cyclic light. *Investigative Ophthalmology and Visual Science* 16:188–190.

Organisciak, D. T., Xie, A., Wang, H.-M., Jiang, Y.-L., Darrow, R. M., and Donoso, L. A. (1991). Adaptive changes in visual cell transduction protein levels: effect of light. *Experimental Eye Research* 53:773–779.

O'Steen, W. K., Anderson, K. V., and Shear, C. R. (1974). Photoreceptor degeneration in albino rats: dependency on age. *Investigative Ophthalmology* 13:334–339.

Palczewski, K., and Smith, W. C. (1996). Splice variants of arrestins. *Experimental Eye Research* 63:599–602.

Palczewski, K., Buczylko, J., Ohguro, H., Annan, R. S., Carr, S. A., Crabb, J. W., Kaplan, M. W., Johnson, R. S., and Walsh, K. A. (1994). Characterization of a truncated form of arrestin isolated from bovine rod outer segments. *Protein Science* 3:314–324.

Penn, J. S., and Anderson, R. E. (1987). Effect of light history on rod outer-segment membrane composition in the rat. *Experimental Eye Research* 44:767–778.

Penn, J. S., and Anderson, R. E. (1992). Effects of light history on the rat retina. In: Osborne, N. N., and Chader, G. J., eds. *Progress in Retinal Research, Volume 11.* Pergamon Press, New York. pp. 75–98.

Penn, J. S., and Williams, T. P. (1986). Photostasis: regulation of daily photon-catch by rat retinas in response to various cyclic illuminances. *Experimental Eye Research* 43:915–928.

Penn, J. S., Baker, B. N., Howard, A. G., and Williams, T. P. (1985). Retinal light-damage in albino rats: lysosomal enzymes, rhodopsin and age. *Experimental Eye Research* 41:275–284.

Penn, J. S., Thum, L. A., and Naash, M. I. (1989). Photoreceptor physiology in the rat is governed by the light environment. *Experimental Eye Research* 49:205–215.

Philippov, P. P., Arshavsky, V. Y., and Dizhoor, A. M. (1990). The phosphodiesterase cascade of a photoreceptor cell: mechanisms of activation and quenching. *Soviet Scientific Reviews* 9:243–307.

Pugh, E. N., Jr. and Lamb, T. D. (1993). Amplification and kinetics of the activation steps in phototransduction. *Biochimica et Biophysica Acta* 1141:111–149.

Rao, V. R., Cohen, G. B., and Oprian, D. D. (1994). Rhodopsin mutation G90D and a molecular mechanism for congenital night blindness. *Nature* 367:639–642.

Raport, C. J., Lem, J., Makino, C., Chen, C.-K., Fitch, C. L., Hobson, A., Baylor, D., Simon, M. I., and Hurley, J. B. (1994). Downregulation of cGMP phosphodiesterase induced by expression of GTPase-deficient cone transducin in mouse rod photoreceptors. *Investigative Ophthalmology and Visual Science* 35:2932–2947.

Rapp, L. M., and Williams, T. P. (1980). A parametric study of retinal light damage in albino and pigmented rats. In: Williams, T. P., and Baker, B. N., eds. *The Effects of Constant Light on Visual Processes.* Plenum Press, New York. pp. 135–159.

Reiser, M. A., Williams, T. P., and Pugh, E. N., Jr. (1996). The effect of light history on the aspartate-isolated fast-PIII responses of the albino rat retina. *Investigative Ophthalmology and Visual Science* 37:221–229.

Remé, C. E. (1977). Autophagy in visual cells and pigment epithelium. *Investigative Ophthalmology and Visual Science* 16:807–814.

Restagno, G., Maghtheh, M., Bhattacharya, S., Ferrone, M., Garnerone, S., Samuelly, R., and Carbonara, A. (1993). A large deletion at the 3' end of the rhodopsin gene in an Italian family with a diffuse form of autosomal dominant retinitis pigmentosa. *Human Molecular Genetics* 2:207–208.

Robinson, P. R., Cohen, G. B., Zhukovsky, E. A., and Oprian, D. D. (1992). Constitutively active mutants of rhodopsin. *Neuron* 9:719–725.

Sanyal, S., and Hawkins, R. K. (1986). Development and degeneration of retina in *rds* mutant mice: effects of light on the rate of degeneration in albino and pigmented homozygous and heterozygous mutant and normal mice. *Vision Research* 26:1177–1185.

Schremser, J.-L., and Williams, T. P. (1995a). Rod outer segment (ROS) renewal as a mechanism for adaptation to a new intensity environment. I. Rhodopsin levels and ROS length. *Experimental Eye Research* 61:17–24.

Schremser, J.-L., and Williams, T. P. (1995b). Rod outer segment (ROS) renewal as a mechanism for adaptation to a new intensity environment. II. Rhodopsin synthesis and packing density. *Experimental Eye Research* 61:25–32.

Sieving, P. A., Richards, J. E., Naarendorp, F., Bingham, E. L., Scott, K., and Alpern, M. (1995). Dark-light: model for nightblindness from the human rhodopsin gly-90 → asp mutation. *Proceedings of the National Academy of Sciences of the USA* 92:880–884.

Smith, S. B., Cope, B. K., and McCoy, J. R. (1994a). Effects of dark-rearing on the retinal degeneration of the C57BL/6-mi^{vit}/mi^{vit} mouse. *Experimental Eye Research* 58:77–84.

Smith, W. C., Milam, A. H., Dugger, D., Arendt, A., Hargrave, P. A., and Palczewski, K. (1994b). A splice variant of arrestin: molecular cloning and localization in bovine retina. *Journal of Biological Chemistry* 269:15407–15410.

Stark, W. S., Chen, D.-M., Johnson, M. A., and Frayer, K. L. (1983). The rdgB gene in *Drosophila:* retinal degeneration in several mutant alleles and inhibition by *norpA*. *Journal of Insect Physiology* 29:123–131.

Sung, C.-H., Makino, C., Baylor, D., and Nathans, J. (1994). A rhodopsin gene mutation responsible for autosomal dominant retinitis pigmentosa results in a protein that is defective in localization to the photoreceptor outer segment. *Journal of Neuroscience* 14:5818–5833.

Sweet, R. W., Yokoyama, S., Kamata, T., Feramisco, J. R., Rosenberg, M., and Gross, M. (1984). The product of ras is a GTPase and the T24 oncogenic mutant is deficient in this activity. *Nature* 311:273–275.

Vaney, D. I. (1997). Neuronal coupling in rod—signal pathways of the retina. *Investigative Ophthalmology and Visual Science* 38:267–273.

Vaughan, D. K., and Fisher, S. K. (1989). Cytochalasin D disrupts outer segment disc morphogenesis *in situ* in rabbit retina. *Investigative Ophthalmology and Visual Science* 30:339–342.

Wilden, U. (1995). Duration and amplitude of the light-induced cGMP hydrolysis in vertebrate photoreceptors are regulated by multiple phosphorylation of rhodopsin and by arrestin binding. *Biochemistry* 34:1446–1454.

Wilden, U., and Kuhn, H. (1982). Light-dependent phosphorylation of rhodopsin: number of phosphorylation sites. *Biochemistry* 21:3014–3022.

Williams, D. S., Linberg, K. A., Vaughan, D. K., Fariss, R. N., and Fisher, S. K. (1988). Disruption of microfilament organization and deregulation of disk membrane morphogenesis by cytochalasin D in rod and cone photoreceptors. *Journal of Comparative Neurology* 272:161–176.

Williams, T. P., and Baker, B. N., eds. (1980). *The Effects of Constant Light on Visual Processes.* Plenum Press, New York.

Williams, T. P., and Howell, W. L. (1983). Action spectrum of retinal light-damage in albino rats. *Investigative Ophthalmology and Visual Science* 24:285–287.

Xu, J., Dodd, R. L., Makino, C. L., Simon, M. I., Baylor, D. A., and Chen, J. (1997). Prolonged photoresponses in transgenic mouse rods lacking arrestin. *Nature* 389:505–509.

Yamamoto, S., Sippel, K. C., Berson, E. L., and Dryja, T. P. (1997). Defects in the rhodopsin kinase gene in the Oguchi form of stationary night blindness. *Nature Genetics* 15:175–178.

Yarfitz, S., and Hurley, J. B. (1994). Transduction mechanisms of vertebrate and invertebrate photoreceptors. *Journal of Biological Chemistry* 269:14329–14332.

9

PHOTOSTASIS

Implications for Pineal, Circadian, and Visual Photosensitivity

Ziad Boulos[1] and Michael Terman[2]

[1]Institute for Circadian Physiology
One Alewife Center
Cambridge, Massachusetts 02140
[2]Department of Psychiatry
Columbia University, and
New York State Psychiatric Institute
722 West 168th Street
New York, New York 10032

1. INTRODUCTION

Extensive evidence in rats indicates that variations in the daily lighting environment lead to gradual changes in retinal structure and function and that these changes represent an adaptive process, termed photostasis, whereby the sensitivity of the retina is slowly altered to ensure the absorption of a constant number of photons per day—a process that may have evolved to allow adaptation to seasonal variations in daytime light intensity and duration.

Light input to the mammalian retina serves a number of different functions, which include entrainment of circadian rhythms to the daily illumination cycle and regulation of melatonin production by the pineal gland. Exposure to light at night results in an acute suppression of melatonin levels, and there are indications in a few species that the degree of suppression is a function of the animal's prior lighting environment.

The aim of this chapter is to consider some potential functional consequences of retinal photostasis in mammals. In particular, the suggestion is made that retinal photostatic adjustments underlie parallel changes in the responsiveness of the pineal gland to light and, possibly, that of the circadian timing system. Evidence of seasonal variations in several light-dependent processes is then reviewed in humans, including patients suffering from winter depression, and the possible involvement of retinal photostasis in these effects is discussed.

Photostasis and Related Phenomena, edited by Williams and Thistle
Plenum Press, New York, 1998

2. EFFECTS OF LIGHT HISTORY ON PHOTOSENSITIVITY IN ANIMALS

2.1. Light History and Retinal Physiology: The Photostasis Hypothesis

Data indicating that an animal's ambient light environment affects retinal structure and function were initially obtained in albino rats maintained under daily cycles in which the lights were on for 12 hours and off for 12 hours each day (12L:12D, 50–100 lux) or in constant darkness (12D:12D) for up to 140 days (Organisciak and Noell, 1977). Total rhodopsin per eye was found to be higher in 12D:12D animals than in 12L:12D animals. At least 10 days of 12D:12D were required for rhodopsin to reach a new steady-state level. A lower phospholipid/opsin ratio was also observed in 12D:12D, suggesting an increase in opsin density in rod outer segments (ROSs) relative to the 12L:12D condition. In addition, ROS length, estimated by multiplication of the phospholipid/opsin ratio by total rhodopsin per eye, appeared to be greater in the 12D:12D animals. Direct measurements of ROS length were subsequently reported by Battelle and LaVail (1978), who found a maximum increase of 25% in 12D:12D albino rats. Animals reared in 12D:12D or switched from 12L:12D to 12D:12D also had higher dark-adapted rhodopsin levels; a maximum increase of about 50% was reached after 10 days. Changes in the opposite direction were observed in rats transferred from 12D:12D to 12L:12D. No differences were observed in ROS diameter or in outer segment disc packing. The fact that rhodopsin level increased more than ROS length suggested that opsin density in newly synthesized ROS discs was also increased in 12D:12D.

A limitation of these data is that, under natural conditions, rats are unlikely to be exposed to complete darkness for extended durations, and changes observed under such conditions may therefore reflect retinal pathology rather than physiological adaptation. Futhermore, a number of retinal functions and processes display circadian variations (Remé *et al.*, 1991; Tosini and Menaker, 1996), and group data could therefore be affected by possible damping of some rhythms and by the phase dispersal that would be expected in 12D:12D because of individual differences in free-running period. A study by Penn and Williams (1986) avoided this potential problem by comparing the retinas of albino rats reared from birth in 12L:12D with daytime light intensity ranging from 3 to 800 lux. Both dark-adapted and steady-state rhodopsin levels (the latter measured at the end of the daily light period) decreased with increasing light intensity up to 400 lux (no measurable rhodopsin was found at 800 lux, probably because of permanent light damage), as did ROS length, light absorbance by rhodopsin in ROSs, and outer-nuclear-layer thickness (reflecting a decrease in photoreceptor density).

Penn and Williams also calculated the total daily photon catch by the retina, from rhodopsin extracts in dark-adapted and steady-state whole retinas and from rhodopsin regeneration rates following bleaching *in vivo.* Their estimates showed a nearly constant daily catch of about 10^{16} photons per eye for all light intensities below 800 lux. On the basis of these results, the authors proposed that the observed changes in retinal anatomy and physiology represent an adaptive process, termed photostasis, whereby the sensitivity of the retina is gradually altered to ensure that a constant number of photons is absorbed each day, regardless of the daily amount of light that the animal is exposed to, at least for light intensities that do not cause permanent retinal damage.

The time course of these adaptive changes was examined in albino rats born and reared under 12L:12D at 8 lux, then switched to 400 lux (Williams *et al.*, 1988). Two weeks after the switch, dark-adapted rhodopsin levels were almost 50% lower than in animals that had been kept in 8 lux, but daily photon catch was the same in the two groups (about 10^{16} photons per eye). Another group of rats was born and reared in 12L:12D at 400 lux and then switched to 8 lux. Transverse light absorbance in ROSs was higher after two weeks in 8 lux than after one day, but no further change was observed 12 weeks after the switch, indicating that the process approaches saturation level within two weeks. Noting the similarity between the time course of these photostatic adjustments and that of ROS renewal (older ROS discs are continually shed and replaced by newly synthesized discs, and the entire ROS is renewed in about 10 days; Young, 1976), Williams *et al.* suggested that ROS renewal was the mechanism underlying photostasis and, indeed, that a primary purpose of outer segment renewal is to allow adaptation to gradual changes in ambient lighting conditions, particularly seasonal changes.

More direct evidence linking photostatic adjustments with ROS renewal was reported recently by Schremser and Williams (1995a, b). As in the earlier studies, rhodopsin levels and ROS lengths were reduced in rats reared in 200 lux compared to rats reared in 3 lux, and similar changes were observed after switching the animals from one light intensity to the other (Schremser and Williams, 1995a). Microspectrophotometric measurements also confirmed the increase in transverse absorbance of ROSs in 3-lux animals relative to 200-lux animals. Absorbance was also measured in individual ROSs obtained 5 days after a switch from 3 to 200 lux (about half the time required for complete ROS turnover). When transverse illumination was limited to the tip of the ROS, absorbance was comparable to that measured in animals kept in 3 lux, but when the base of the ROS was illuminated, absorbance was similar to that in 200-lux animals (Schremser and Williams, 1995b). Thus, these results imply that discs synthesized after a switch from low to high light intensity carry fewer rhodopsin molecules; these discs are added to the base of the ROS, gradually replacing the older, more densely packed discs. A new steady state is reached at the end of the ROS turnover period, reflecting the animal's new lighting environment.

Other adaptive changes reflecting an animal's light history include changes in ERG thresholds and amplitudes, both *in vivo* (Penn *et al.,* 1987, 1989) and *in vitro* (Reiser *et al.,* 1996), in ROS membrane fatty-acid composition and cholesterol content (Penn and Anderson, 1987), in retinal levels of several antioxidants, and in retinal susceptibility to light damage (Penn *et al.,* 1985, 1987). Changes have also been observed in the levels of two proteins involved in phototransduction, transducin and arrestin (Organisciak *et al.,* 1991), as well as in the levels of mRNAs encoding these and two other transduction proteins, opsin and cGMP-phosphodiesterase (Farber *et al.,* 1991).

If retinal sensitivity is a function of total daily light exposure, as suggested by Penn and Williams (1986), then it should reflect not only prior light intensity but also prior light duration, or day length. This prediction was borne out in albino rats born and reared under 16L:8D or 8L:16D, with daytime light intensity kept at 200 lux in both conditions (Parker and Williams, 1995). Animals reared under short days had longer ROSs, more nuclei in the outer nuclear layer, and approximately five times more rhodopsin per eye than animals reared under long days. In both groups, however, total daily light absorption was approximately 10^{16} photons per eye. These observations support Penn and Williams' suggestion that photostasis allows adaptation to seasonal changes in ambient lighting conditions, which, at temperate latitudes, include prominent changes in day length.

The results summarized above were all obtained in albino rats. Rhodopsin levels were measured in the eyes of pigmented rats by Battelle and LaVail (1978), who found a 7–9% increase after 10 days in darkness over levels in 12L:12D (<150 lux)—substantially less than the approximately 50% increase found in albino rats. In pigmented animals, changes in pupillary diameter provide a much more effective means of regulating the amount of light reaching the retina than they do in albinos, and a given change in ambient light intensity therefore results in a smaller change in retinal illumination. Despite these differences, recent data from pigmented rats indicate that daily retinal light absorption is also maintained at a constant level, as it is in albinos (Williams, this volume).

2.2. Light Input Pathways to the Circadian System and Pineal Gland

Light falling on the mammalian retina is transmitted along several specialized pathways, one of which, the retinohypothalamic tract (RHT), conveys ambient light information to the primary circadian pacemaker located in the suprachiasmatic nucleus (SCN) of the hypothalamus. Light input to the SCN resets the pacemaker in a phase-dependent manner, thereby entraining the many SCN-dependent behavioral and physiological rhythms to the local day-night cycle (Rusak and Boulos, 1981; Johnson *et al.,* 1988; Morin, 1994).

Among the rhythmic functions driven by the SCN is the synthesis of melatonin by the pineal gland. In both diurnal and nocturnal animals, melatonin levels in the pineal gland and in the circulation are high at night and fall to low levels during the day. This rhythm persists in darkness but not in bright constant illumination. In a number of mammalian species, the duration of the nocturnal increase in melatonin varies with nighttime duration and thus provides the organism with an internal temporal cue that can serve to synchronize reproductive, metabolic, and other seasonally changing activities to the annual cycle in day length (Goldman and Elliott, 1988). Entrainment of the daily melatonin rhythm by light involves a multisynaptic pathway that is still not fully characterized, but that includes the RHT, the SCN, parvocellular regions of the hypothalamic paraventricular nucleus (PVN), sympathetic preganglionic neurons in the thoracic spinal cord, and the superior cervical ganglia that innervate the pineal gland (Moore, 1996).

Pineal and blood melatonin levels are also acutely suppressed by exposure to light at night; the degree of suppression depends on the intensity and spectral characteristics of the light stimulus (Reiter, 1985; Brainard *et al.,* 1986, 1988; Bronstein *et al.,* 1987; McIntyre *et al.,* 1989; Nelson and Takahashi, 1991). The pathway mediating this direct effect of light on pineal melatonin is generally held to be the same as that mediating entrainment of the melatonin rhythm and thus to involve the SCN. This conclusion is based on data from studies in rodents showing that both the rhythm in melatonin production and the pineal-dependent response of the reproductive system to day length are abolished by lesions of the SCN and of the PVN, as well as by horizontal knife cuts dorsal to the SCN (Klein and Moore, 1979; Klein *et al.,* 1983; Pickard and Turek, 1983; Lehman *et al.,* 1984; Eskes and Rusak, 1985; Hastings and Herbert, 1986). Early neuroanatomical data had pointed to the SCN as the sole recipient of RHT terminals (e.g., Moore and Lenn, 1972), which implied that the SCN was necessary for the generation and entrainment of the melatonin rhythm, as well as for mediating the direct effects of light on melatonin production.

Subsequent studies, using more sensitive anatomical tracing techniques, showed that the RHT innervates several hypothalamic structures in addition to the SCN, including, in particular, the sub-paraventricular zone immediately dorsal to the SCN and, in some species, the PVN as well (for review, see Morin, 1994). These projections would also be severed by SCN lesions and by horizontal knife cuts dorsal to the SCN, and a role for an extra-SCN projection in relaying light input to the pineal gland cannot, therefore, be ruled out. Recently, however, Moore *et al.* (1995) showed that many of the RHT terminals in the sub-paraventricular zone are collaterals of axons passing through the SCN, suggesting at least partial overlap between the retinal components of the pathways mediating circadian entrainment and the acute effects of light on melatonin. Common retinal components are also indicated by the similarity between the spectral sensitivity of the circadian system and that of melatonin suppression in rodents (Takahashi *et al.,* 1984; Reiter, 1985; Bronstein *et al.,* 1987; Boulos, 1995).

2.3. Effects of Light History on Melatonin Suppression

Evidence that an animal's prior light environment can affect the responses of its pineal gland to light was described by Lynch *et al.* (1981), who compared urinary melatonin rhythms in rats exposed for 17 days to daily cycles of 12 hours of bright light (45–110 $\mu W/cm^2$) and 12 hours of dim light (0.1–0.3 $\mu W/cm^2$), or 12 hours of dim light and 12 hours of darkness. All animals were then exposed to constant dim light for 15 days. Both groups showed high melatonin levels during the dimmer halves of their respective light cycles and low levels during the brighter halves, indicating that both cycles entrained the animals' rhythms (Fig. 1). When transferred to constant dim light, however, the rats that had been exposed to the bright-dim cycle showed clear free-running rhythms in melatonin excretion, whereas those that had been exposed to the dim-dark cycle showed significantly lower melatonin levels and highly damped rhythms, indicating greater melatonin suppression.

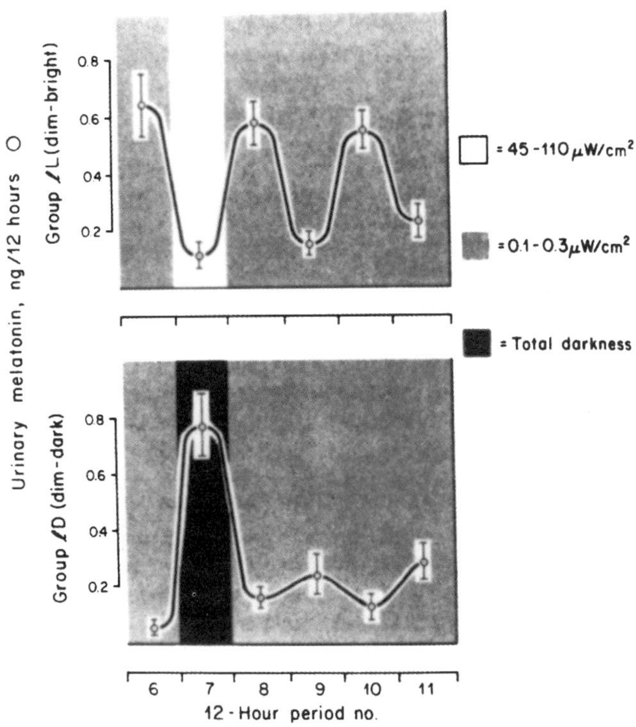

Figure 1. Urinary melatonin levels in two groups of albino rats on the last day of exposure to a daily cycle of 12 hours of bright light and 12 hours of dim light (top panel) or of 12 hours of dim light and 12 hours of darkness (bottom panel), followed by two days under constant dim light. Melatonin levels were measured in consecutive 12-hours intervals. Reproduced from Lynch *et al.* (1981) with permission of S. Karger AG, Basel.

Reiter *et al.* (1983b) examined the effects of light on pineal melatonin levels in laboratory-reared and wild-caught 13-lined ground squirrels, *Spermophilus tridecemlineatus.* The wild-caught animals were kept in the laboratory for one week before they were used in the experiment. These rodents are diurnally active, and, in a natural environment, would be exposed to high light intensities during the day. In contrast, the laboratory-reared animals were exposed to daytime intensities at least two orders of magnitude lower than that of sunlight. When kept on a daily 14L:10D cycle, both groups showed the typical entrained pattern: high melatonin levels at night and low levels during the day. As seen in Figure 2, exposure to 400 $\mu W/cm^2$ (about 1300 lux) at night completely suppressed the nocturnal melatonin peak in the laboratory-reared animals but had no effect in the wild-caught animals.

A relative insensitivity of pineal melatonin to the acute, suppressive effects of light was also observed in wild-caught rodents from two other diurnal species, eastern chipmunks, *Tamias striatus* (Reiter *et al.,* 1982), and Richardson's ground squirrels, *Spermophilus richardsonii* (Reiter *et al.,* 1983a). In contrast, the threshold for pineal melatonin suppression in wild-caught cotton rats, *Sigmodon hispidus,* a nocturnal species, was only between 0.01 and 0.005 $\mu W/cm^2$ (Thiele *et al.,* 1983)—a little lower than the threshold observed by Brainard *et al.* (1982) in laboratory-reared Syrian hamsters (between 0.185 and 0.019 $\mu W/cm^2$). In their natural environment, many nocturnal species avoid exposure to bright sunlight by spending most of the daytime hours in light-excluding burrows, emerging at dusk and regaining their shelters at dawn. Thus, the light histories of wild-caught

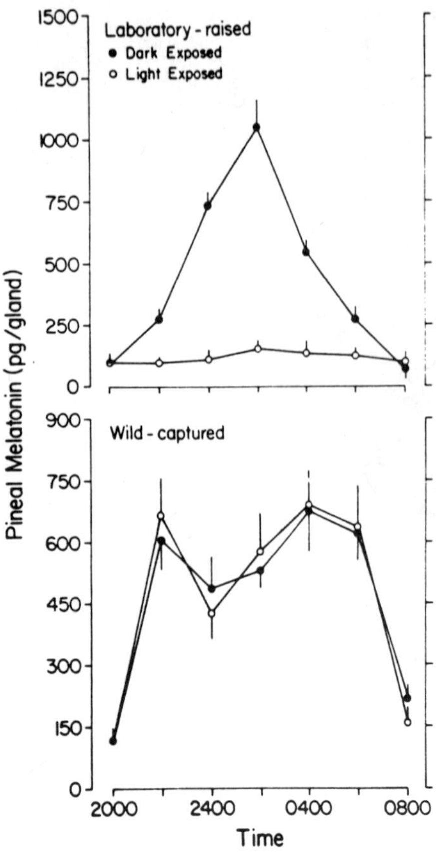

Figure 2. Pineal melatonin levels in laboratory-reared (top panel) and wild-captured (bottom panel) 13-lined ground squirrels exposed to either darkness or moderately bright light (400 $\mu W/cm^2$) at night. Light exposure completely suppressed the nocturnal rise in melatonin in laboratory-reared animals but had no effect on wild-caught animals. Reproduced from Reiter *et al.* (1983b) with permission of Elsevier Science Inc.

and laboratory-reared animals are likely to be much more similar for nocturnal-fossorial species than they are for diurnal species.

In summary, the results described in this and the preceding section indicate that long-term exposure to a dim environment increases, and to a bright environment decreases, both retinal and pineal photosensitivity. Whether the latter effect is a consequence of the former or a separate phenomenon has yet to be determined, but the fact that changes in pineal responsiveness to light were observed in rats after 17 days in an altered photic environment (Lynch *et al.,* 1981) is consistent with the documented time course of retinal photostatic adjustments.

2.4. Possible Implications for Circadian Photosensitivity

The likelihood that the pathways mediating circadian and pineal photoreception include common retinal elements suggests that the sensitivity of the circadian system to light also depends on the animal's prior lighting environment. As yet, however, there is very little information relevant to that issue.

Shimomura and Menaker (1994) compared light-induced phase shifts of the circadian activity rhythm of Syrian hamsters kept in darkness for various periods ranging from 1 to 49 days after extended exposure to cyclic lighting. Both wild-type and homozygous *tau* mutant hamsters were studied (the *tau* mutation results in a shortening of the free-running period from about 24 hours in wild-type animals to about 20 hours in homozygotes; Ralph and Menaker, 1988). In both groups, phase shifts induced by 60-min light pulses increased with time in darkness: in wild-type hamsters, maximum phase delays were obtained between 1 and 4 weeks after the beginning of darkness, whereas in the mutants, the phase delays continued to increase throughout the 7 weeks of continuous darkness. In the latter case, however, the increase in circadian photosensitivity could be at least partially accounted for by gradual changes in circadian pacemaker properties unrelated to retinal photostasis (Shimomura and Menaker, 1994).

Thus, the hypothesis that retinal photostatic mechanisms influence circadian photosensitivity remains to be tested. A potentially useful approach would be to compare light-induced phase shifts in groups of animals previously exposed to light-dark cycles with the same photoperiod duration but with different daytime light intensities.

2.5. Photoreceptors Mediating Circadian and Pineal Responses to Light

The identity of the photoreceptors mediating the effects of light on the mammalian circadian system and pineal gland is still uncertain. Although spectral sensitivity functions for light-induced circadian phase shifts and melatonin suppression in nocturnal rodents resemble the action spectrum for rhodopsin, data obtained from mice with inherited retinal degeneration (Foster *et al.,* 1991, 1993; Colwell and Foster, 1992; Provencio *et al.,* 1994) and from rodless transgenic mice (Foster *et al.,* 1995; Argamaso *et al.,* 1995) indicate that rods are not necessary for normal circadian photoreception. Several studies suggest that, in some nocturnal rodents, circadian and pineal photoreception may instead be mediated by two classes of cones, one with peak sensitivity near 500 nm and another in the ultraviolet range, at around 360 nm (Brainard *et al.,* 1994; García-Fernández *et al.,* 1995; Provencio and Foster, 1995; Yoshimura and Ebihara, 1996). This conclusion is consistent with recent neuroanatomical data indicating that the retinal components mediating circadian photoreception in rats include cones, cone bipolar cells, and a distinct subset of Type III ganglion cells that project in the RHT (Moore *et al.,* 1995; Speh and Moore, 1996).

Although evidence of retinal photostasis has thus far been limited to rods, the dependence of pineal photosensitivity on prior lighting conditions in a diurnal rodent with a cone-dominated retina (Reiter *et al.*, 1983b) raises the possibility that cone photosensitivity depends similarly on the animal's lighting history (on the assumption, of course, that the observed changes in pineal response reflect changes in retinal photoreception rather than in subsequent processing of the light signal). Certainly, the mechanism underlying photostasis in rods, outer segment renewal, is also present in cones, although the turnover period of cone outer segments and the manner of incorporation of newly synthesized protein into cone outer segment discs may not be the same as those of rods (Young, 1976; Anderson *et al.*, 1978; Remé *et al.*, 1991).

3. SEASONAL VARIATIONS IN HUMAN PHOTOSENSITIVITY

In human rods, direct measurement of rhodopsin content as a function of light history is infeasible, and indirect approaches (e.g., reflectometry) have yet to be tried, but there have been inferential approaches that might draw on a photostatic "explanation." After a several-week period of dim light exposure, for example, one might expect to find increases in melatonin suppression, scotopic visual sensitivity, and ERG responses. The human duplex retina is of course designed for both scotopic and photopic light reception, but the lack of data for cone photostasis in animals makes it difficult to predict how photopic light sensitivity would be affected by protracted exposure to dim or bright light.

The seasonal variation in outdoor light provides an annual course of bright light exposure that may affect physiological function for the population as a whole (Lacoste and Wirz-Justice, 1989). Across the seasons, midday light is always "bright" at temperate latitudes, a factor that might prevent sizeable seasonal variation in rhodopsin content, but incremental winter night length, lower midday maxima, and adverse weather conditions that discourage outdoor activity could result in a relative light deprivation, with distinct psychiatric consequences in a significant proportion of the population, especially at more extreme latitudes.

3.1. Seasonal Affective Disorder

The winter depression of seasonal affective disorder (SAD) is usually presaged in autumn, after the equinox, by increased fatigue, difficulty awakening (often with hypersomnia), and a specific craving for carbohydrate-rich foods (which, if indulged, can lead to recurrent annual weight gain of up to 30 lb). By December or January, mood plummets, and patients experience decreased work productivity, activities, libido, and concentration and increased anxiety. In severe cases, suicidal thinking is common. Prevalence estimates for SAD are still imprecise. Beyond clinically severe cases, subsyndromal winter depression—short of Major Depressive Disorder, yet still very bothersome—occurs more than twice as often as SAD. Taken together, the total U.S. population prevalence is approximately 25%. Prevalence depends on latitude. For example, less than 2% of the population in the southernmost tier of states reports symptoms consistent with SAD.

In late spring and summer, the clinical picture reverses. Most patients with SAD become euthymic; i.e., their moods regain an "even keel." A significant subgroup, however, shows a bipolar switch to hypomania—less frequently, mania—which is characterized by euphoria, increased libido, rapid thinking and speech, increased energy despite decreased sleep, and creative spurts. Although these symptoms may sound wonderful, hypomania often has a distinct downside, including subjective speediness, difficulty concentrating, aggressive behavior, impulsivity, and serious risk-taking.

3.2. Light Treatment of Winter Depression

Beyond the correlation of mood state with the seasonal changes in light availability, the success of artificial light therapy points to a functional connection with retinally transduced light. As first shown by Rosenthal *et al.* (1984), clinical improvement—including complete remission of winter symptoms—can be obtained within a week of daily bright-light exposure (2500 lux for 2 hours or longer) in the morning and evening. Subsequent studies showed that evening light exposure had no incremental benefit over a single daily morning exposure (Terman *et al.,* 1989) and that session durations as short as 30 min were effective when light intensity was increased to 10,000 lux (Terman *et al.,* 1990).

Undoubtedly, cone transduction is involved at these high intensities, which fall within the daylight range (>750 lux) and are likely to exceed rod saturation levels, but rod function cannot be excluded. Indeed, relatively dim simulated dawn signals presented in the bedroom to the sleeping patient have been shown to result in clinical improvement similar to that obtained with postawakening bright light (Terman and Schlager, 1990; Terman, in press). Rod activity is likely to predominate throughout most of these simulated dawn twilights, which are incremented gradually from 0.001 lux to ~300 lux at the level of the eyes, but are further attenuated by closed eyelids. In contrast, standard bright light treatment on dark winter mornings may serve to end nighttime scotopic function abruptly through a sudden bleach.

Although the central mechanisms of action of light therapy are still unknown, they may involve the RHT projection to the SCN pacemaker rather than the classical visual projections. One hypothesis states that early-morning light exposure phase-advances the circadian clock in winter, when otherwise it would drift later, tracking winter sunrise. If patients with SAD showed abnormally delayed rhythms in winter (at least some surely do), the corrective phase advance would be therapeutic (Sack *et al.,* 1990). Recent controlled clinical trials support this explanation: morning light exceeded evening light in clinical efficacy, and the response to morning light was proportional to the phase advance of the melatonin rhythm (Terman and Terman, 1995, 1996).

3.3. Photostasis in SAD?

At least two conflicting hypotheses have been proposed. The first (Remé *et al.,* 1990) assumes that retinal photosensitivity normally increases in winter in response to low-light conditions, maintaining a constant daily photon catch, as shown in albino rats. Patients with SAD may show deficient photostasis, thus failing to absorb sufficient light to forestall the depressive syndrome. Light therapy serves to stimulate synthesis of new photopigment, thereby restoring normal sensitivity. Prior to light therapy, therefore, patients should show subsensitivity in retinal response to light.

The second hypothesis (Beersma, 1990) assumes that patients with SAD show exaggerated photostatic adjustment to dim light in comparison with normals. Thus, the SCN interprets dim indoor light in the evening as if it were an extension of outdoor daylight, similar to summer, and fails to trigger a normal photoperiodic response. Light therapy serves to decrease rhodopsin availability through photostatic adjustment and thus restores the SCN's ability to detect the shortening of the photoperiod in autumn and winter. Before treatment, the patients should therefore show supersensitivity in retinal response to light.

Experimental investigations of seasonal variation in light sensitivity, in both healthy subjects and patients with SAD, have yet to produce a preponderance of results in favor of either hypothesis. Just as with animal studies, photostatic adjustment can be inferred from differences in light-induced melatonin suppression resulting from recent histories of exposure to relatively dim (winter) or intense (summer) light. Other experiments have attempted to localize changes in light sensitivity to the retina, where seasonal variations would more closely reflect changes in photoreceptor physiology.

3.4. Light-Induced Melatonin Suppression

An antarctic locale (75° S latitude) provided extreme seasonal contrast in ambient lighting for a study of melatonin suppression in two groups of healthy men (Owen and Arendt, 1992). Each group was exposed to two light levels (300 lux and 2200 lux) at one of two times of night (0100–0200 hours or 0500–0600 hours), both in winter, when daylight reaches only 500 lux, and in summer, when the sun never sets and daylight can exceed 100,000 lux. In winter, both light intensities caused significant suppression of plasma melatonin levels at both times of night, whereas in summer, significant suppression was observed only during exposure to 2200 lux at 0100–0200 hours. These results must be considered preliminary, because of the small number of subjects studied (only four subjects in one group and three in the other completed the study) and the fact that baseline melatonin levels varied considerably at the two times of year, but they are consistent with the animal data described above in showing greater suppression after exposure to dim than to bright light. Recently, Lewy *et al.* (1997) found autumn and winter increases in melatonin suppression at 45° N latitude, but only in females. The sexes did not differ in spring and summer, and males showed no significant seasonal pattern.

There are also indications that patients with SAD show heightened sensitivity to light in winter, as assessed by melatonin suppression. In one study, patients, but not normal controls, showed a significant seasonal variation in their sensitivity to 2000 lux presented at 0030–0130 hours, showing greater melatonin suppression than controls in winter and less in summer (Thompson *et al.*, 1990). In a case study (McIntyre *et al.*, 1990), exposure to only 200 lux resulted in clear melatonin suppression (67%) while the patient was depressed in winter and reduced suppression (37%) after two weeks of successful bright-light therapy. The results of both studies are compatible with Beersma's (1990) hypothesis of supersensitivity in SAD and a pattern of rod-like photostatic adjustment to either seasonal or artificial daily light exposure. Other investigators, however, have failed to find differences in melatonin suppression between patients with SAD and normal controls (Murphy *et al.*, 1993).

3.5. Electrooculogram (EOG) Studies

The EOG reflects an electrical potential between the front and back of the eye produced primarily by the cornea, lens, and pigment epithelium. The potential decreases during dark adaptation and increases during light adaptation. The Arden ratio compares the EOG under light- and dark-adapted conditions, primarily indexing changes in the pigment epithelium, where photoreceptor discs are shed and degraded. As a result, the EOG may indirectly reflect photoreceptor function.

In an initial investigation of the EOG in SAD, Lam *et al.* (1991) compared Arden ratios during winter in a large group of depressed patients and normal controls (N = 38). The patients showed a significant but small reduction in the Arden ratio. More than twice as many patients as controls fell into the clinically borderline range, but pathology was not apparent. Group differences were insufficient to categorize subjects according to diagnosis. The result is consistent with the hypothesis of deficient photostasis in SAD (Remé *et al.,* 1990), which might be corrected by supplemental light exposure.

Ozaki *et al.* (1993) recorded EOGs in patients and controls before and after bright light treatment in winter. As in the study of Lam *et al.,* the Arden ratio was slightly lower in depressed patients. A week of light treatment did not serve to increase the ratio. The authors therefore suggested that the EOG provides a subtle trait marker for SAD, but treatment duration may have been insufficient to promote such an ocular change. A subsequent study by the same authors (Ozaki *et al.,* 1995) showed lower Arden ratios in patients than in controls in winter, but the two groups did not differ in summer. These findings were attributable to a significant seasonal variation in the controls, who had higher Arden ratios in winter than in summer, but not in the patients, whose ratios did not differ at the two times of year. As noted by the authors, however, limitations of the EOG technique and the substantial overlap observed between the Arden ratios of patients and controls preclude any strong conclusions regarding visual sensitivity in SAD and its potential role in pathogenesis.

3.6. Electroretinogram (ERG) Studies

The ERG provides a more direct measure of activity in the retina's photoreception apparatus, but the results of three studies are equivocal. Lam *et al.* (1992) measured dark-adapted ERGs using a flashing bright-light stimulus—which elicits a mixed rod/cone response—in a large group of patients and controls (N = 48). There were no group differences in b-wave amplitude, but a significant diagnosis-×-sex interaction was observed; female patients (N = 18) showed a small reduction in amplitude relative to matched controls, while male patients (N = 6) showed a small increase. Many more women than men report SAD—a ratio of about 5:1—and the result might suggest a slight retinal subsensitivity in the more vulnerable sex. Oren *et al.* (1993) used a pattern ERG, following a model for tests of retinal change in Parkinson's disease, but for patients with SAD and normal controls (N = 24), there were no group differences in b-wave amplitude or latency across a range of spatial and temporal stimulus frequencies.

Recently, Hébert *et al.* (1997) measured scotopic ERGs to diffuse flashes of blue-green light, under dilation, in 12 subjects with subsyndromal SAD (S-SAD) and 9 normal controls (not matched for age or sex). The ERG thresholds were virtually identical in summer, whereas in winter subjects with S-SAD showed a small (~0.14 log unit) but significant increase of the threshold relative to the controls. Seasonal variation in the ERG threshold was found only for the subjects with S-SAD. Interestingly, they also showed a large positive correlation ($r = 0.52$) between the percentage change in summer-winter thresholds and a global seasonality score for winter symptoms (including fatigue, hypersomnia, appetite and weight gain, dysphoric mood, and social withdrawal). These results, and the data for females with SAD (Lam *et al.,* 1992), are compatible with the hypothesis of a shortfall in photostatic adjustment of rods in response to reduced light exposure in winter (Remé *et al.,* 1990). The small effect sizes, however, are unlikely to be clinically significant.

3.7. Other Ocular and Visual System Measures

As part of a comprehensive study at the National Institute of Mental Health, Oren *et al.* (1993) obtained winter data for patients with SAD and for controls. Comparisons included measures of pupil size, color vision (anomaloscope and 100-hue test), intraocular pressure (also tested after light treatment), and pattern visual evoked cortical potentials (amplitude and latency to P100). No group differences were found for any of the measures, leading the authors to doubt that SAD is linked to abnormal ophthalmic function.

3.8. Dark Adaptation Studies

An annual cycle of dark-adapted absolute threshold was first documented by Sweeney *et al.* (1960) in three healthy men, tested weekly. The size of a dim stimulus was varied around the point of minimum detectability. Thresholds gradually decreased from August to March and increased again in June and July, leading to the conclusion that scotopic vision becomes increasingly sensitive with reduced exposure to sunlight in winter. A complication of this study was that the initial high-threshold result for August was not replicated to the same extent a year later. Interestingly, however, in the two weeks preceding the second August tests, there was a precipitous drop in sunlight exposure (as gauged from subjects' logs and degree of tanning)—and apparently an appropriate photostatic response. Thus, one might hypothesize that indoor/outdoor lifestyles—which may vary with profession, habit, and mood state—produce variations in ambient light exposure that may bear little or no relation to the seasonal cycle in natural light availability. Indeed, when ambient light exposure is measured by ambulatory photometry, urbanites show little daily exposure to light levels above 1000 lux, even in a city as sunny and warm as San Diego (Savides *et al.,* 1986; Campbell *et al.,* 1988; Eastman, 1990).

Two recent studies obtained complete dark-adaptation curves for larger groups of patients with SAD and for age- and sex-matched normal controls. Both studies used the classic Goldmann-Weekers method, in which the threshold response to a small white stimulus target is tracked for 30 min after an intense *Ganzfeld* bleach. Tests were without dilation or use of an artificial pupil.

In the first study (Oren *et al.,* 1991), patients and controls (N = 20) were tested in winter, while the patients were depressed. Surprisingly, the patients showed significantly faster recovery across the range of 2 to 20 min, although the final scotopic threshold at 30 min did not differ. Thus, in distinction from the EOG and ERG data discussed earlier, patients with SAD appeared supersensitive, in that they reached their final threshold about 5 min earlier than the controls. When the investigators tested a larger group of 50 subjects, however, the dark-adaptation curves for patients and controls were indistinguishable, and the hypothesis of supersensitivity in SAD was dropped. Furthermore, no differences were found for subgroups of patients and controls who were tested in both winter and summer—a failure to replicate the seasonal pattern found by Sweeney *et al.* (1960).

Terman and Terman used a similar experimental design, comparing (a) patients and controls in winter and summer, (b) seasonal variation in each group separately, and (c) patients before and after bright-light therapy (preliminary results were reported by Terman and Terman, 1991). The analysis identified each subject's cone-rod breakpoint (between about 3 and 7 min post bleach) and derived separate curves for photopic and scotopic dark adaptation. As in the initial report by Oren *et al.* (1991), patients with SAD showed significantly faster scotopic recovery than normal controls but converged on a common final threshold (see Fig. 3).

The most robust results, however, involved cone sensitivity, which was not assessed in earlier studies of seasonal variation, either in patients with SAD or in normal controls. As shown in Figure 3, patients showed a significantly lower cone final threshold than did controls (~0.4 log unit), indicating "supersensitivity." Patients also showed significantly

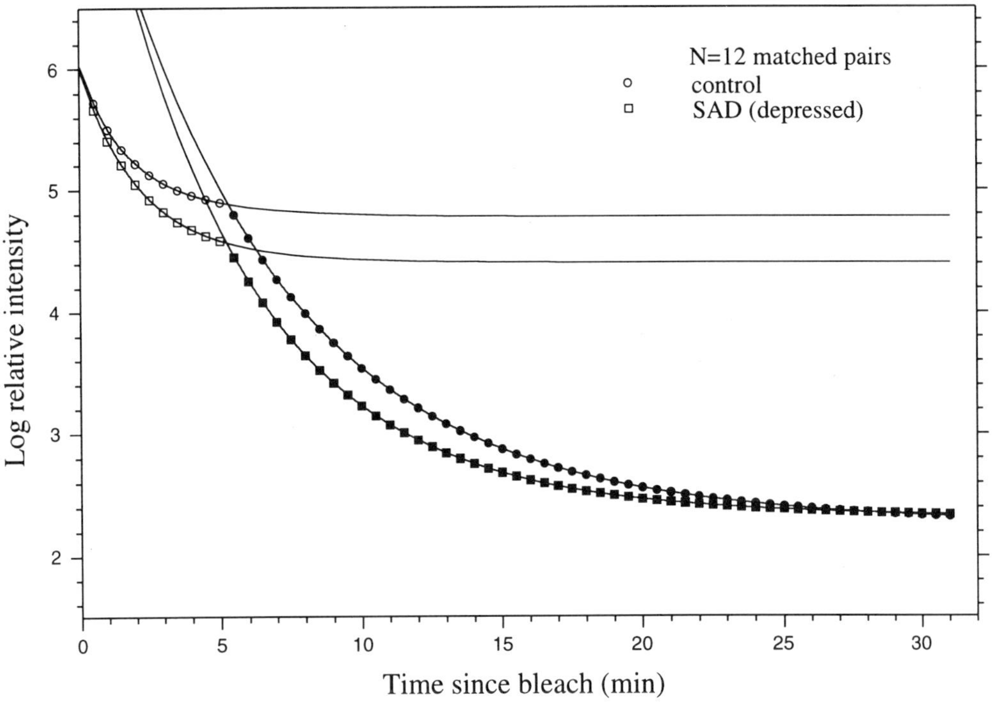

Figure 3. Averaged photopic and scotopic dark adaptation curves for patients with seasonal affective disorder, measured in winter while they were depressed, and in age- and sex-matched normal controls. Extrapolated portions are based on best-fitting negative exponential functions of the form $y = a + b^{-cx}$, where a is the final threshold, a + b is the initial postbleach threshold, and c is a constant that describes the recovery rate.

lower cone—but not rod—threshold in summer than in winter (~0.3 log unit). A similar but smaller decrease in cone threshold (~0.2 log unit) followed 7–10 days of morning light therapy in winter. The differences were much greater in patients with blue eyes: the seasonal effect on photopic threshold was nearly double (~0.6 log unit). Such magnified response may be attributable to increased light transmission through the iris. (For a clinical discussion about blue-eyed patients with SAD, see Terman, in press.)

3.9. Summary of Human Photosensitivity Studies

Evidence of long-term adaptation to dim or bright light conditions is not uniform and appears to vary with the measure used (melatonin suppression, ERG, EOG, psychophysical thresholds, dark-adaptation rates), clinical status (SAD or healthy), and gender. Problems have included small sample sizes, tests without 24-hour control of lighting conditions or specific knowledge of light exposure history, inadequate matching for age and gender, and lack of standard laboratory controls (e.g., mydriasis). Replication attempts have been few and not necessarily confirmatory.

In healthy subjects, an increase in light-induced melatonin suppression in winter was observed in males in Antarctica (Owen and Arendt, 1992), but at temperate latitudes such supersensitivity in winter was found in females only (Lewy *et al.,* 1997) or not at all (Thompson *et al.,* 1990). The scotopic absolute detection threshold decreased in winter for a small group of male subjects (Sweeney *et al.,* 1960), but neither sex showed a summer-winter difference in two larger studies (Oren *et al.,* 1993; Terman and Terman, unpublished data). EOG ratios were found to be higher in winter than in summer (Ozaki *et al.,* 1995), but ERG thresholds did not vary across the seasons (Hébert *et al.,* 1997).

Patients with SAD may provide a special—or exaggerated—case. In winter, they have shown increased melatonin suppression relative to controls in one study (Thompson *et al.,* 1990), although not in another (Murphy *et al.,* 1993). Oculoretinal measures have tended to show an opposite pattern of results, suggestive of photostatic deficiency. Two of three studies showed reduced ERG b-wave response in winter (Lam *et al.,* 1992; Hébert *et al.,* 1997); in one (Lam *et al.,* 1992), this response was restricted to females. Likewise, three studies found wintertime subsensitivity in EOG response in patients relative to normal controls (Lam *et al.,* 1991; Ozaki *et al.,* 1993, 1995). One group could not detect seasonal variations in dark adaptation or differences between patients and controls (Oren *et al.,* 1993), but Terman and Terman found a lower photopic threshold in depressed patients in winter and resultant supersensitivity relative to controls. Although their scotopic absolute threshold was not lower in winter, patients did show an accelerated rate of scotopic dark adaptation, which suggests photostatic enhancement. A puzzle remains, however: patients also showed significant summertime increases in both photopic and scotopic sensitivity, an effect not predicted by the principles of photostasis.

4. CONCLUSIONS

The ability to adapt to variations in illumination by changes in photosensitivity is a basic property of vertebrate retinas; its purpose is to maintain a more or less constant response in the face of changes in illumination level (Shapley and Enroth-Cugell, 1984). Much of the relevant research has focused on retinal mechanisms with relatively short time constants, allowing rapid adaptation to momentary illumination changes. Retinal photostasis, however, represents a much more gradual phenomenon—with a time course of several days rather than seconds or minutes—allowing adaptation to slow changes in ambient illumination, such as may be encountered across the seasons at temperate and more extreme latitudes.

The site(s) of adaptation also differ in the two cases: the more rapid forms of light adaptation have been documented in several classes of retinal neurons, including rods in some vertebrates, but they do not appear to take place in mammalian rods (Shapley and Enroth-Cugell, 1984); in contrast, photostasis has thus far only been demonstrated in rod photoreceptors in rats. An important, and as yet unanswered, question is whether cone photosensitivity is similarly dependent on prior lighting conditions and, if so, whether a cone-dominated or duplex retina also absorbs a constant number of photons per day, regardless of the animal's photic environment.

Functional consequences of photostasis beyond the retina remain to be explored. We have suggested changes in pineal and in circadian photosensitivity as two potential consequences, but these would require that cones be capable of long-term photostatic adjustments, because of the likelihood that both pineal and circadian responses to light are cone-mediated.

Seasonal variations in light availability provide a natural laboratory for studies of photic sensitivity and photostatic response. In humans, however, such variations are greatly attenuated by contemporary urban lifestyles, which minimize exposure to natural daylight. Measures of ocular/retinal response to light show mixed evidence of photostatic adjustment in humans. Bright-light transduction is cone-dependent, and adaptation to ambient lighting conditions may differ from that of rods. These issues have direct practical significance, given that seasonal reduction in light availability can induce psychiatric depressive episodes, but research has yet to demonstrate a clear relationship between mood state and retinal, hypothalamic, or pineal adaptation to the lighting environment.

LITERATURE CITED

Anderson, D. H., Fisher, S. K., and Steinberg, R. H. (1978). Mammalian cones: disc shedding, phagocytosis, and renewal. *Investigative Ophthalmology and Visual Science* 17:117–133.

Argamaso, S. M., Froelich, A. C., McCall, M. A., Nevo, E., Provencio, I., and Foster, R. G. (1995). Photopigments and circadian systems of vertebrates. *Biophysical Chemistry* 56:3–11.

Battelle, B.-A., and LaVail, M. M. (1978). Rhodopsin content and rod outer segment length in albino rats: modification by dark adaptation. *Experimental Eye Research* 26:487–497.

Beersma, D. G. M. (1990). Do winter depressives experience summer nights in winter? *Archives of General Psychiatry* 47:879–880.

Boulos, Z. (1995). Wavelength dependence of light-induced phase shifts and period changes in hamsters. *Physiology and Behavior* 57:1025–1033.

Brainard, G. C., Podolin, P. L., Rollag, M. D., Northrup, B. E., Cole, C., and Barker, F. M. (1986). The influence of light irradiance and wavelength on pineal physiology of mammals. In: Klein, D. C., and O'Brien, P., eds. *Pineal-Retinal Relationships.* Academic Press, New York. pp. 253–277.

Brainard, G. C., Lewy, A. J., Menaker, M., Fredrickson, R. H., Miller, S. T., Weleber, R. G., Cassone, V., and Hudson, D. (1988). Dose-response relationship between light irradiance and the suppression of plasma melatonin in human volunteers. *Brain Research* 454:212–218.

Brainard, G. C., Barker, F. M., Hoffman, R. J., Stetson, M. H., Hanifin, J. P., Podolin, P. L., and Rollag, M. D. (1994). Ultraviolet regulation of neuroendocrine physiology in rodents. *Vision Research* 34:1521–1533.

Bronstein, D. M., Jacobs, G. H., Haak, K. A., Neitz, J., and Lytle, L. D. (1987). Action spectrum of the retinal mechanism mediating nocturnal light-induced suppression of rat pineal gland N-acetyltransferase. *Brain Research* 406:352–356.

Campbell, S. S., Kripke, D. F., Gillin, J. C., and Hrubovcak, J. C. (1988). Exposure to light in healthy elderly subjects and Alzheimer's patients. *Physiology and Behavior* 42:141–144.

Colwell, C. S., and Foster, R. G. (1992). Photic regulation of Fos-like immunoreactivity in the suprachiasmatic nucleus of the mouse. *Journal of Comparative Neurology* 324:135–142.

Eastman, C. (1990). Natural summer and winter sunlight exposure patterns in seasonal affective disorder. *Physiology and Behavior* 48:611–616.

Eskes, G. A., and Rusak, B. (1985). Horizontal knife cuts in the suprachiasmatic area prevent hamster gonadal responses to photoperiod. *Neuroscience Letters* 61:261–266.

Farber, D. B., Seager-Danciger, J., and Organisciak, D. T. (1991). Levels of mRNA encoding proteins of the cGMP cascade as a function of light environment. *Experimental Eye Research* 53:781–786.

Foster, R. G., Argamaso, S., Coleman, S., Colwell, C. S., Lederman, A., and Provencio, I. (1993). Photoreceptors regulating circadian behavior: a mouse model. *Journal of Biological Rhythms* 8(Suppl.):S17–S23.

Foster, R. G., Froelich, A., Argamaso-Hernan, S. M., and McCall, M. A. (1995). Rodless transgenic mice show increased circadian responses to light. *Investigative Ophthalmology and Visual Science* 36(Suppl.):S424.

Foster, R. G., Provencio, I., Hudson, D., Fiske, S., De Grip, W., and Menaker, M. (1991). Circadian photoreception in the retinally degenerate mouse (*rd/rd*). *Journal of Comparative Physiology A* 169:39–50.

García-Fernández, J. M., Jimenez, A. J., and Foster, R. G. (1995). Persistence of cone photoreceptors within the dorsal retina of aged retinally degenerate mice (*rd/rd*): implications for circadian organization. *Neuroscience Letters* 187:33–36.

Goldman, B. D., and Elliott, J. A. (1988). Photoperiodism and seasonality in hamsters: role of the pineal gland. In: Stetson, M. H., ed. *Processing of Environmental Information in Vertebrates.* Springer Verlag, New York. pp. 203–218.

Hastings, M. H., and Herbert, J. (1986). Neurotoxic lesions of the paraventriculo-spinal projection block the nocturnal rise in pineal melatonin synthesis in the Syrian hamster. *Neuroscience Letters* 69:1–6.

Hébert, M., Guillemette, J., Dumont, M., and LaChapelle, P. (1997). Decrease in retinal sensitivity associated with seasonal mood disorder. *Society for Light Treatment and Biological Rhythms Abstracts* 9:9.

Johnson, R. F., Moore, R. Y., and Morin, L. P. (1988). Loss of entrainment and anatomical plasticity after lesions of the hamster retinohypothalamic tract. *Brain Research* 460:297–313.

Klein, D. C., and Moore, R. Y. (1979). Pineal N-acetyltransferase and hydroxyindole-O-methyltransferase: control by the retinohypothalamic tract and the suprachiasmatic nucleus. *Brain Research* 174:245–262.

Klein, D. C, Smoot, R., Weller, J. L., Higa, S., Markey, S. P., Creed, G. J., and Jacobowitz, D. M. (1983). Lesions of the paraventricular nucleus area of the hypothalamus disrupt the suprachiasmatic-spinal cord circuit in the melatonin rhythm generating system. *Brain Research Bulletin* 10:647–652.

Lacoste, V., and Wirz-Justice, A. (1989). Seasonal variations in normal subjects: An update of variables current in depression research. In: Rosenthal, N. E., and Blehar, M. C., eds. *Seasonal Affective Disorders and Phototherapy.* Guilford Press, New York. pp. 167–229.

Lam, R. W., Beattie, C. W., Buchanan, A., Remick, R. A., and Zis, A. P. (1991). Low electrooculographic ratios in patients with seasonal affective disorder. *American Journal of Psychiatry* 148:1526–1529.

Lam, R. W., Beattie, C. W., Buchanan, A., and Mador, J. A. (1992). Electroretinography in seasonal affective disorder. *Psychiatry Research* 43:55–63.

Lehman, M. A., Bittman, E. L., and Newman, S. W. (1984). Role of the hypothalamic paraventricular nucleus in neuroendocrine responses to daylength in the golden hamster. *Brain Research* 308:25–32.

Lewy, A. J., Cutler, N. L., Hughes, R. J., and Sack, R. L. (1997). Women are more sensitive than men to 500 lux light suppression of melatonin (LSM) in fall and winter. *Sleep Research* 26:731.

Lynch, H. J., Rivest, R. W., Ronsheim, P. M., and Wurtman, R. J. (1981). Light intensity and the control of melatonin secretion in rats. *Neuroendocrinology* 33:181–185.

McIntyre, I. M., Norman, T. R., Burrows, G. D., and Armstrong, S. M. (1989). Human melatonin suppression by light is intensity dependent. *Journal of Pineal Research* 6:149–156.

McIntyre, I. M., Norman, T. R., and Burrows, G. D. (1990). Melatonin supersensitivity to dim light in seasonal affective disorder. *Lancet* 335:488.

Moore, R. Y. (1996). Neural control of the pineal gland. *Behavioural Brain Research* 73:125–130.

Moore, R. Y., and Lenn, N. J. (1972). A retinohypothalamic projection in the rat. *Journal of Comparative Neurology* 146:1–14.

Moore, R. Y., Speh, J. C., and Card, J. P. (1995). The retinohypothalamic tract originates from a distinct subset of retinal ganglion cells. *Journal of Comparative Neurology* 352:351–366.

Morin, L. P. (1994). The circadian visual system. *Brain Research Reviews* 67:102–127.

Murphy, D. G. M., Murphy, D. M., Abbas, M., Palaidou, E., Binnie, C., Arendt, J., Campos Costa, D., and Checkley, S. A. (1993). Seasonal affective disorder: response to light as measured by electroencephalogram, melatonin suppression, and cerebral blood flow. *British Journal of Psychiatry* 163:327–331.

Nelson, D. E., and Takahashi, J. S. (1991). Comparison of visual sensitivity for suppression of pineal melatonin and circadian phase-shifting in the golden hamster. *Brain Research* 554:272–277.

Oren, D. A., Joseph-Vanderpool, J. R., and Rosenthal, N. E. (1991). Adaptation to dim light in depressed patients with seasonal affective disorder. *Psychiatry Research* 36:187–193.

Oren, D. A., Moul, D. E., Schwartz, P. J., Alexander, J. R., Yamada, E. M., and Rosenthal, N. E. (1993). An investigation of ophthalmic function in winter seasonal affective disorder. *Depression* 1:29–37.

Organisciak, D. T., and Noell, W. K. (1977). The rod outer segment phospholipid/opsin ratio of rats maintained in darkness or cyclic light. *Investigative Ophthalmology and Visual Science* 16:188–190.

Organisciak, D. T., Xie, A., Wang, H.-M., Jiang, Y.-L., Darrow, R. M., and Donoso, L. A. (1991). Adaptive changes in visual cell transduction protein levels: effects of light. *Experimental Eye Research* 53:773–779.

Owen, J., and Arendt, J. (1992). Melatonin suppression in human subjects by bright and dim light in Antarctica: time and season-dependent effects. *Neuroscience Letters* 137:181–184.

Ozaki, N., Rosenthal, N. E., Moul, D. E., Schwartz, P. J., and Oren, D. A. (1993). Effects of phototheraopy on electrooculographic ratio in winter seasonal affective disorder. *Psychiatry Research* 49:99–107.

Ozaki, N., Rosenthal, N. E., Myers, F., Schwartz, P. J., and Oren, D. A. (1995). Effects of season on electro-oculographic ratio in winter seasonal affective disorder. *Psychiatry Research* 59:151–155.

Parker, C., and Williams, T. P. (1995). The effects of day length on photostasis. *Investigative Ophthalmology and Visual Science* 36(Suppl.):S515.

Penn, J. S., and Anderson, R. E. (1987). Effect of light history on rod outer-segment membrane composition in the rat. *Experimental Eye Research* 44:767–778.

Penn, J. S., and Williams, T. P. (1986). Photostasis: regulation of daily photon-catch by rat retinas in response to various cyclic illuminances. *Experimental Eye Research* 43:915–928.

Penn, J. S., Howard, A. G., and Williams, T. P. (1985). Light damage as a function of "light history" in the albino rat. In: LaVail, M. M., Hollyfield, J. G., and Anderson, R. E., eds. *Retinal Degeneration: Experimental and Clinical Studies.* Alan R. Liss, New York. pp. 439–447.

Penn, J. S., Naash, M. I., and Anderson, R. E. (1987). Effect of light history on retinal antioxidants and light damage susceptibility in the rat. *Experimental Eye Research* 44:779–788.

Penn, J. S., Thum, L. A., and Naash, M. I. (1989). Photoreceptor physiology in the rat is governed by the light environment. *Experimental Eye Research* 49:205–215.

Pickard, G. E., and Turek, F. W. (1983). The hypothalamic paraventricular nucleus mediates the photoperiodic control of reproduction but not the effects of light on the circadian rhythm of activity. *Neuroscience Letters* 43:67–72.

Provencio, I., and Foster, R. G. (1995). Circadian rhythms in mice can be regulated by photoreceptors with cone-like characteristics. *Brain Research* 694:183–190.

Provencio, I., Wong, S., Lederman, A., Argamaso, S. M., and Foster, R. G. (1994). Visual and circadian responses to light in aged retinally degenerate mice. *Vision Research* 34:1799–1806.

Ralph, M. R., and Menaker, M. (1988). A mutation of the circadian system in golden hamsters. *Science* 241:1225–1227.

Reiser, M. A., Williams, T. P., and Pugh, E. N., Jr. (1996). The effect of light history on the aspartate-isolated fast-PIII responses of the albino rat retina. *Investigative Ophthalmology and Visual Science* 37:221–229.

Reiter, R. J. (1985). Action spectra, dose-response relationships, and temporal aspects of light's effects on the pineal gland. *Annals of the New York Academy of Sciences* 453:215–230.

Reiter, R. J., King, T. S., Richardson, B. A., Hurlbut, E. C., Karasek, M. A., and Hansen, J. T. (1982). Failure of room light to inhibit pineal N-acetyltransferase activity and melatonin content in a diurnal species, the eastern chipmunk (*Tamias striatus*). *Neuroendocrinology Letters* 4:1–6.

Reiter, R. J., Hurlbut, E. C., Brainard, G. C., Steinlechner, S., and Richardson, B. A. (1983a). Influence of light irradiance on hydroxyindole-O-methyltransferase activity, serotonin-N-acetyltransferase activity, and radioimmunoassayable melatonin levels in the pineal gland of the diurnally active Richardson's ground squirrel. *Brain Research* 288:151–157.

Reiter, R. J., Steinlechner, S., Richardson, B. A., and King, T. S. (1983b). Differential response of pineal melatonin levels to light at night in laboratory-raised and wild-captured 13-lined ground squirrels (*Spermophilus tridecemlineatus*). *Life Sciences* 32:2625–2629.

Remé, C. E., Terman, M., and Wirz-Justice, A. (1990). Are deficient retinal photoreceptor renewal mechanisms involved in the pathogenesis of winter depression? *Archives of General Psychiatry* 47:878–879.

Remé, C. E., Wirz-Justice, A., and Terman, M. (1991). The visual input stage of the mammalian circadian pacemaking system. I. Is there a clock in the mammalian eye? *Journal of Biological Rhythms* 6:5–29.

Rosenthal, N. E., Sack, D. A., Gillin, J. C., Lewy, A. J., Goodwin, F. K., Davenport, Y., Mueller, P. S., Newsome, D. A., and Wehr. T. A. (1984). Seasonal affective disorder: a description of the syndrome and preliminary findings with light therapy. *Archives of General Psychiatry* 41:72–80.

Rusak, B., and Boulos, Z. (1981). Pathways for photic entrainment of mammalian circadian rhythms. *Photochemistry and Photobiology* 34:267–273.

Sack, R. L., Lewy, A. J., White, D. M., Singer, C. M., Fireman, M. J., and Vandiver, R. (1990). Morning versus evening light treatment for winter depression: evidence that the therapeutic effects of light are mediated by circadian phase shifts. *Archives of General Psychiatry* 47:343–351.

Savides, T. J., Messin, S., Senger, C., and Kripke, D. F. (1986). Natural light exposure of young adults. *Physiology and Behavior* 38:571–574.

Schremser, J.-L., and Williams, T. P. (1995a). Rod outer segment (ROS) renewal as a mechanism for adaptation to a new intensity environment. I. Rhodopsin levels and ROS length. *Experimental Eye Research* 61:17–24.

Schremser, J.-L., and Williams, T. P. (1995b). Rod outer segment (ROS) renewal as a mechanism for adaptation to a new intensity environment. II. Rhodopsin synthesis and packing density. *Experimental Eye Research* 61:25–32.

Shapley, R., and Enroth-Cugell, C. (1984). Visual adaptation and retinal gain controls. *Progress in Retinal Research* 3:263–346.

Shimomura, K., and Menaker, M. (1994). Light-induced phase shifts in *tau* mutant hamsters. *Journal of Biological Rhythms* 9:97–110.

Speh, J. C., and Moore, R. Y. (1996). The circadian timing (CTS) receives light information through an "ON" type cone pathway. *Society for Research on Biological Rhythms Abstracts* 5:126.

Sweeney, E. J., Kinney, J. S., and Ryan, A. (1960). Seasonal changes in scotopic sensitivity. *Journal of the Optical Society of America* 50:237–240.

Takahashi, J. S., DeCoursey, P. J., Bauman, L., and Menaker, M. (1984). Spectral sensitivity of a novel photoreceptive system mediating entrainment of mammalian circadian rhythms. *Nature* 308:186–188.

Terman, J. S., Terman, M., Schlager, D., Rafferty, B., Rosofsky, M., Link, M. J., Gallin, P. F., and Quitkin, F. M. (1990). Efficacy of brief, intense light exposure for treatment of winter depression. *Psychopharmacology Bulletin* 26:3–11.

Terman, M. (in press). Light on sleep. In: Schwartz, W. J., ed. *Scientific Foundations of Sleep Therapies.* Karger, Basel.

Terman, M., and Schlager, D. S. (1990). Twilight therapeutics, winter depression, melatonin, and sleep. In: Montplaisir, J., and Godbout, R., eds. *Sleep and Biological Rhythms.* Oxford University Press, New York. pp. 113–128.

Terman, M., and Terman, J. S. (1991). Seasonal variation in detection of dim light by SAD patients and controls. *Society for Light Treatment and Biological Rhythms Abstracts* 3:12.

Terman, M., and Terman, J. S. (1995). Phase shifts in melatonin and sleep under light therapy for winter depression. *Society for Light Treatment and Biological Rhythms Abstracts* 7:15.

Terman, M., and Terman, J. S. (1996). A multi-year controlled trial of bright light and negative ions. *Society for Light Treatment and Biological Rhythms Abstracts* 8:1.

Terman, M., Terman, J. S., Quitkin, F. M., McGrath, P. J., Stewart, J. W., and Rafferty, B. (1989). Light therapy for seasonal affective disorder: a review of efficacy. *Neuropsychopharmacology* 2:1–22.

Thiele, G., Holtorf, A., Steinlechner, S., and Reiter, R. J. (1983). The influence of different light irradiances on pineal N-acetyltransferase activity and melatonin levels in the cotton rat, *Sigmodon hispidus. Life Sciences* 33:1543–1547.

Thompson, C., Stinson, D., and Smith, A. (1990). Seasonal affective disorder and season-dependent abnormalities of melatonin suppression by light. *Lancet* 336:703–706.

Tosini, G., and Menaker, M. (1996). Circadian rhythms in cultured mammalian retina. *Science* 272:419–421.

Williams, T. P., Penn, J. S., Bush, R. A., and Makino, C. L. (1988). Renewal of rod outer segment and regulation of daily photon catch by the rat retina. In: *Proceedings of the Yamada Conference XXI.* Yamada Science Foundation, Kyoto, Japan. pp. 255–260.

Yoshimura, T., and Ebihara, S. (1996). Spectral sensitivity of photoreceptors mediating phase-shifts of circadian rhythms in retinally degenerate CBA/J (*rd/rd*) and normal CBA/N (+/+) mice. *Journal of Comparative Physiology A* 178:797–802.

Young, R. W. (1976). Visual cells and the concept of renewal. *Investigative Ophthalmology* 15:700–725.

10

CIRCADIAN REGULATION OF VERTEBRATE PHOTORECEPTORS

Rhythms in the Gating of Cationic Channels

Stuart E. Dryer, Theresa D'Souza, and Michelle A. Reiser

Program in Neuroscience
Department of Biological Science
Florida State University
Tallahassee, Florida 32306-4075

1. CIRCADIAN OSCILLATORS AND VISUAL-SYSTEM FUNCTION

Even though ambient light intensity can be as much as 12 orders of magnitude greater during the day than during the night, visual systems can detect contrasts in the face of these enormous changes in background illumination. Some of this behavior can be attributed to acute adaptive phenomena, but it is now clear that circadian oscillators also play an important role in regulating photoreceptor responses to large daily changes in ambient illumination (reviewed by Cahill and Besharse, 1995).

Circadian oscillators are biological clocks with a period of close to 24 hours that drive many behavioral, physiological, and biochemical rhythms. Circadian oscillators continue to generate rhythmicity under constant conditions devoid of any external time cues (i.e. they can free-run) but are entrained by external cues such as light-dark cycles. Because of circadian oscillators, a dark-adapted retina during the night behaves differently from a dark-adapted retina during the day, even in environments devoid of external time cues. For example, psychophysical and physiological measurements in fully dark-adapted animals have indicated an increase in light sensitivity during the night in several vertebrate species, including mammals (Terman and Terman, 1985; Bassi and Powers, 1986, 1987; White and Hock, 1992; Shaw *et al.,* 1993; and others). This overall increase in light sensitivity is associated with several biochemical and morphological changes, the nature and relative importance of which depend on the species being studied. The fundamental biological significance of control by intrinsic circadian oscillators, as opposed to acute adaptive processes, is that they allow complex changes in retinal function, some of which require coordinated changes in gene expression, to anticipate changes in ambient illumination. Intrinsic circadian oscillators are present in most vertebrate photoreceptors, including those of the retina as well as the nonmammalian pineal gland.

Photostasis and Related Phenomena, edited by Williams and Thistle
Plenum Press, New York, 1998

Circadian oscillators may also play a role in retinal dysfunction. Fain and Lisman (1993) have proposed that they provide a common pathway mediating retinal degeneration caused by constant light, vitamin A deficiency, and certain forms of retinitis pigmentosa (RP). Certain details of their theory are controversial, in particular the idea that mutations of the opsin chromophore-binding site, such as K296E, cause chronic excitation of the visual cascade (Robinson *et al.*, 1994; but see Li *et al.*, 1995). Nevertheless, a primary or secondary role for circadian oscillators in some forms of retinal degeneration remains plausible, as discussed in detail by Cahill and Besharse (1995).

2. CIRCADIAN REGULATION OF RETINAL DOPAMINE AND MELATONIN SECRETION

A number of biochemical processes in the vertebrate retina are regulated by intrinsic circadian oscillators. The most extensively studied is the synthesis and secretion of melatonin. Retinal and pineal melatonin secretion is elevated during the night in all species that have been examined to date. Melatonin is extremely lipophilic and is secreted by simple diffusion across the plasma membrane, in a manner similar to that of steroid hormones. Thus, the rate of melatonin secretion is simply determined by its rate of synthesis. The two most important enzymes regulating retinal melatonin synthesis are tryptophan hydroxylase (TPH) and serotonin N-acetyltransferase (NAT), and it is the activity of these enzymes, especially the latter, that determines the overall rate of melatonin secretion. These enzymes are expressed in the photoreceptor layer, which also contains a circadian oscillator (Green *et al.*, 1994; Bernard *et al.*, 1997). The rhythmic secretion of melatonin in cultured retina persists under constant-dark conditions in many vertebrates, indicating control by an intrinsic circadian oscillator (Besharse and Iuvone, 1983; Tosini and Menaker, 1996). In *Xenopus*, this phenomenon can be observed in a cultured photoreceptor layer in the absence of inner retinal elements (Cahill and Besharse, 1993), indicating that those cells contain the circadian oscillator as well as the melatonin biosynthetic enzymes. Light produces two distinct effects on retinal melatonin secretion—acute inhibition of melatonin secretion and entrainment of the photoreceptor circadian oscillator (Cahill and Besharse, 1991, 1993). Consistent with these effects, the activity of retinal NAT and TPH and the transcription of their genes are under direct circadian control (Green *et al.*, 1994; Bernard *et al.*, 1997). In chickens, the activity of NAT appears to be regulated by a combination of transcriptional and posttranslational mechanisms (Bernard *et al.*, 1997). The neurohormone dopamine (DA) is also under circadian control in the vertebrate retina. DA secretion from subpopulations of amacrine and/or interplexiform neurons (Witkovsky and Schutte, 1991) appears to be elevated during the circadian day (Wirz-Justice *et al.*, 1984; Kolbinger *et al.*, 1990; McCormak and Burnside, 1993). The photoreceptor circadian oscillator also controls the expression of a limited number of other genes, including the cone visual pigment iodopsin (Pierce *et al.*, 1993), *c-fos* (Yoshida *et al.*, 1993), and a novel leucine-zipper protein known as nocturnin (Green and Besharse, 1996). The function of nocturnin is unknown, but it is probably either a component of the circadian oscillator or a fairly proximal output of the photoreceptor circadian oscillator.

Circadian rhythms in retinal melatonin and dopamine secretion appear to drive other circadian rhythms in the retina. For example, Wang and Mangel (1996a) have demonstrated that goldfish L-type cone horizontal cells are driven almost exclusively by red (625 nm) cones during the day but are driven primarily by rods during the night. These changes take place even in constant dark conditions, indicating control by a circadian oscillator, and are mediated at least in part by melatonin secretion from photoreceptors (Wang and Mangel, 1996b). Melatonin secreted from photoreceptors appears to be a paracrine signal that produces a number of dark-adaptive effects (Cahill and Besharse, 1995), including cone elongation (Pierce and Besharse, 1985) and retinal pigment epithelium (RPE) melanin-granule aggregation (Kemali *et al.,* 1987). Melatonin may also cause "dark priming" of light-induced rod-disc shedding (Besharse and Dunis, 1983; Besharse *et al.,* 1988). In contrast, DA appears to be a retinal paracrine signal that produces several light-adaptive effects in direct opposition to those of melatonin (reviewed by Cahill and Besharse, 1995; Besharse and Iuvone, 1992). Thus, the effects of DA include cone contraction and dispersion of RPE melanin granules (Pierce and Besharse, 1985; Deary and Burnside, 1986, 1989; McCormack and Burnside, 1992). DA also enhances cone input to horizontal cells (Witkovsky *et al.,* 1988) and stimulates formation of dendritic spines on horizontal cells (Weiler *et al.,* 1988). In addition, DA can shift the phase of the intrinsic circadian oscillator of *Xenopus* photoreceptors, producing a phase-response curve identical to that of light (Cahill and Besharse, 1993). DA also exerts some effects in retina that are not obviously related to light adaptation (Mangel and Dowling, 1985). Melatonin and DA are mutually inhibitory. DA inhibits the secretion of melatonin from retinal photoreceptors in several species including the chick (Zawilska and Iuvone, 1989), whereas melatonin inhibits retinal DA release (Dubocovich, 1983).

3. THE AVIAN PINEAL GLAND IS A PHOTORECEPTIVE CIRCADIAN OSCILLATOR

The avian pineal gland is a classic model system for cellular studies of vertebrate circadian oscillators (reviewed by Takahashi *et al.,* 1989). Pineal cells of chicks exhibit circadian rhythms in melatonin secretion in dissociated cell culture that can persist under constant-dark conditions for at least 10 cycles (Barrett and Takahashi, 1995). Chick pineal cells are also directly photosensitive. As with retinal photoreceptors, light produces two effects on chick pineal photoreceptors—acute inhibition of melatonin secretion and entrainment of the intrinsic circadian oscillator. There is evidence that the two effects of light utilize different transduction cascades that are differentially sensitive to blockade by pertussis toxin (Zatz, 1988, 1994). Avian pineal cells express several "retinal" proteins, including all of the fundamental components of a phototransduction cascade. Thus, chick pineal cells express a visual pigment known as pinopsin (Okano *et al.,* 1994; Max *et al.,* 1995), a molecule that shares significant homology with cone opsins. In addition, some avian pineal cells are labeled by monoclonal antibodies directed against rhodopsin (Foster *et al.,* 1987; Araki *et al.,* 1992) and iodopsin (Araki *et al.,* 1992). Together with recent studies examining the effects of vitamin A depletion (Zatz, 1994), these results are consistent with the hypothesis that the two different effects of light utilize transduction cascades initiated by different photopigments. Avian pineal cells also express transducin (van Veen *et al.,* 1986; Foster *et al.,* 1987), a light-sensitive phosphodiesterase (Takahashi *et al.,* 1989), recoverin (Bastianelli and Pochet, 1994), visinin (Gotto *et al.,* 1989), and rod arrestin (S-antigen) (Collin *et al.,* 1986; Foster *et al.,* 1987). Our laboratory has shown that chick pineal cells express two distinct types of cGMP-activated cationic channels biophysically similar to those of retinal photoreceptors (Dryer and Henderson, 1991, 1993), a conclusion recently confirmed by molecular cloning (Bonigk *et al.,* 1996). The properties of these channels will be discussed in more detail below. Chick pinealocytes contain a 9+0 sensory cilium associated with a rudimentary outer segment (Bischoff, 1969; Boya and Zamorano, 1975; Calvo and Boya, 1979). These results underscore that chick pineal cells are simply another class of photoreceptor with many fundamental similarities to retinal photoreceptors, especially cones. Electrophysiological responses to light have not been described in avian pineal cells but have been observed in pinealocytes of several lower vertebrates. These consist of hyperpolarizing responses in lampreys (Pu and Dowling, 1981; Tamotsu and Morita, 1986) and teleosts (Meissl and Ekstrom, 1988) and are similar to those of retinal photoreceptors, albeit somewhat slower, possibly reflecting a greater diffusional space for signaling molecules in the photoreceptive outer segment (Lamb and Pugh, 1992). Thus, the broad outlines of phototransduction in the nonmammalian pinealocyte appear at least qualitatively similar to those of retinal photoreceptors.

4. PROPERTIES OF CIRCADIAN OSCILLATORS IN PINEAL AND RETINA

Entrainment and phase resetting of circadian oscillators have been extensively studied in chick pineal cells (extensive literature reviewed by Takahashi *et al.,* 1989; note especially the many papers by M. Zatz and coworkers cited therein). Indeed, chick pineal photoreceptors have served as a model for the design of circadian experiments in retinal cells in various species. To the extent that similar studies have been performed in retina and chick pineal, the results are generally similar. Minor differences may be due to species differences. Thus, regulation of melatonin secretion by light and by second messengers is strikingly similar in chick retina and chick pineal (see below), and identical melatonin biosynthetic enzymes are expressed in both tissues, although at substantially higher levels in the pineal. Moreover, the phase-response curves for light are similar in *Xenopus* retina and chick pineal (Cahill and Besharse, 1991; Takahashi *et al.,* 1989). It should also be noted that the abnormal period of free-running circadian rhythms in isolated retinas from *tau*-mutant hamsters is identical to the period of overall behavioral rhythms in the same animals (Tosini and Menaker, 1996). This result suggests a fundamentally similar molecular organization of intrinsic circadian oscillators in brain and in retinal and pineal photoreceptors.

The activities of TPH and NAT and the rate of melatonin secretion in the chick retina and pineal gland are positively regulated by cyclic AMP (Takahashi *et al.,* 1989; Iuvone *et al.,* 1991; Florez and Takahashi, 1996; and many others). Thus, agents that increase cAMP or mimic the actions of cAMP evoke increases in NAT activity and melatonin secretion in both preparations. Conversely, neurohormones that inhibit adenylate cyclase inhibit melatonin secretion. Melatonin secretion from the chick pineal and retina is also positively regulated by intracellular free Ca^{2+}. In both preparations, KCl-evoked membrane depolarization or application of BAY K 8644 stimulates melatonin secretion. These effects are blocked by inorganic and dihydropyridine L-type Ca^{2+} channel blockers. There is also evidence that Ca^{2+} and cAMP interact in avian photoreceptors; large increases in intracellular free Ca^{2+} increase cAMP levels in cultured chick retinal cells (Gan *et al.,* 1995) and pineal cells (Zatz, 1992a) and shift the dose-response curve for 8-Br-cAMP to the left (Nikaido and Takahashi, 1996). However, decreasing extracellular Ca^{2+} from 3 mM to 0.5 mM has no effect on cAMP levels but still shifts the 8-Br-cAMP dose-response curve to the right (Nikaido and Takahashi, 1996). In other words, the effects of intracellular Ca^{2+} on melatonin secretion can occur independently of effects on cyclic AMP.

5. OUTPUT PATHWAYS OF THE CIRCADIAN OSCILLATOR IN AVIAN PINEAL PHOTORECEPTORS

The concentration of cyclic AMP exhibits a free-running circadian rhythm in cultured chick pineal cells and is both in phase and highly correlated with the secretion of melatonin (reviewed by Takahashi *et al.,* 1989). This pattern initially suggested that the circadian oscillator causes a proximal regulation of cAMP levels, which in turn control NAT activity. A recent study by Zatz (1992b) has disproved this theory in its simplest form. In those experiments, high concentrations of cAMP analogs or forskolin were shown to produce a saturating effect on melatonin secretion, such that higher concentrations evoked no additional stimulation of melatonin release. That melatonin secretion continued to oscillate under those conditions indicates that the circadian clock does not act *exclusively* through cAMP to stimulate melatonin secretion. The fact that intracellular Ca^{2+} can also stimulate melatonin secretion suggested another possibility, namely that circadian oscillations in intracellular free Ca^{2+} play a role in circadian regulation of melatonin secretion. There is strong evidence in support of this theory. Membrane-permeant cAMP analogs and adenylate cyclase activators such as forskolin are more effective at stimulating melatonin secretion during the night, even in cells free-running under constant-dark conditions (Takahashi *et al.,* 1989; Nikaido and Takahashi, 1996), but only if Ca^{2+} is present in the extracellular medium. Certain interpretations of these data remain controversial but there is general agreement that Ca^{2+} influx is increased during the circadian night and that this increase plays a role in circadian-clock regulation of melatonin secretion. Only a portion of this effect can be attributed to increased nighttime activity of L-type Ca^{2+} channels (Nikaido and Takahashi, 1996), indicating enhanced nighttime activity of other Ca^{2+} transport mechanisms. We have recently identified a cationic channel, known as I_{LOT}, that is permeable to Ca^{2+} and under circadian control in cultured chick pineal cells (D'Souza and Dryer, 1996, and below). Gating of this channel is only observed during the night, even in cells free-running in constant darkness. Therefore, this newly discovered ionic channel may be responsible for the increased Ca^{2+} influx during the circadian night and therefore represents a plausible mechanism for clock regulation of melatonin secretion and/or other physiological processes. I_{LOT} channels are also present in chick retinal photoreceptors (see below).

6. METHODS FOR STUDYING CIRCADIAN REGULATION OF IONIC CHANNELS IN CHICK PINEAL AND RETINAL PHOTORECEPTORS

The free-running circadian rhythm in melatonin secretion from chick pineal cells persists for at least 10 cycles in dissociated cell culture under constant-dark conditions if the cells are maintained at 39–40°C. Rhythms dampen after 3–4 cycles at 37°C (Barrett and Takahashi, 1995).

Chick pineal glands were obtained from chicks 1–2 days after hatching (P1–2), and cells were dissociated and plated onto poly-D-lysine-coated glass coverslips and maintained at 37°C in CO_2 incubators equipped with cool fluorescent lights and timers for entrainment *in vitro.* Cells were entrained to a 12 hour:12 hour light-dark (12L:12D) cycle for 4 days, at which point they are switched to constant darkness at the expected time of lights-off. With this experimental design, it is possible to record from cells at different phases of the light-dark cycle, or to record from cells at various circadian times after they have been switched to constant dark conditions, and thus to determine whether rhythms were driven by the light-dark cycle or by the intrinsic circadian oscillator. Experiments performed during normal 12L:12D cycles were carried out on day 4 (in the case of pineal cells) or day 5 (in the case of retinal cells) *in vitro*. In most of our studies to date, patch-clamp recordings have been made under visible illumination during the light phase (ZT4–6) or under infrared illumination during the dark phase (ZT16–18). Recordings from free-running cells have been made under infrared illumination on day 2 of constant darkness (daytime, CT4–6; and nighttime, CT16–18). (ZT stands for Zeitgeber time; CT for circadian time, usually expressed relative to the original entraining light-dark cycle. ZT0 = time of lights-on in cells maintained on 12L:12D cycles; ZT12 = time of lights-off. CT0 = time of expected lights-on in free-running cells; CT12 = time of expected lights-off.) Patch-clamp recordings were made according to standard methods at room temperature (21–23 °C). The recording apparatus included an inverted-stage microscope equipped with Hoffman modulation-contrast optics housed in a dark room. Cells were protected by a second layer of light shielding consisting of heavy black cloth tightly covering the Faraday cage and all microscope light sources. We obtained infrared illumination by passing a 790-nm cutoff filter into the light path of the microscope. An infrared-sensitive video camera and a video monitor allowed for electrode placement in the absence of visible light. Unfortunately, it is not possible to make recordings from a single cell that last long enough for reliable observations of circadian changes in channel gating, so it was necessary to sample populations of cells at different circadian times. In the case of I_{LOT}, we typically monitored a single cell for 10 minutes before deciding whether or not an active channel was present.

7. VOLTAGE- AND LIGAND-ACTIVATED IONIC CHANNELS AND INTRACELLULAR CALCIUM STORES IN CHICK PINEAL CELLS

As noted above, isolated chick pineal cells express dihydropyridine-sensitive L-type Ca^{2+} channels that can be readily detected in whole-cell recordings (Harrison and Zatz, 1989; Henderson and Dryer, 1992) and fura-2 imaging experiments (D'Souza and Dryer, 1994). It should be noted that pharmacological studies suggest that L-type Ca^{2+} channels play only a small role in regulating increases in Ca^{2+} influx during the circadian night (Takahashi *et al.*, 1989; Nikaido and Takahashi, 1996). Moreover, their gating should depend primarily on changes in membrane potential evoked by other ionic channels. For example, cGMP-activated cationic channels are also expressed in chick pineal cells. These channels are an essential component of phototransduction cascades in rods and cones. Two different types of cGMP-activated channels can be detected in inside-out patches excised from acutely isolated chick pineal cells (Dryer and Henderson, 1991, 1993). The predominant form found in pineal cells can be activated by cGMP (ED_{50} = 35 μM, Hill slope = 2.8) but not by 500 μM cAMP. Higher concentrations of cyclic AMP (> 1.5 mM) can cause some activation of these channels, and that cyclic nucleotide appears to function as a partial agonist (Dryer and Henderson, 1993). The unitary conductance of the most abundant form of these channels is 22 pS at −60 mV, and there is some rectification in the current-voltage relationship in symmetrical NaCl in the absence of divalent cations. These channels exhibit rapid voltage-dependent blockade in the presence of millimolar concentrations of divalent cations, especially Mg^{2+}, applied to either side of the patch membrane. The second variant of this channel is similar except that it has a slightly higher unitary conductance of 30 pS (Dryer and Henderson, 1993). U. B. Kaupp and coworkers have recently cloned two distinct cGMP-activated channel α-subunit splice variants (cone type) from the chick pineal gland (Bonigk *et al.,* 1996). The density of cGMP-activated channels in pineal cells is much lower than that in rod or cone outer segments, as determined by electrophysiology (Dryer and Henderson, 1991) and immunocytochemistry (Bonigk *et al.,* 1996) but is comparable to that found in rod inner segments. Application of phosphodiesterase inhibitors causes activation of cGMP-activated channels within 1–3 minutes in intact acutely isolated chick pineal cells under visible illumination (D'Souza and Dryer, 1995). This activity is immediately suppressed by patch excision. In contrast, application of forskolin for up to 30 minutes does not stimulate cGMP-activated channels. The presence of cGMP-activated channels and other essential components of a "retinal" phototransduction cascade suggests a mechanism for daily rhythms in intracellular Ca^{2+} in chick pineal cells maintained on light-dark cycles as well as for the acute inhibitory effects of light on melatonin secretion, but we have not seen any evidence for rhythms in the gating of cGMP-activated channels in chick pineal cells free-running under constant dark conditions (D'Souza and Dryer, unpublished data). Some other mechanism is therefore probably responsible for the free-running circadian rhythms in Ca^{2+} influx observed in constant darkness. Evidence for such a mechanism is presented below.

8. AN UNUSUAL NON-SELECTIVE CATIONIC CHANNEL UNDER DIRECT CIRCADIAN CONTROL IN CHICK PINEAL PHOTORECEPTORS

I_{LOT} (for "long open-time" channel) is an unusual nonselective cationic channel that we have found to be under circadian control in cultured chick pineal photoreceptors (D'Souza and Dryer, 1996). We have also detected this channel in cultured chick retinal photoreceptors, where preliminary results suggest that it is also rhythmic (see below). To the best of our knowledge, I_{LOT} represents the first case of a plasma-membrane ionic channel that is regulated by a vertebrate intrinsic circadian oscillator. I_{LOT} is spontaneously active in cell-attached patch recordings made during the night and can be easily identified because of the very long open time of the unitary currents in pineal cells (Fig. 1A, B) and in retinal cells (Fig. 2, discussed below). The gating of these channels is not voltage dependent and is not altered by application of negative or positive pressure to the recording pipette (D'Souza and Dryer, 1996). I_{LOT} channels occasionally stay open without apparent interruptions for seconds at a time. These long-lived events account for a considerable fraction of the total integral current accumulated over a period of minutes. Bursts of I_{LOT} openings are often separated by quiescent periods of several seconds. I_{LOT} can also be detected during the night in cell-attached patches when recording pipettes are filled with 80 mM $CaCl_2$ in the absence of other cations (Fig. 1C). Although unitary current amplitudes and open-channel noise are reduced under those conditions, the characteristic long open times still allow for unambiguous identification of these channels. I_{LOT} channels therefore have a finite permeability to Ca^{2+}.

I_{LOT} can also be detected in inside-out patches from pineal cells excised into Ca^{2+}-free external salines (Fig. 3). Spontaneous activity is not significantly changed by patch excision (D'Souza and Dryer, 1996). The unitary slope conductance of I_{LOT} under these recording conditions is 40 pS, and there is very little rectification in the current-voltage relationship. I_{LOT} channels can be detected at −60 mV and reverse at close to 0 mV under

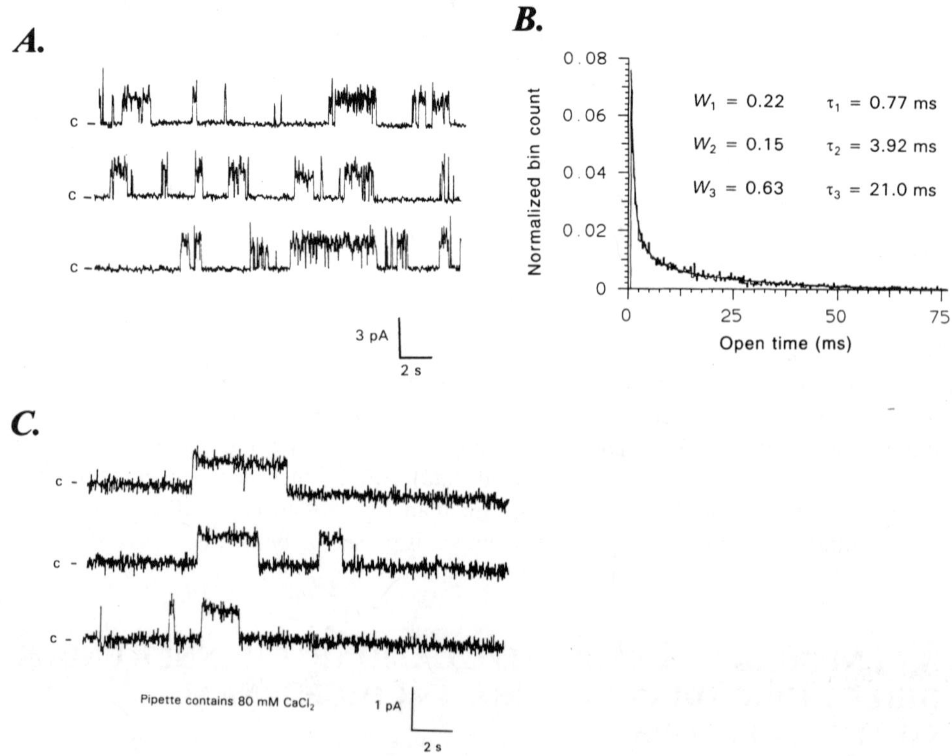

Figure 1. Characteristics of I_{LOT} channels in cultured chick pineal photoreceptors. Recordings were made under infrared illumination 4 hours after lights-off (ZT16) from cells maintained for 5 days on 12L:12D cycle. Currents moving out of the recording electrode are shown as outward. A. I_{LOT} unitary currents in a cell-attached patch. The recording electrode was filled with normal extracellular saline (containing 5.4 mM $CaCl_2$), and the patch was hyperpolarized by 80 mV from the resting potential. Current level of the closed state is indicated to the left of the traces. Note that many of the bursts lasted for several seconds. B. Open-time histogram of a single I_{LOT} channel in a different cell-attached patch constructed from 10 minutes of continuous data. Normalized histogram is fitted with a probability distribution function containing three exponential components with time constants and weights indicated. C. I_{LOT} unitary currents in a cell-attached patch. Conditions were as in A except that the recording pipette was filled with a solution consisting of 80 mM $CaCl_2$, 10 mM HEPES, pH 7.4. Modified from D'Souza and Dryer (1996).

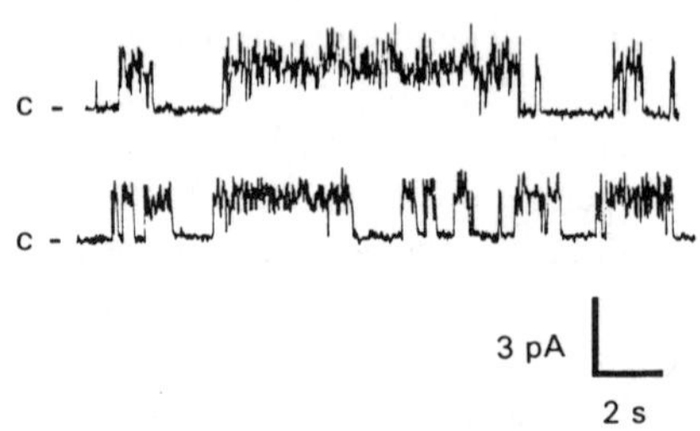

Figure 2. Example of an I_{LOT} channel in a cell-attached patch on a cultured chick retinal photoreceptor identified on the basis of its large oil droplet. Chick retinal cells were dissociated on E18 and maintained *in vitro* for 5 days on a 12L:12D cycle. This recording was made at ZT16 on the fifth day in culture. Recording pipette was filled with normal extracellular saline and was hyperpolarized 80 mV from the resting potential.

conditions where the calculated chloride equilibrium potential (E_{Cl}) is −59 mV. These results indicate that I_{LOT} is a nonselective cationic channel whose gating does not depend on continued exposure to soluble cytosolic messengers. Consistent with this conclusion, patch excision followed by bath application of 1–10 μM Ca^{2+} does not cause activation of I_{LOT} in quiescent inside-out patches. Thus, I_{LOT} gating is not inhibited by cytosolic messengers. Active I_{LOT} channels are not blocked by 1 mM $MgCl_2$. We do not know whether patch excision alters the gating of I_{LOT} channels in cultured retinal cells. The properties of I_{LOT} channels are in some ways reminiscent of those of gap-junctional hemichannels, which in a few cases have been shown to exhibit similar kinetics (DeVries and Schwartz, 1992). However, the gap-junctional hemichannels described to date are closed in the presence of physiological concentrations of external Ca^{2+} (for example, see DeVries and Schwartz, 1992), whereas I_{LOT} channels are functional even in the presence of 80 mM external $CaCl_2$ (D'Souza and Dryer, 1996). Moreover, the gating of all gap-junctional hemichannels described to date is quite voltage dependent (DeVries and Schwartz, 1992; Suchyna *et al.,* 1993; Verselis *et al.,* 1994; Ebihara *et al.,* 1995; Trexler *et al.,* 1996; Steiner and Ebihara, 1996), whereas gating of I_{LOT} channels is independent of voltage over a 100-mV range (D'Souza and Dryer, 1996). That being said, functional gap junctions are found in rat pineal cells (Saez *et al.,* 1991) as well as in various retinal cells (DeVries and Schwartz, 1992; McMahon and Brown, 1994; Lu and McMahon, 1996; McMahon and Matson, 1996). At this time we certainly cannot exclude that I_{LOT} is an unusual form of gap-junctional hemichannel comprised of connexins that is functional in the absence of a cell partner in physiological external Ca^{2+}. In that case, ionic influx from the external compartment would occur for every hemichannel that was not perfectly paired with a hemichannel from a partner cell, and the distinction between an unopposed gap-junction hemichannel and other types of plasma-membrane ionic channels would be more semantic than real.

Active I_{LOT} channels are readily detected in cultured chick pineal cells with normal recording pipettes during the night, when melatonin secretion is highest, but are almost never observed during the day. This pattern is seen in cells maintained on a 12L:12D cycle, as well as in cells free-running in constant darkness for 2 days (Table 1). These results indicate that the gating of I_{LOT} is under direct circadian control. Note that I_{LOT} gating detected during the night is not inhibited by 10-minute pulses of 100-lux visible light (Fig. 4).

9. THE MECHANISMS CONTROLLING I_{LOT} GATING ARE UNKNOWN

I_{LOT} provides a mechanism for coupling of the electrophysiological properties of a photoreceptor cell to the output of its clock, but factors that typically regulate ion-channel gating do not affect I_{LOT}. Thus, gating of I_{LOT} is not voltage, stretch, or Ca^{2+} dependent (or inhibited by external Ca^{2+}) and does not require continued contact with soluble cytosolic messengers (Fig. 3). Moreover, I_{LOT} is not activated in quiescent cells by patch excision, application of 10 μM melatonin, acute activation of adenylate cyclase, depletion of internal Ca^{2+} stores, or application of NO donors such as sodium nitroprusside. It should be noted that NO donors inhibit gap-junction hemichannels in retinal horizontal cells (DeVries and Schwartz, 1992). However, the nighttime appearance of I_{LOT} requires protein synthesis, in free-running pineal cells as well as in cells maintained on 12L:12D cycles. In these experiments, chick pineal cells were treated with the translational inhibitor anisomycin (1 μM) at CT10 and ZT10. In control experiments, we found that I_{LOT} channels are not active at the time of anisomycin treatment (0/50 cells in both cases). Active I_{LOT} channels were observed at the normal frequency in control cells at CT18 (16/47 cells) and ZT18 (9/45 cells) but were not observed in anisomycin-treated cells at CT18 (0/56 cells) or ZT19 (0/85 cells). These results are highly statistically significant ($p < 0.0005$, G-test, Sokal and Rohlf, 1981, chapter 17). Thus, protein synthesis is required for the normal nighttime expression of functional I_{LOT} channels, even in light-dark cycles, but it is not clear what proteins are being synthesized. The circadian oscillator could plausibly control I_{LOT} gating by at least three different classes of mechanisms: (1) Gating of plasma-membrane I_{LOT} channels may be controlled by covalent modification, possibly by direct phosphorylation/dephosphorylation of the channel molecules or closely associated auxiliary subunits. According to this scheme, the biosynthesis of regulatory kinases, phosphatases, or other protein-modifying enzymes is under direct circadian control. (2) I_{LOT} gating may be regulated by other substances whose biosynthesis is under circadian control and dependent on protein synthesis. Moreover, we propose that these substances remain associated with the plasma membrane during patch excision. Examples of such substances could include lipids or G-proteins. (3) I_{LOT} channel molecules or a portion thereof may only be present in the plasma membrane during the night. Biosynthesis and/or plasma-membrane insertion of the channel is under circadian control and dependent on protein synthesis. These theories are not mutually exclusive, and one could certainly conceive of other schemes of varying complexity. In the absence of molecular tools for studying I_{LOT} expression, these hypotheses are not easy to test conclusively, but one has to start somewhere. A more detailed model is presented below.

10. I_{LOT} CAN BE DETECTED IN RETINAL PHOTORECEPTORS

As discussed above, several groups, including ours, have looked for and found "retinal" proteins in chick pineal photoreceptors. Conversely, we have detected I_{LOT} in cultured chick retinal photoreceptors (Fig. 2). In these experiments, we dissociated retinas relatively late in embryonic development, at embryonic days 18–19 (E18–19), and recorded after 4–7 days *in vitro* (because robust free-running circadian rhythms in retinal melatonin secretion are not detected until E20 *in vivo* and reach maximum amplitude at P3; Iuvone, 1990). The culture conditions are the same as for pineal cells except that the culture medium contains 40 ng/ml basic fibroblast growth factor, which we find improves the survival and health of the photoreceptors. With this culture system, we can maintain presumptive cone photoreceptors (easily identified by their oblong shape, prominent oil droplets, and peanut hemaglutinin staining) for up to two weeks *in vitro*. Survival rates are dramatically lower for cells dissociated later than E19.

Figure 3. I_{LOT} channels in excised inside-out patches. Recordings were made under infrared illumination from cultured chick pineal cells about 4 hours after lights-off (ZT16). A. I_{LOT} channels observed after patch excision. Note characteristic long open times of the unitary currents. Patch potential was −80 mV. B. Current-voltage relationship of I_{LOT} channels. Slope conductance was 40 pS. The bath saline contained 155 mM KCl, 10 mM EGTA, and 10 mM HEPES, pH 7.4, and the recording pipette contained 150 mM NaCl, 10 mM EGTA, and 10 mM HEPES, pH 7.4. C. Unitary currents of normal amplitude were observed at −60 mV but not at 0 mV when bath salines contained 15 mM Cl^- and the recording pipette contained 155 mM Cl^-. The chloride equilibrium potential under these conditions was −59 mV. Modified from D'Souza and Dryer (1996).

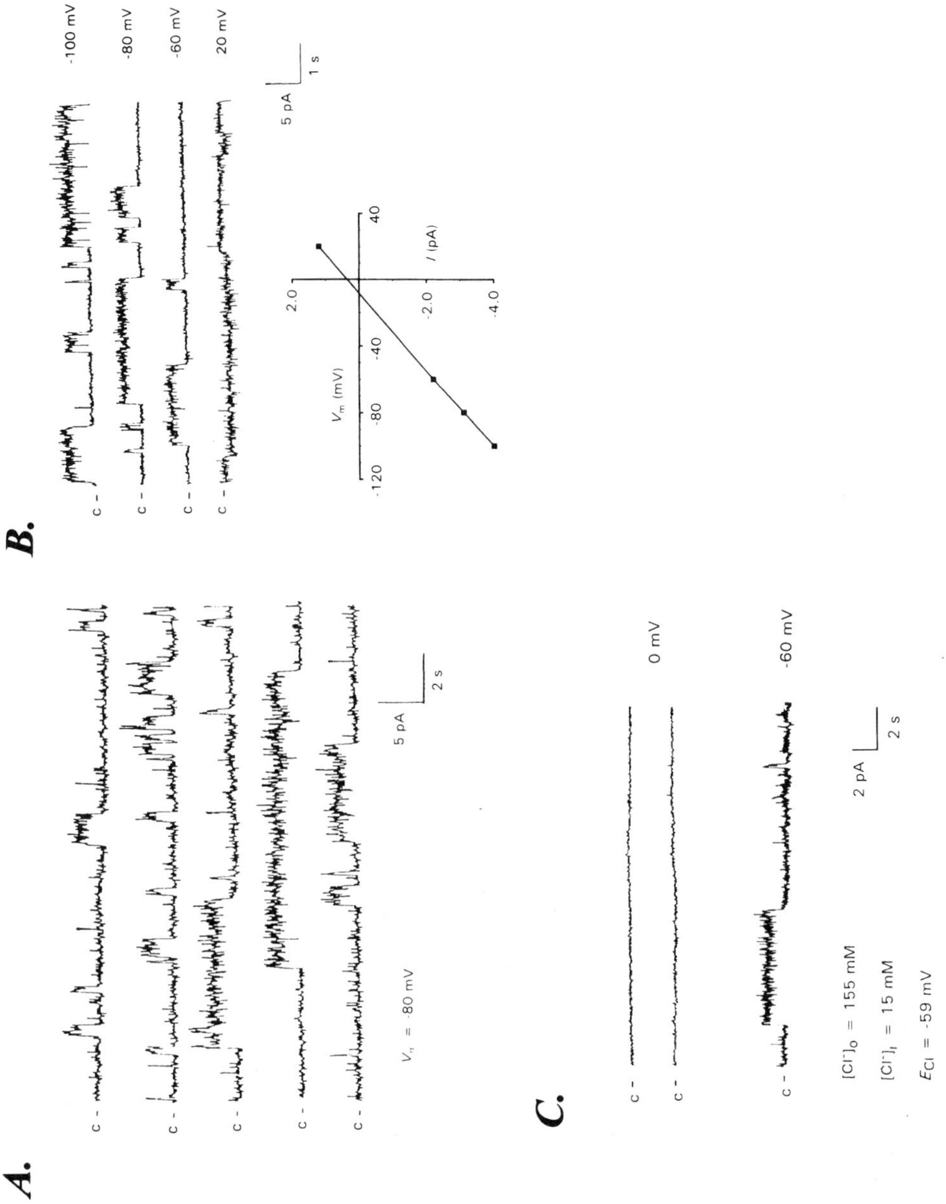
A.
B.
C.
C -
-100 mV
-80 mV
-60 mV
20 mV
5 pA
1 s
V_m (mV)
-120
-80
-40
40
2.0
-2.0
-4.0
I (pA)
V_H = -80 mV
5 pA
2 s
0 mV
-60 mV
2 pA
2 s
$[Cl^-]_o$ = 155 mM
$[Cl^-]_i$ = 15 mM
E_{Cl} = -59 mV

Table 1. Circadian rhythms in the gating of I_{LOT} channels

Activity of I_{LOT} channels in chick pineal cells during 12L:12D cycle[a]		Activity of I_{LOT} channels in free-running chick pineal cells in constant darkness[c]	
Phase in light-dark cycle	Number of cells with active I_{LOT} channels	Phase in circadian cycle	Number of cells with active I_{LOT} channels
Light (n = 78 cells)	0	Day (n = 50 cells)	0
Dark (n = 75 cells)	19 ($p < 0.005$)[b]	Night (n = 72 cells)	16 ($p < 0.025$)[b]

[a]Chick pineal cells were cultured for 5 days with continuous 12L:12D cycles. Cell-attached patch recordings were made on the fifth day *in vitro*. Cells in the light phase were sampled 4–6 hours after lights-on. Cells in the dark phase were sampled 4–6 hours after lights-off.

[b]G-test for homogeneity of two independent samples.

[c]Cultured chick pineal cells were entrained to 12L:12D cycles for 5 days. On the fifth day, cells were switched to constant darkness at the expected lights-off. Cell-attached patch recordings were made in the dark on the second day of constant-dark conditions. Cells in the day were sampled 4–6 hours after *expected* lights-on. Cells in the night were sampled 4–6 hours after *expected* lights-off.

We have detected I_{LOT} channels in eight out of 14 cell-attached patches on cultured retinal photoreceptors at ZT16–20 (night) (Fig. 2). These channels have very long open-time kinetics indistinguishable from those of I_{LOT} channels of pineal cells, and their gating is not voltage dependent. Moreover, I_{LOT} channels have been seen in only four out of 27 retinal cells at ZT6–10 (day) ($p < 0.05$, G-test; see Sokal and Rohlf, 1981, chapter 17). These data, although very preliminary, suggest that I_{LOT} gating also exhibits a daily rhythm in retinal cells, at least when they are maintained on 12L:12D cycles. The most widely used methods for culturing chick retinal photoreceptors are similar to those of Adler (1986), in which retinas are dissociated at E6–8 and kept *in vitro* for several days. Under these culture conditions there are free-running circadian rhythms in iodopsin gene expression (Pierce *et al.,* 1993), and 8-Br-cAMP and forskolin can evoke robust stimulation of melatonin secretion, but free-running circadian rhythms in melatonin secretion are

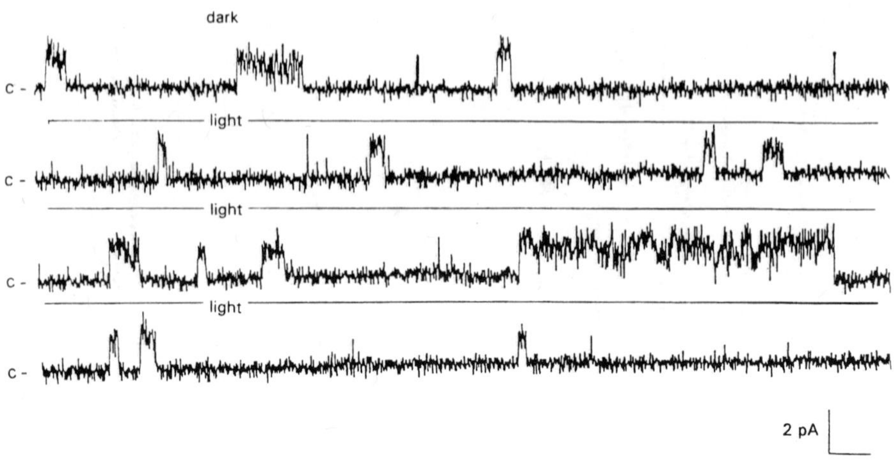

Figure 4. Acute application of 100-lux visible light does not inhibit the gating of I_{LOT} channels in cell-attached patch recordings made under infrared illumination at ZT16. The patch was hyperpolarized by 80 mV from the resting potential.

not observed (J. Takahashi, pers. comm.). Some essential mechanism for coupling circadian clock output to melatonin secretion is apparently missing at these early developmental stages. If the working model described below is correct, I_{LOT} would be a plausible candidate.

11. A WORKING MODEL OF I_{LOT} FUNCTION IN RETINAL AND PINEAL PHOTORECEPTORS

The model shown in Figure 5 is deliberately simplified so as to restrict attention to the physiological role of I_{LOT} in the context of melatonin secretion. We propose that similar processes take place in chick pineal cells and the cell bodies and inner segments of retinal photoreceptors. The model comprises a single cell containing an intrinsic circadian oscillator that can be entrained by light (and also by dopamine in the case of retinal photoreceptors). The most proximal outputs of the oscillator are assumed to be the transcriptional/translational control of a relatively small number of genes. In the diagram, the synthesis of various protein kinases and/or phosphatases and I_{LOT} are shown as alternative clock outputs (A. and B.), although we do not conceive of these alternatives as mutually exclusive or all inclusive. Upon translation,

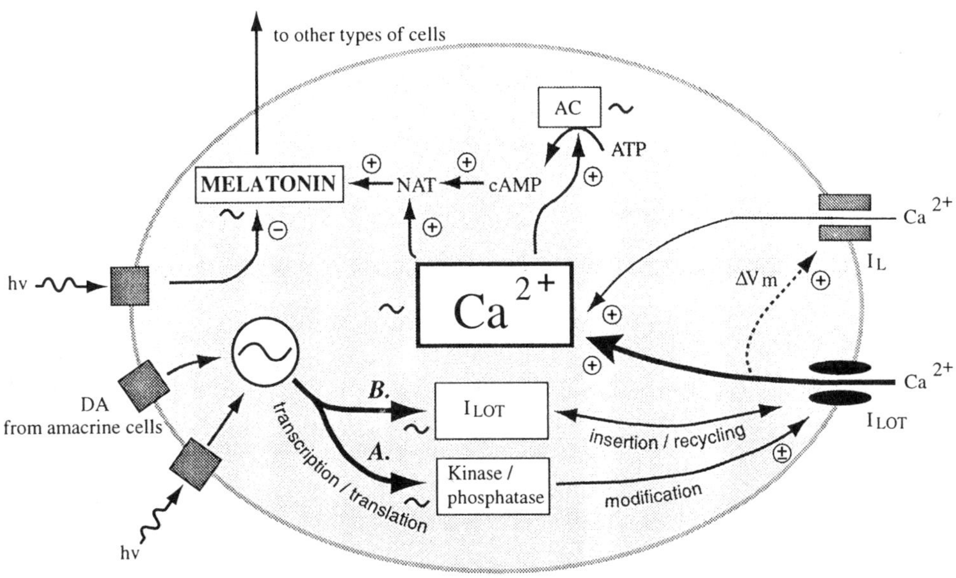

Figure 5. A working model of I_{LOT} function in photoreceptors. The model proposes that I_{LOT} plays a central role in coupling melatonin secretion to the output of the intrinsic circadian oscillator. Rhythmic changes in I_{LOT} gating require protein synthesis. Two possibilities are shown here. In A, the clock regulates the expression of kinases and/or phosphatases that in turn regulate I_{LOT} gating by direct modification of the plasma membrane channels. In B, the clock regulates the synthesis of I_{LOT} channels directly. These newly synthesized channels then pass through the Golgi apparatus and are inserted into the plasma membrane at the beginning of the night. They are removed from the plasma membrane at the end of the night. Here we propose that the majority of the nighttime Ca^{2+} influx is through I_{LOT} channels, but activation of I_{LOT} channels also causes membrane depolarization leading to activation of L-type Ca^{2+} channels, which amplify the signal by allowing for some additional Ca^{2+} influx. We also show that the circadian oscillator is entrained by light, as well as by dopamine in the case of retinal photoreceptors. Light also causes acute suppression of melatonin secretion, possibly by inhibiting cGMP-activated channels not shown in the diagram. Additional details are in the text.

newly synthesized I_{LOT} channels pass through the Golgi apparatus and are inserted into the plasma membrane. This process is depicted as reversible, as the I_{LOT} rhythm may reflect cycling of the channels in and out of the plasma membrane. In addition, we propose that the clock-regulated kinases and/or phosphatases cause direct modification of the I_{LOT} channels, thereby regulating their gating. For reasons of graphical simplicity, the diagram implies that these are cytosolic enzymes, but they could also be membrane bound, and they could certainly be the endpoint of a complex biochemical cascade. We propose that activation of I_{LOT} results in substantial and physiologically significant Ca^{2+} influx. In addition, we propose that I_{LOT} activation causes membrane depolarization, leading to activation of L-type Ca^{2+} channels, which serves to amplify the signal by causing additional Ca^{2+} influx. The resulting increase in intracellular free Ca^{2+} causes a direct increase in NAT activity, as well as a secondary cAMP-mediated increase in NAT. The increase in NAT activity is responsible for enhanced nighttime melatonin secretion. Light causes acute inhibition of melatonin secretion (probably in part because of inhibition of cGMP-activated channels not shown in the figure). This model is supported by some data from our own and other laboratories as described above. For example, we know that the nighttime expression of I_{LOT} requires the synthesis of some protein or proteins, as it is blocked by anisomycin. We know that I_{LOT} channels are permeable to Ca^{2+} ions, although the available data do not allow us to make any statements as to the amount of Ca^{2+} influx that could be expected at any particular membrane potential or external Ca^{2+} concentration. I_{LOT} is a nonselective cationic channel, and its activation will cause some membrane depolarization. Therefore I_{LOT} activation could also lead to L-type Ca^{2+} channel activation.

This model is focused on control of melatonin secretion simply because that is the best-studied biochemical output of photoreceptor circadian oscillators, but it should be emphasized that I_{LOT} may be important (perhaps even more important) for other physiological processes in retinal photoreceptors. For example, I_{LOT}-mediated Ca^{2+} influx into the synaptic terminals of photoreceptors could potentially have significant effects on the release of neurotransmitter onto second-order neurons. It is already known that cGMP-activated channels can directly regulate exocytosis in salamander cones (Rieke and Schwartz, 1994). I_{LOT} could have a similar function regardless of whether or not it proves to be under circadian control in retinal cells. I_{LOT} could also affect photoreceptor physiology independent of effects on Ca^{2+} influx by simply short circuiting the membrane, thereby altering the efficacy of gap-junctional transmission. Finally, we cannot completely exclude the possibility that I_{LOT} is itself a component of gap junctions, in which case electrical coupling of photoreceptors to each other or to other cell types may be under direct circadian control.

ACKNOWLEDGMENTS

This research was supported by AFOSR Grant F-49620.

LITERATURE CITED

Adler, R. (1986). Developmental predetermination of the structural and molecular polarization of photoreceptor cells. *Developmental Biology* 117:520–527.

Araki, M., Fukada, Y., Shichida, Y., Yoshizawa, T., and Tokunaga, F. (1992). Differentiation of both rod and cone types of photoreceptors in the *in vivo* and *in vitro* developing pineal glands of the quail. *Developmental Brain Research* 65:85–92.

Barrett, R. K., and Takahashi, J. S. (1995). Temperature compensation and temperature entrainment of the chick pineal cell circadian clock. *Journal of Neuroscience* 15:5681–5692.

Bassi, C. J., and Powers, M. K. (1986). Daily fluctuations in the detectability of dim lights by humans. *Physiology and Behavior* 38:871–877.

Bassi, C. J., and Powers, M. K. (1987). Circadian rhythm in goldfish visual sensitivity. *Investigative Ophthalmology and Visual Science* 28:1811–1815.

Bastianelli, E., and Pochet, R. (1994). Calbindin-D28k, calretinin, and recoverin immunoreactivities in developing chick pineal gland. *Journal of Pineal Research* 17:103–111.

Bernard, M., Iuvone, P. M., Cassone, V. M., Roseboom, P. D., Coon, S. L., and Klein, D. C. (1997). Avian melatonin synthesis: photic and circadian regulation of serotonin *N*-methyltransferase mRNA in the chicken pineal gland and retina. *Journal of Neurochemistry* 68:213–224.

Besharse, J. C., and Dunis, D. A. (1983). Methoxyindoles and photoreceptor metabolism: activation of rod shedding. *Science* 219:1341–1343.

Besharse, J. C., and Iuvone, P. M. (1983). Circadian clock in *Xenopus* eye controlling retinal serotonin *N*-acetyltransferase. *Nature* 305:133–135.

Besharse, J. C., and Iuvone, P. M. (1992). Is dopamine a light-adaptive or a dark-adaptive modulator in retina? *Neurochemistry International* 20:193–199.

Besharse, J. C., Iuvone, P. M., and Pierce, M. E. (1988). Regulation of rhythmic photoreceptor metabolism: a role for postreceptoral neurons. In: Osborne, N. N., and Chader, G. J., eds. *Progress in Retinal Research, Vol. 7.* Pergamon Press, Oxford. pp. 21–61.

Bischoff, M. B. (1969). Photoreceptoral and secretory structures in the avian pineal organ. *Journal of Ultrastructure Research* 28:16–26.

Bonigk, W., Muller, F., Middendorff, R., Weyand, I., and Kaupp, U. B. (1996). Two alternatively spliced forms of the cGMP-gated channel α-subunit from cone photoreceptor are expressed in the chick pineal organ. *Journal of Neuroscience* 16:7458–7468.

Boya, J., and Zamorano, L. (1975). Ultrastructural study of the pineal gland of the chicken (*Gallus gallus*). *Acta Anatomica* 92:202–226.

Cahill, G. M., and Besharse, J. C. (1991). Resetting the circadian clock in cultured *Xenopus* eyecups: regulation of retinal melatonin rhythms by light and D2 dopamine receptors. *Journal of Neuroscience* 11:2959–2971.

Cahill, G. M., and Besharse, J. C. (1993). Circadian clock functions localized in *Xenopus* retinal photoreceptors. *Neuron* 10:573–577.

Cahill, G. M., and Besharse, J. C. (1995). Circadian rhythmicity in vertebrate retinas: regulation by a photoreceptor oscillator. *Progress in Retinal Eye Research* 14:267–291.

Calvo, J., and Boya, J. (1979). Ultrastructural study of the embryonic development of the pineal gland of the chicken (*Gallus gallus*). *Acta Anatomica* 103:39–73.

Collin, J. P., Mirshahi, M., Brisson, P., Falcon, J., Guerlotte, J., and Faure, J. P. (1986). Pineal-retinal molecular relationships: distribution of "S-antigen" in the pineal complex. *Neuroscience* 19:657–666.

D'Souza, T., and Dryer, S. E. (1994). Intracellular free Ca^{2+} in dissociated cells of the chick pineal gland: regulation by membrane depolarization, second messengers, and neuromodulators, and evidence for release of intracellular Ca^{2+} stores. *Brain Research* 656:85–94.

D'Souza, T., and Dryer, S. E. (1995). Effects of phosphodiesterase inhibitors and forskolin on cyclic GMP-activated channels in intact isolated cells of the chick pineal gland. *Neurochemistry International* 27:527–533.

D'Souza, T., and Dryer, S. E. (1996). A novel cationic channel regulated by a vertebrate intrinsic circadian oscillator. *Nature* 382:165–167.

Deary, A., and Burnside, B. (1986). Dopaminergic regulation of cone retinomotor movement in isolated teleost retinas: I. Induction of cone contraction is mediated by D2 receptors. *Journal of Neurochemistry* 46:1006–1021.

Deary, A., and Burnside, B. (1989). Light-induced dopamine release from teleost retinas acts as a light-adaptive signal to the retinal pigment epithelium. *Journal of Neurochemistry* 53:870–878.

DeVries, S. H., and Schwartz, E. A. (1992). Hemi-gap-junction channels in solitary horizontal cells of the catfish retina. *Journal of Physiology (London)* 445:201–230.

Dryer, S. E., and Henderson, D. (1991). A cyclic GMP-activated channel in dissociated cells of the chick pineal gland. *Nature* 353:756–758.

Dryer, S. E., and Henderson, D. (1993). Cyclic GMP-activated channels of the chick pineal gland: effects of divalent cations, pH, and cyclic AMP. *Journal of Comparative Physiology [A]* 172:271–279.

Dubocovich, M. L. (1983). Melatonin is a potent modulator of dopamine release in retina. *Nature* 306:782–784.

Ebihara, L., Berthoud, V., and Beyer, E. (1995). Distinct behavior of connexin 56 and connexin 46 gap junctional channels can be predicted from the behavior of their hemi-gap-junction channels. *Biophysical Journal* 68:1796–1803.

Fain, G. L., and Lisman, J. E. (1993). Photoreceptor degeneration in vitamin A deprivation and retinitis pigmentosa: the equivalent light hypothesis. *Experimental Eye Research* 57:335–340.

Florez, J. C., and Takahashi, J. S. (1996). Regulation of tryptophan hydroxylase by cyclic AMP, calcium, norepinephrine, and light in cultured chick pineal cells. *Journal of Neurochemistry* 67:242–250.

Foster, R. G., Korf, H. W., and Schalken, J. J. (1987). Immunocytochemical markers revealing retinal and pineal but not hypothalamic photoreceptor systems in the Japanese quail. *Cell and Tissue Research* 248:161–167.

Gan, J., Alonso-Gomez, A. L., Avendano, G., Johnson, B., and Iuvone, P. M. (1995). Melatonin biosynthesis in photoreceptor-enriched chick retinal cell cultures: a role of cyclic AMP in the K^+-evoked, Ca^{2+}-dependent induction of serotonin *N*-acetyltransferase activity. *Neurochemistry International* 27:147–155.

Gotto, K., Miki, N., and Kondo, H. (1989). An immunohistochemical study of pinealocytes of chicks and some other lower vertebrates by means of visinin (retinal cone-specific protein)-immunoreactivity. *Archives of Histology and Cytology* 52:451–458.

Green, C. B., and Besharse, J. C. (1996). Identification of a novel vertebrate circadian clock-regulated gene encoding the protein nocturnin. *Proceedings of the National Academy of Sciences of the USA* 93:14884–14888.

Green, C. B., Cahill, G. M., and Besharse, J. C. (1994). Tryptophan hydroxylase expression is regulated by a circadian clock in *Xenopus laevis* retina. *Journal of Neurochemistry* 62:2420–2428.

Harrison, N. L., and Zatz, M. (1989). Voltage-dependent calcium channels regulate melatonin output from cultured chick pineal cells. *Journal of Neuroscience* 9:2462–2467.

Henderson, D., and Dryer, S. E. (1992). Voltage- and Ca^{2+}-activated ionic currents in acutely isolated cells of the chick pineal gland. *Brain Research* 572:182–189.

Iuvone, P. M. (1990). Development of melatonin synthesis in chicken retina: regulation of serotonin *N*-acetyltransferase activity by light, circadian oscillators, and cyclic AMP. *Journal of Neurochemistry* 54:1562–1568.

Iuvone, P. M., Gan, J., and Avendano, G. (1991). K^+-evoked depolarization stimulates cyclic AMP accumulation in photoreceptor-enriched retinal cell cultures: role of calcium influx through dihydropyridine-sensitive calcium channels. *Journal of Neurochemistry* 57:615–621.

Kemali, M., Kemali, D., Lovero, N., Maj, N., and Milici, N. (1987). Lithium and melatonin: morphological modifications induced in frog retina pigment screening. *Pharmacopsychiatria (Stuttgart)* 20:224–226.

Kolbinger, W., Kohler, K., Oetting, H., and Weiler, R. (1990). Endogenous dopamine and cyclic events in the fish retina: HPLC assay of total content, release, and metabolic turnover during differing light/dark cycles. *Visual Neuroscience* 5:417–428.

Lamb, T. D., and Pugh, E. N., Jr. (1992). A quantitative account of the activation steps involved in phototransduction in amphibian photoreceptors. *Journal of Physiology (London)* 449:719–758.

Li, T., Franson, W. K., Gordon, J. W., Berson, E. L., and Dryja, T. P. (1995). Constitutive activation of phototransduction by K296E opsin is not a cause of photoreceptor degeneration. *Proceedings of the National Academy of Sciences of the USA* 92:3551–3555.

Lu, C., and McMahon, D. G. (1996). Gap junction channel gating at bass retinal electrical synapses. *Visual Neuroscience* 13:1049–1057.

McCormack, C. A., and Burnside, B. (1992). A role for endogenous dopamine in circadian regulation of retinal cone movements. *Experimental Eye Research* 55:511–520.

McCormack, C. A., and Burnside, B. (1993). Light and circadian modulation of teleost retinal tyrosine hydroxylase activity. *Investigative Ophthalmology and Visual Science* 34:1853–1860.

McMahon, D. H., and Brown, D. R. (1994). Modulation of gap-junction channel gating at zebrafish retinal electrical synapses. *Journal of Neurophysiology* 72:2257–2268.

McMahon, D. G., and Matson, M. P. (1996). Horizontal cell electrical coupling in the giant danio: synaptic modulation by dopamine and synaptic maintenance by calcium. *Brain Research* 718:89–96.

Mangel, S. C., and Dowling, J. E. (1985). Responsiveness and receptive field size of carp horizontal cells are reduced by prolonged darkness and dopamine. *Science* 229:1107–1109.

Max, M., McKinnon, P. J., Seidenman, K. J., Barrett, R. K., Appelbury, M. L., Takahashi, J. S., and Margolskee, R. F. (1995). Pineal opsin: a nonvisual opsin expressed in chick pineal. *Science* 267:1502–1506.

Meissl, H., and Ekstrom, P. (1988). Photoreceptor responses to light in the isolated pineal organ of the trout, *Salmo gairdneri*. *Neuroscience* 25:1071–1076.

Nikaido, S. S., and Takahashi, J. S. (1996). Calcium modulates circadian variation in cAMP-stimulated melatonin in chick pineal cells. *Brain Research* 716:1–10.

Okano, T., Yoshizawa, T., and Fukada, Y. (1994). Pinopsin is a chicken pineal photoreceptive molecule. *Nature* 372:94–97.

Pierce, M. E., and Besharse, J. C. (1985). Circadian regulation of retinomotor movements. *Journal of General Physiology* 86:671–689.

Pierce, M. E., Sheshberadaran, H., Zhang, Z., Fox, L. E., Appelbury, M. L., and Takahashi, J. S. (1993). Circadian regulation of iodopsin gene expression in embryonic photoreceptors in retinal cell culture. *Neuron* 10:579–584.

Pu, G. A., and Dowling, J. E. (1981). Anatomical and physiological characteristics of pineal photoreceptor cell in the larval lamprey, *Petromyzon marinus*. *Journal of Neurophysiology* 46:1018–1038.

Rieke, F., and Schwartz, E. A. (1994). A cGMP-gated current can control exocytosis at cone synapses. *Neuron* 13:863–873.

Robinson, P. R., Buczylko, J., Ohguro, H., and Palczewski, K. (1994). Opsins with mutations at the site of chromophore attachment constitutively activate transducin but are not phosphorylated by rhodopsin kinase. *Proceedings of the National Academy of Sciences of the USA* 91:5411–5415.

Saez, J. C., Berthoud, V. M., Kadle, R., Traub, O., Nicholson, B. J., Bennett, M. V. L., and Dermietzel, R. (1991). Pinealocytes in rats: connexin identification and increase in coupling caused by norepinephrine. *Brain Research* 568:265–275.

Shaw, A. P., Collazo, C. R., Easterling, K., Young, C. D., and Karwoski, C. J. (1993). Circadian rhythm in the visual system of the lizard *Anolis carolinensis*. *Journal of Biological Rhythms* 8:107–124.

Sokal, R. R., and Rohlf, F. J. (1981). *Biometry*. Freeman, New York.

Steiner, E., and Ebihara, L. (1996). Functional characterization of canine connexin 45. *Journal of Membrane Biology* 150:153–161.

Suchyna, T. M., Xu, L.-X., Gao, F., Fourtner, C. R., and Nicholson, B. J. (1993). Identification of a proline residue as a transduction element in voltage gating of gap junctions. *Nature* 365:847–850.

Takahashi, J. S., Murakami, N., Nikaido, S. S., Pratt, B. L., and Robertson, L. M. (1989). The avian pineal, a vertebrate model system of the circadian oscillator: cellular regulation of circadian rhythms by light, second messengers and macromolecular synthesis. *Recent Progress in Hormone Research* 45:279–352.

Tamotsu, S., and Morita, Y. (1986). Photoreception in pineal organs of larval and adult lampreys, *Lampetra japonica*. *Journal of Comparative Physioliology [A]* 159:1–5.

Terman, M., and Terman, J. (1985). A circadian pacemaker for visual sensitivity? *Annals of the New York Academy of Sciences* 453:147–161.

Tosini, G., and Menaker, M. (1996). Circadian rhythms in cultured mammalian retina. *Science* 272:419–421.

Trexler, E. B., Bennett, M. V. L., Bargiello, T. A., and Verselis, V. K. (1996). Voltage gating and permeation in a gap junction hemichannel. *Proceedings of the National Academy of Sciences of the USA* 93:5836–5841.

van Veen, T., Ostholm, T., Gierschik, P., Spiegel, A., Somers, R., Korf, H. W., Klein, D. C. (1986). Alpha-transducin immunoreactivity in retinae and sensory pineal organs of adult vertebrates. *Proceedings of the National Academy of Sciences of the USA* 83:912–916.

Verselis, V. K., Ginter, C. S., and Bargiello, T. A. (1994). Opposite voltage gating polarities of two closely related connexins. *Nature* 368:348–351.

Wang, Y., and Mangel, S. C. (1996a). A circadian clock regulates rod and cone input to fish retinal cone horizontal cells. *Proceedings of the National Academy of Sciences of the USA* 93:4655–4660.

Wang, Y., and Mangel, S. C. (1996b). Melatonin acts as a circadian clock regulator of rod and cone pathways in fish retina. *Society for Neuroscience Abstracts* 22:792.12 (abstract).

Weiler, R., Kohler, K., Kursch, M., and Wagner, H. J. (1988). Glutamate and dopame modulate synaptic plasticity in horizontal cell dendrites of fish retina. *Neuroscience Letters* 87:205–209.

White, M. P., and Hock, P. A. (1992). Effects of continuous darkness on ERG correlates of disc shedding in the rabbit retina. *Experimental Eye Research* 54:173–180.

Wirz-Justice, A., Da Prada, M., and Remé, C. (1984). Circadian rhythm in rat retinal dopamine. *Neuroscience Letters* 45:21–25.

Witkovsky, P., and Schutte, M. (1991). Organization of dopaminergic neurons in vertebrate retinas. *Visual Neuroscience* 7:113–124.

Witkovsky, P., Stone, S., and Besharse, J. C. (1988). Dopamine modifies the balance of rod and cone inputs to horizontal cells of the *Xenopus* retina. *Brain Research* 449:332–336.

Yoshida, K., Kayamura, K., and Imaki, J. (1993). Differential expression of *c-fos* mRNA in rat retinal cells: regulation by light/dark cycle. *Neuron* 10:1049–1054.

Zatz, M. (1992a). Agents that affect calcium influx can change cyclic nucleotide levels in cultured chick pineal cells. *Brain Research* 583:304–307.

Zatz, M. (1992b). Does the circadian pacemaker act through cyclic AMP to drive the melatonin rhythm in chick pineal cells? *Journal of Biological Rhythms* 7:301–311.

Zatz, M. (1994). Photoendocrine transduction in cultured chick pineal cells: IV. What do vitamin A depletion and retinaldehyde addition do to the effects of light on the melatonin rhythm? *Journal of Neurochemistry* 62:2001–2011.

Zatz, M. (1988). Two mechanisms of photoendocrine transduction in cultured chick pineal cells: pertussis toxin blocks the acute but not the phase-shifting effects of light on the melatonin rhythm. *Brain Research* 453:63–71.

Zawilska, J. B., and Iuvone, P. M. (1989). Catecholamine receptors regulating serotonin *N*-acetyltransferase activity and melatonin content of chick retina and pineal gland: D2-dopamine receptors and alpha-2 adrenergic receptors in pineal gland. *Journal of Pharmacology and Experimental Therapeutics* 250:86–92.

11

PHOTORECEPTOR DEVELOPMENT AND PHOTOSTASIS

Anne B. Fulton,[1] Ronald M. Hansen,[1] Janice Dodge,[2] and
Theodore P. Williams

[1]Departments of Ophthalmology
Children's Hospital and Harvard Medical School
Boston, Massachusetts
[2]Department of Biological Science
Florida State University
Tallahassee, Florida 32306-4340

1. INTRODUCTION

In photostasis, the rhodopsin content of the retina and the rod outer segment (ROS) lengths change in response to the light of the animals' habitat. Specifically, for rats maintained in a habitat with dim, cyclic light, rhodopsin content is higher and ROS's longer than for those maintained in bright cyclic light (Penn and Williams, 1986; Reiser *et al.*, 1996). Alterations in ROS shedding and renewal (Young, 1976) and in rhodopsin distribution in the ROS appear to underlie the observed ROS adaptations to lighting habitat (Schremser and Williams, 1995a, b).

In common with photostasis, developmental processes alter ROS length and the rhodopsin content of the retina. Maturation of rod outer segment (ROS) length and rhodopsin content in rat retina are summarized by the growth curves in Figure 1, illustrating that the developmental increase in rhodopsin content, determined by extraction from whole, dark-adapted retinas, lags behind that of ROS length. The age at which rhodopsin is about half that of adults is about 18 to 19 days. The age at which ROS lengths are half those in adults is about 13 days. Despite the appearance that the ROS grow and then "fill up" with rhodopsin, Dodge *et al.* (1996a) have found no evidence of apo-opsin, that is, opsin without 11-*cis* chromophore, in immature rat ROS. Ratto *et al.* (1991), on the other hand, suggested apo-opsin was present in immature ROSs.

Phototransduction in the ROS requires proteins in addition to opsin. The mRNAs for the proteins in the transduction cascade, and the proteins themselves, are detectable early in the course of ROS development (Bok *et al.*, 1984; Bowes *et al.*, 1988; Broekhuyse and Kuhlmann, 1989; Cantera *et al.*, 1990; Colombaioni and Strettoi, 1993; DeLeeuw *et al.*, 1990; Eisenfeld *et al.*, 1985; Gonzalez-Fernandez *et al.*, 1984; Lee *et al.*, 1990; Lolley and

Photostasis and Related Phenomena, edited by Williams and Thistle
Plenum Press, New York, 1998

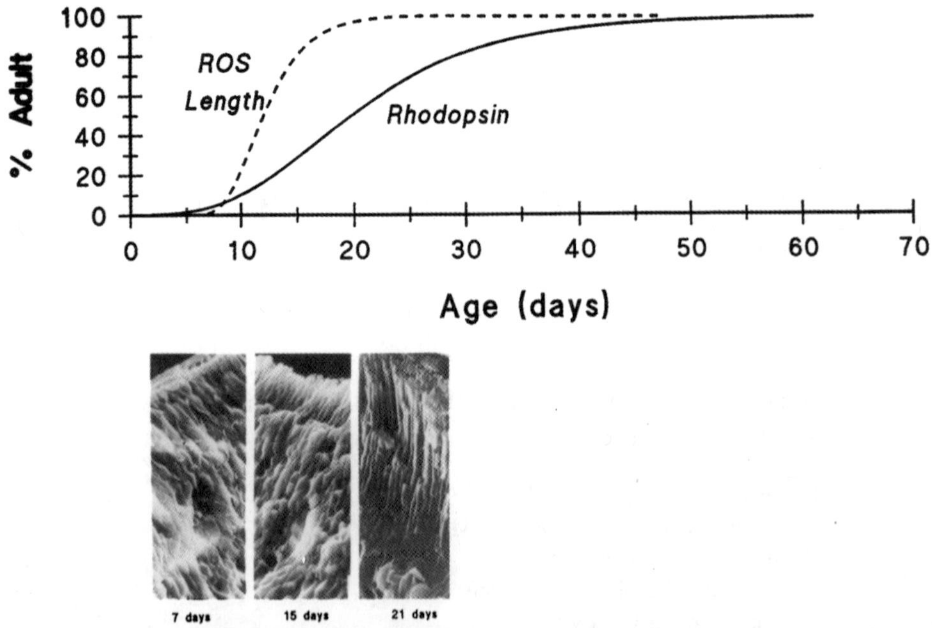

Figure 1. The scanning electron micrographs show that rat rod outer segments (ROS) are quite short at age 7 days, longer at 15 days, and as long as those in adults at 21 days. The smooth curves are logistic growth curves (Fulton *et al.*, 1995) that summarize the developmental increase in ROS lengths and rhodopsin content. Data for the ROS length curve are taken from Dowling and Sidman (1962), Bonting *et al.* (1961), and Ratto *et al.* (1991). The rhodopsin values were taken from Dowling and Sidman (1962), Bonting *et al.* (1961), Fulton and Baker (1984), and Fox and Rubenstein (1989). Each set of data was fitted by the growth curve and normalized to the maximum adult value, and then the combined, normalized values were fitted with the summarizing growth curve.

Farber, 1980; Nir *et al.*, 1984; van Veen *et al.*, 1989). The isomerase (Bernstein *et al.*, 1987) responsible for the *cis-trans* conversion of the chromophore is also present at young ages and appears to increase over the ages when ROS length increases (Bridges, 1989). The developmental increase in some of the ROS-specific proteins are summarized by the growth curves in Figure 2. No protein, including the protein carrier for retinal, intraretinal binding protein (IRBP), has a slower developmental course than does rhodopsin. Transducin, phosophodiesterase, and arrestin have developmental courses antedating that of rhodopsin. Thus, developmental deficiencies in these proteins appear unlikely to limit the processes involved in the activation or deactivation of phototransduction.

The function of the ROS also suggests that the molecular compositions of the discs in immature and mature rods are similar. For example, the gain of the activation of phototransduction is the same in immature and mature ROS (Ratto *et al.*, 1991). Furthermore, in a study (Fulton *et al.*, 1995) that tracked the maturation of ROS function using ERG procedures and a model (Hood and Birch, 1994) of rod phototransduction based on that of Lamb and Pugh (1992), the developmental increase in the gain of the rod cell response was the same as that for the increase of rhodopsin content (Fig. 3). Interestingly, for both the isolated ROS (Ratto *et al.*, 1991) and the intact retina (Fulton *et al.*, 1995), the developmental increase in the saturated amplitude of the rod cell response lagged behind that of ROS length growth. The result for the intact retina is shown in Figure 3.

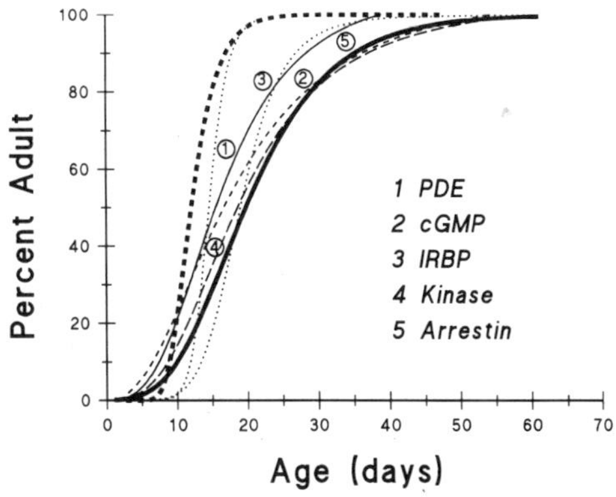

Figure 2. Plotted along with the ROS-length (heavy dashes) and rhodopsin (heavy solid) curves are the logistic growth curves that summarize the developmental increases in phosphodiesterase (PDE), cGMP, intraretinal binding protein (IRBP), kinase, and arrestin. The data for PDE and cGMP are taken from Lolley and Farber (1980) and Colombaioni and Strettoi (1993), those for IRBP from Gonzalez-Fernandez *et al.* (1984), those rhodopsin kinase from Ho *et al.* (1986), and those for arrestin from Broekhuyse and Kuhlmann (1989). In each case a logistic growth curve was fit to the data to determine the age at which 50% of the adult value was reached. In every case, that age was less than 19.2 days, the age at which rhodopsin is 50% of the adult value.

Without evidence of developmental deficiency in particular ROS proteins, and with evidence that the molecular compositions of the discs must be the same in infants and adults, how is it that the developmental increase in rhodopsin is delayed compared to ROS length (Fig. 1)? Using the microspectrophotometer in T. P. W.'s laboratory, we examined this issue by measuring rhodopsin concentration along the ROS (Dodge *et al.,* 1996a). The 13-day-old ROS's were quite short, but it was possible to get the beam of the microspectrophotometer, transversely, through the middle of one. The rhodopsin absorbance was

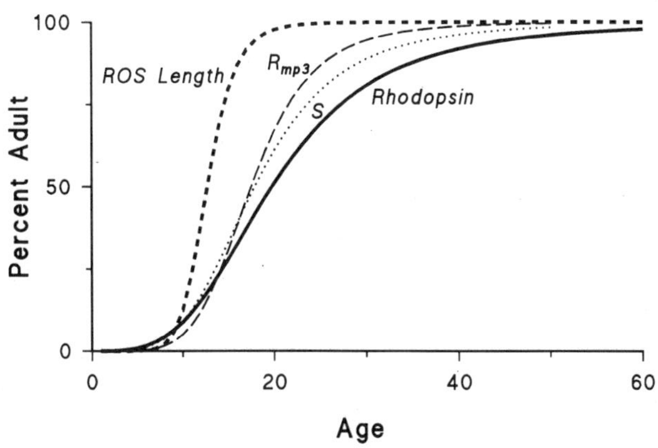

Figure 3. Developmental increase in R_{mp3}, the saturated amplitude of the rod cell responses, and S, the sensitivity or gain of the processes involved in the activation of phototransduction. The courses of R_{mp3} and S previously reported (Fulton *et al.,* 1995) are shown along with the ROS length and rhodopsin curves replotted from Figure 1. The age at which amplitude was half the maximum value was at 17.5 days (95% confidence interval 15 to 20 days). The ROS lengths are at half the maximum value at 13 days. Sensitivity, S, is half maximal at 17.9 days (16.2 to 19.6 days).

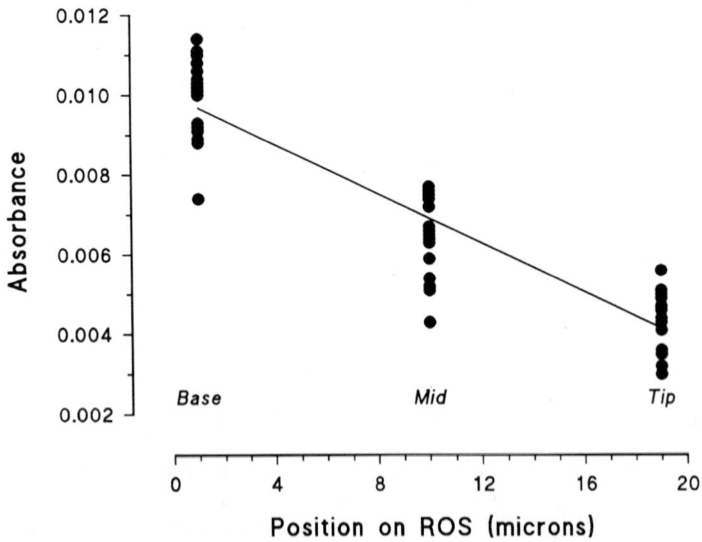

Figure 4. Absorbance as a function of position on the outer segment in 19-day-old rats. At age 19 days, when ROS length is the same as that of adults but rhodopsin content is about 50% that of adults, the absorbance at the base does not differ from that of adults, but it is significantly lower at the midpoint and tip of the ROS. On the assumptions of an ROS length of 20 m and microspectrophotometric measurements 1, 10, and 19 m from the tip, a straight line has been fitted (least squares criteria) to the data ($n = 50$). The line $y = -3.07 \times 10^{-4} x + 0.001$) is a reasonable description of the gradient ($r^2 = 0.85$, $t = 16.4$, $p < 0.01$).

about half that in the adult ROS. The 19-day-old ROS were about as long as those in the adult retina. Transverse absorbances were measured at the base, the midpoint, and the tip of the ROS. In the 19-day-olds, there was a gradient (Fig. 4). At the tip, where the oldest discs are located, the absorbance was low and nearly the same as that found at the midpoint in the 13-day-olds. The absorbance at the midpoint was higher. The absorbance at the base, near the inner segment, was the same as that in the 34-day-olds and the adult ROS. Absorbances were uniform along the length of the 34-day-old and adult ROS. These absorbances, measured transversely, gave a good accounting of the rhodopsin content of the whole retinas (Dodge *et al.,* 1996a).

Because the measuring beam in the microspectrophotometer covers many discs, we cannot determine whether the low absorbances at the tip of the 19-day-old ROS are due to fewer rhodopsin molecules per disc or to fewer discs per unit length, but electron micrographs of immature mouse ROSs (De Robertis, 1960), which have a developmental course virtually identical to that of the rat, show discs that are widely separated and irregularly oriented. In the mature mammalian ROS, the disc membranes, including the newly formed ones at the ROS base, are regularly and closely spaced (Steinberg *et al.,* 1980).

Therefore, because the physical properties of the immature ROS show the gradient in absorbance, whereas the gain of the molecular processes in the disc are the same in the mature and immature ROS (Ratto *et al.,* 1991; Fulton *et al.,* 1995), we suspect that the infant and adult differ in the physical spacing of the discs rather than in their molecular composition. This interpretation also accommodates the observation that development of the saturated amplitude of the rod photoresponse lags behind the developmental course of

ROS length (Ratto *et al.,* 1991; Fulton *et al.,* 1995). The amplitude reflects the number of channels available for closure by light. In mature rods, at least in the toad (Baylor *et al.,* 1979), the number of channels per unit length is uniform along the length of the ROS. If the mature ROS has 36 discs m^{-1} (Pugh and Lamb, 1993), if the tip of the 19-day-old ROS has only half the absorbance and 18 discs m^{-1}, if the gradient from 18 at the tip to 36 at the base is linear (Fig. 4), and if the length of the 19-day-old and adult ROSs are 20 m, then the total number of discs in the 19-day-old is 518, or 72% of the 720 discs in the adult. If it is assumed that the number of channels per micron is proportional to the number of discs, then in the 19-day-old the saturated amplitude of the rod photoresponse should be 72% of that in adults. Because the beam of the microspectrophotometer is about 2 m long as it traverses the ROS, 72% can be considered an overestimate. Saturated amplitudes of 61% of maximum were reported in 19-day-olds (Fulton *et al.,* 1995), so the agreement with expectation is reasonable.

We decided to find out whether light altered the developmental course of rhodopsin increase in infant rats. Put into different words: Is photostasis operative in the immature rat? Would the light level in the habitat affect rhodopsin content in the young retinas "programmed" for a developmental increase in rhodopsin content? In adults, the level of light in the habitat has a differential effect on the rhodopsin content (Penn and Williams, 1986; Schremser and Williams, 1995a). Light levels known to cause significant changes in the rhodopsin content of adults were used. Both albino and pigmented rats were studied.

2. METHODS

Both albino (Sprague-Dawley) and pigmented (Long-Evans) rats were born and maintained in controlled lighting, on a cycle of 12 hours of light and 12 hours of darkness (12L:12D) each day. Lights came on at 0800 hours. The infants were maintained with their mothers in 1-lux, 200-lux, and, in addition for Long-Evans rats, 3500-lux habitats. Retinas were obtained for rhodopsin extractions at ages 13, 19, and 35 days and from adults (60 to 90 days old). A total of 84 albino rats, median of 11 (range 6 to 13) in each age group, and 64 pigmented rats, median of 6 (range 2 to 10) in each group, were studied. One eye of each rat was used for the rhodopsin studies and analyses presented here. Fellow eyes were used for other biochemical and microspectrophotometric studies (Dodge *et al.,* 1996b).

In dim red light whole retinas were removed from the dark-adapted rats as previously described (Dodge *et al.,* 1996b). Each retina was put into a separate centrifuge tube. To each centrifuge tube 1 ml of distilled water was added, and the contents were stirred vigorously with a stainless-steel spatula. The samples were incubated in the dark for 1 hour and then centrifuged at 12,000 g for 15 min. The supernatant was discarded, and 1.0 ml of 1% Emulphogene (Sigma) in 50 mM Tris-acetate buffer, pH 6.9, was added to each pellet. The pellets were disrupted with a spatula.

The tubes were kept in the dark for 1 hour at 20°C, then spun again at 12,000 g. An aliquot of the supernatant was removed, and absorbance spectra were measured from 350 nm to 700 nm first in the dark and then after exposure to bright, white light for 5 min. From these pre- and postexposure spectra, a difference spectrum was calculated. For each sample the absorbance of rhodopsin at its λ_{max} was obtained from its difference spectrum. The number of nanomoles of rhodopsin present in each retina was calculated according to the Beer-Lambert law and with an extinction coefficient of 42,000 M^{-1} cm^{-1}.

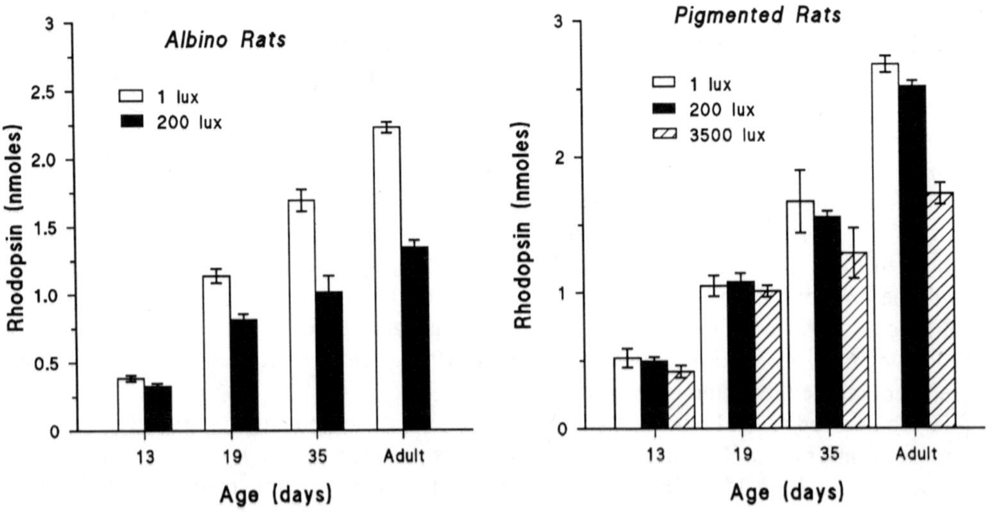

Figure 5. The mean (±SEM) nanomoles of rhodopsin extracted from retinas of albino (left panel) and pigmented (right panel) rats. Rhodopsin content increased with age. At all ages albinos in the 200-lux habitat had less rhodopsin than those in the 1-lux habitat. At all ages pigmented rats in the 3500-lux habitat had less rhodopsin than those in the 1-lux and 200-lux habitats.

3. RESULTS

In both albino and pigmented rats, the rhodopsin content increased with age and varied with lighting condition. The mean rhodopsin values for each group are summarized in Figure 5. At all ages, rats in the brightest habitats, 200 lux for the albinos and 3500 lux for the pigmented rats, had lower amounts of rhodopsin than were found in the 1-lux group. The effect of light on the rhodopsin content was similar at all ages. Among albinos, the mean rhodopsin values of the 200-lux group were 80%, 72%, 60%, and 61% of that of the 1-lux group for 13-, 19-, and 35-day-old and adult groups, respectively. Among the pigmented rats, the mean rhodopsin values of the 3500-lux group were 81%, 83%, 72%, and 65% of the 1-lux group for 13-, 19-, and 35-day-old and adult groups. For albino rats, rhodopsin content varied significantly with both age ($F = 251$; df 3,77; $p < 0.01$) and light level ($F = 162$; df 1,77; $p < 0.01$). For pigmented rats, rhodopsin content also varied significantly with both age ($F = 26$; df 2,66; $p < 0.01$) and light level ($F = 254$; df 3,66; $p < 0.01$). Thus both age and light affect total rhodopsin content. No significant interaction of age and light was found in albino or pigmented rats.

Rhodopsin values for individual albino rats, normalized to the adult mean for group, are shown in Figure 6. A logistic growth curve fitted to the data for the rats in the 1-lux habitat indicated that the age at which rhodopsin content was half the adult value was 20.2 days (95% confidence interval 17.5–22.9 days). The result for the 200-lux group was 18.2 days (16.3–20.1 days). The courses of developmental increase in rhodopsin with age are therefore indistinguishable for rats reared in the two lighting habitats. The normalized values for the pigmented rats are shown in Figure 7. For the pigmented rats, the developmental courses for rhodopsin increase were similar in 1-lux, 200-lux, and 3500-lux habitats. Growth-curve analysis indicates that the age at which 50% adult values are reached is 16.8 days (95% confidence interval 14.4–19.2 days) for the 3500-lux group; 21.7 days

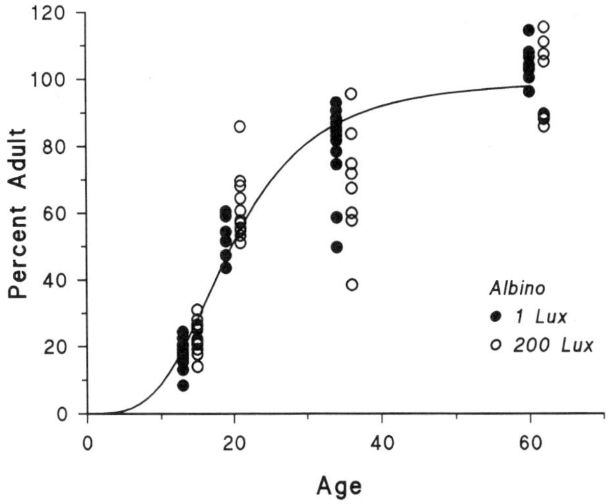

Figure 6. Rhodopsin development for albinos in 1- and 200-lux habitats. The points plot normalized values for individual rats. The smooth curve is the rhodopsin growth curve shown in Figure 1.

(19.3–24.1 days) for the 200-lux group, and 21.2 days (19.8–22.6 days) for the 1-lux group. The data for 35-day-old rats (Fig. 7) for the 1- and 200-lux groups were omitted from the analysis.

4. DISCUSSION

The effect of light history on rhodopsin content is similar in mature and immature retinas. In both pigmented and albino rats, in all lighting conditions, a developmental increase in rhodopsin content was found. The developmental courses are similar in all conditions. The major share of the developmental increase in ROS length takes place over about 10 days, whereas rhodopsin increase has a more protracted time course (Fig. 1). In

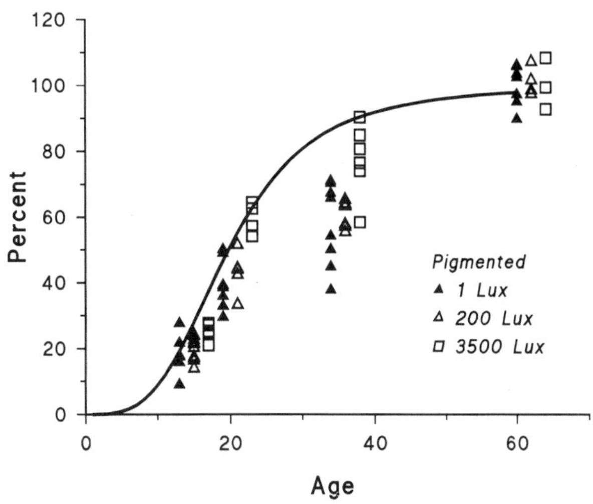

Figure 7. Rhodopsin development for pigmented rats in 1-lux, 200-lux, and 3500-lux habitats. The points plot normalized values for individual rats. The smooth curve is the rhodopsin growth curve shown in Figure 1.

the photostasis experiments of Schremser and Williams (Schremser and Williams, 1995a), in which mature rats were switched from bright to dim habitats, or in companion experiments, from dim to bright habitats, the time course of ROS length changes bore no simple relationship to the rhodopsin concentrations in those ROS.

The discs of the ROS are turned over continually as newly synthesized discs are added at the base and older discs are shed from the tips (Young, 1976). In the mature animal, in stable lighting conditions, the rate of synthesis and shedding are equal, and ROS lengths and rhodopsin content remain constant from day to day. In the developing retina, disc synthesis exceeds shedding, so ROS length (LaVail, 1973) and rhodopsin content increase. The developmental course for shedding (Tamai and Chader, 1979), assessed as the number of phagosomes per unit length of retinal pigment epithelium, is summarized in Figure 8. The course of increase in the number of phagosomes is indistiguishable in two lighting conditions differing by 3 log units, and the developmental course for the phagosomes is similar to that for rhodopsin (Fig. 1).

In mature rats switched from bright to dim habitats, the number of phagosomes is decreased in approximate concert with increases in rhodopsin content. Disc shedding decreased modestly in mature rats moved from dim to bright habitats. Such rats show a relatively large decrease in rhodopsin content. Furthermore, 5 days after the switch from dim to bright habitat, rhodopsin absorbance was less at the base than at the tip of the ROS (Schremser and Williams, 1995b), indicating a gradient opposite to that seen in the immature, 19-day-old ROS (Fig. 4). Light-induced alterations in rhodopsin synthesis, leading to a low number of rhodopsin molecules per disc membrane, have been hypothesized (Schremser and Williams, 1995b), but we also note that these results are not inconsistent with fewer discs per unit length of ROS. Thus, factors involved in disc assembly and the physical flattening of the disc membranes as they are added to the base of the outer segment may also be considered among the mechanisms underlying photostasis.

ACKNOWLEDGMENTS

This research was supported by the Massachusetts Lions Eye Research Foundation and EY10597.

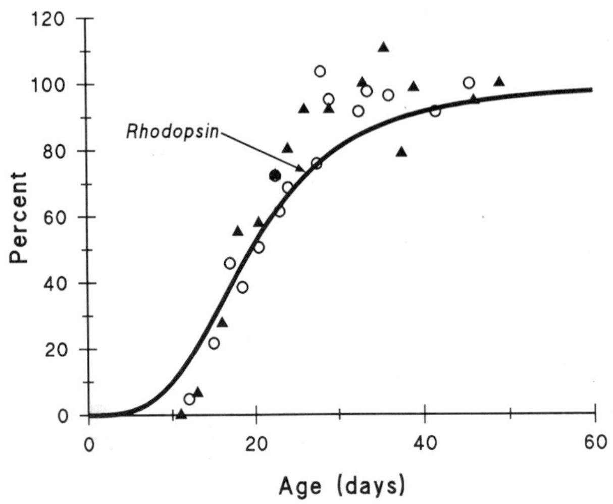

Figure 8. Percent adult phagosomes per unit length of retinal pigment epithelium as a function of age during development. The points plot data from Tamai and Chader (1979). The open symbols are for rats maintained in 40 to 50 foot candles, and the filled symbols are for rats maintained in <0.5 foot candles. The courses of phagosome increase are similar in the two lighting conditions, which differ by 3 log units. The smooth curve is the rhodopsin curve replotted from Figure 1.

LITERATURE CITED

Baylor, D. A., Lamb, T. D., and Yau, K.-W. (1979). The membrane current of single rod outer segments. *Journal of Physiology (London)* 288:589–611.

Bernstein, P. S., Law, W. C., and Rando, R. R. (1987). Isomerization of all-*trans*-retinoids to 11-*cis*-retinoide in vitro. *Proceedings of the National Academy of Sciences of the USA* 84:1849–1853.

Bok, D., Ong, D., and Chytil, F. (1984). Immunocytochemical localization of cellular retinol binding protein in the rat retina. *Investigative Ophthalmology and Visual Science* 25:877–883.

Bonting, S. L., Caravaggio, L. L., and Gouras, P. (1961). The rhodopsin cycle in the developing vertebrate retina: I. Relation of rhodopsin content, electroretinogram and rod structure in the rat. *Experimental Eye Research* 1:14–24.

Bowes, C., van Veen, T., and Farber, D. (1988). Opsin, G-protein and 48 kDa protein in normal and rd mouse retinas: developmental expression of mRNAS and proteins and light/dark cycling of mRNAs. *Experimental Eye Research* 47:369–390.

Bridges, C. (1989). Distribution of retinol isomerase in vertebrate eyes and its emergence during retinal development. *Vision Research* 29:1711–1717.

Broekhuyse, R. M., and Kuhlmann, E. D. (1989). Assay of s-antigen immunoreactivity in mammalian retinas in relation to age ocular dimension and retinal degeneration. *Japanese Journal of Ophthalmology* 33:243–250.

Cantera, R., von Schantz, M., Chader, G., Ehinger, B., Sanyal, S., and van Veen, T. (1990). Postnatal development of photoreceptor-specific proteins in mice with hereditary retinal degeneration. *Experimental Biology* 48:305–312.

Colombaioni, L., and Strettoi, E. (1993). Appearance of cGMP-phosphodiesterase immunoreactivity parallels the morphological differentiation of photoreceptor outer segments in the rat retina. *Visual Neuroscience* 10:395–402.

De Robertis, E. (1960). Some observations on the ultrastructure and morphogenesis of photoreceptors. *Journal of General Physiology* 43:1–13.

DeLeeuw, A., Gaur, V., Saari, J., and Milam, A. (1990). Immunolocalization of cellular retinol-, retinaldehyde- and retinoic acid-binding proteins in rat retina during pre- and postnatal development. *Journal of Neurocytology* 19:253–264.

Dodge, J., Fulton, A. B., Parker, C., Hansen, R. M., and Williams, T. P. (1996a). Rhodopsin in immature rod outer segments. *Investigative Ophthalmology and Visual Science* 37:1951–1956.

Dodge, J., Parker, C., Reynaud, X., Williams, T., and Fulton, A. (1996b). The role of rhodopsin in ERG's attenuated by an early history of oxygen exposure. *Investigative Ophthalmology and Visual Science* 37:S799.

Dowling, J. E., and Sidman, R. L. (1962). Inherited retinal dystrophy in the rat. *Journal of Cell Biology* 14:73–109.

Eisenfeld, A., Bunt-Milam, A., and Saari, J. (1985). Localization of retinoid-binding proteins in developing rat retina. *Experimental Eye Research* 41:299–304.

Fox, D. A., and Rubenstein, S. D. (1989). Age related changes in retinal sensitivity, rhodopsin content and rod outer segment length in hooded rats following low level lead exposure during development. *Experimental Eye Research* 48:237–249.

Fulton, A. B., and Baker, B. N. (1984). The relation of retinal sensitivity and rhodopsin in developing rat retina. *Investigative Ophthalmology and Visual Science* 25:647–651.

Fulton, A. B., Hansen, R. M., and Findl, O. (1995). The development of the rod photoresponse from dark adapted rats. *Investigative Ophthalmology and Visual Science* 36:1038–1045.

Gonzalez-Fernandez, F., Landers, R., Glazebrook, P., Fong, S., Liou, G., Lam, D., and Bridges, C. (1984). An extracellular retinol-binding glycoprotein in the eyes of mutant rats with retinal dystrophy: development, localization, and biosynthesis. *Journal of Cell Biology* 99:2092–2098.

Ho, A. K., Somers, R. L., and Klein, D. C. (1986). Development and regulation of rhodopsin kinase in rat pineal retina. *Journal of Neurochemistry* 46:1176–1179.

Hood, D. C., and Birch, D. G. (1994). Rod phototransduction in retinitis pigmentosa: estimation and interpretation of parameters derived from the rod a-wave. *Investigative Ophthalmology and Visual Science* 35:2948–2961.

Lamb, T. D., and Pugh, E. N., Jr. (1992). A quantitative account of the activation steps involved in phototransduction in amphibian photoreceptors. *Journal of Physiology (London)* 449:719–758.

LaVail, M. M. (1973). Kinetics of rod outer segment renewal in the developing mouse retina. *Journal of Cell Biology* 58:650–661.

Lee, R., Lieberman, B., and Lolley, R. (1990). Retinal accumulation of the phosducin/t and transducin complexes in developing normal mice and in mice and dogs with inherited retinal degeneration. *Experimental Eye Research* 51:325–333.

Lolley, R. N., and Farber, D. B. (1980). Cyclic GMP metabolic defects in inherited disorders of rd mice and RCS rats. *Neurochemistry International* 1:427–440.

Nir, I., Cohen, D., and Papermaster, D. (1984). Immunocytochemical localization of opsin in the cell membrane of developing rat retinal photoreceptors. *Journal of Cell Biology* 98:1788–1795.

Penn, J. S., and Williams, T. P. (1986). Photostasis: regulation of daily photon-catch by rat retinas in response to various cyclic illuminances. *Experimental Eye Research* 43:915–928.

Pugh, E. N., Jr. and Lamb, T. D. (1993). Amplification and kinetics of the activation steps in phototransduction. *Biochimica et Biophysica Acta* 1141:111–149.

Ratto, G. M., Robinson, D. W., Yan, B., and McNaughton, P. A. (1991). Development of the light response in neonatal mammalian rods. *Nature* 351:654–657.

Reiser, M. A., Williams, T. P., and Pugh, E. N., Jr. (1996). The effect of light history on the aspartate isolated fast P III response of the albino rat retina. *Investigative Ophthalmology and Visual Science* 37:221–229.

Schremser, J.-L., and Williams, T. P. (1995a). Rod outer segment (ROS) renewal as a mechanism for adaptation to a new intensity environment. I. Rhodopsin levels and ROS length. *Experimental Eye Research* 61:17–24.

Schremser, J.-L., and Williams, T. P. (1995b). Rod outer segment (ROS) renewal as a mechanism for adaptation to a new intensity environment. II. Rhodopsin synthesis and packing density. *Experimental Eye Research* 61:25–32.

Steinberg, R. H., Fisher, S. K., and Anderson, D. H. (1980). Disc morphogenesis in vertebrate photoreceptors. *Journal of Comparative Neurology* 190:501–518.

Tamai, M., and Chader G. J. (1979). The early appearance of disc shedding in the rat retina. *Investigative Ophthalmology and Visual Science* 18:913–917.

van Veen, T., Cantera, R., Narfstrom, K., Nilsson, S., Sanyal, S., Wiggert, B., and Chader, G. (1989). Postnatal development of photoreceptor proteins in mutant mice and Abyssinian cats with retinal degeneration. In: LaVail, M. M., Anderson, R. E., and Hollyfield, J. G., eds. *Inherited and Environmentally Induced Retinal Degenerations.* Liss, New York. pp. 275–289.

Young, R. W. (1976). Visual cells and the concept of renewal. *Investigative Ophthalmology* 15:700–725.

12

PHOTOSTASIS AND BEYOND

Where Adaptation Ends

Ch. E. Remé, R. Bush, F. Hafezi, A. Wenzel, and Ch. Grimm

University Eye Clinic Zürich
Zürich, Switzerland

1. LIGHT AS A PHOTORECEPTOR MODULATOR

Both disc shedding, the degradative part of the renewal cycle of photoreceptor outer segments, and autophagy, the degradation of cellular organelles in inner segments, follow a circadian rhythm in mammals (LaVail, 1976; Remé *et al.,* 1985). The light history of a given animal (Penn and Williams, 1986), however, defines the amplitude of that rhythm and the overall "sensitivity" of the system to light exposure (Remé *et al.,* 1991). Furthermore, a gradual increase of the light signal simulating a natural twilight transition (Terman, in press) can trigger disc shedding at very low light levels (Bush *et al.,* 1990), raise the upper limit of entrainment to light-dark cycles in hamsters (Boulos *et al.,* 1996), and entrain human circadian rhythms (Wirz-Justice *et al.,* in press).

We have previously suggested that structure and some functions of photoreceptors can be modulated by acute or chronic light exposures that differ from those normally encountered within an animal's light-dark cycle (Remé *et al.,* 1991). On a structural level, such modification would be executed by disc shedding and autophagy. Other light-dependent changes, performed on a molecular level, will not be discussed here (Penn and Anderson, 1987; Organisciak *et al.,* 1991). Thus, disc shedding and autophagy not only follow a circadian rhythm but in addition may be evoked or their amplitude modified by light exposures (Figs. 1, 2). The magnitude of such evoked responses varies with time of day of the exposure. This variation persists under constant conditions; i.e., the sensitivity of the system shows a circadian rhythm (Fig. 2) (Remé *et al.,* 1985, 1986).

2. ADAPTATION TO LIGHT EXPOSURE AND PHYSIOLOGICAL DARK-LIGHT TRANSITIONS

One can thus envision the disc-shedding and autophagy responses evoked by "unscheduled" and/or high-intensity light exposures as an adaptive attempt to adjust to an ap-

Photostasis and Related Phenomena, edited by Williams and Thistle
Plenum Press, New York, 1998

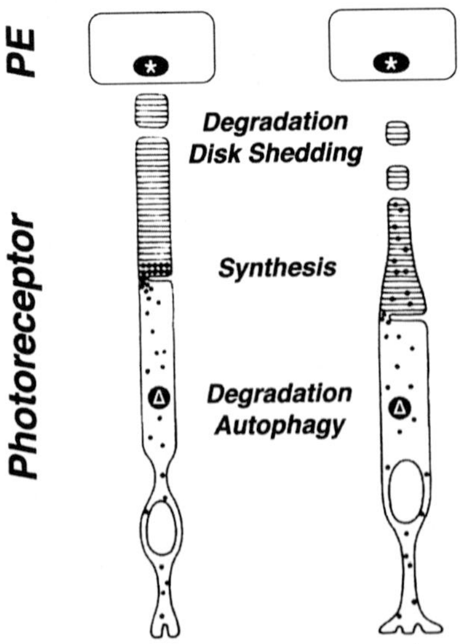

Figure 1. Schematic drawing depicting the renewal of rods and cones. Phagosomes (*) are shed from outer segments and degraded in the pigment epithelium (PE). In inner segments, autophagic degradation (Δ) of organelles occurs. In rods, new disc membranes are formed at the base of the outer segment, whereas in cones molecular renewal takes place throughout the entire outer segment.

parently new and potentially dangerous light environment. A disc-shedding burst in addition to the one following the abrupt light onset in the laboratory light-dark cycle would reduce the length of rod outer segments, the rhodopsin content, and thus the photon absorption of a given rod (Schremser and Williams, 1995). In the case of autophagy, the situation is less obvious: what would the system gain by increasing the degradation of its metabolic machinery unless the photon absorption could directly or indirectly be reduced? That it may in fact be reduced is demonstrated in recent experiments in which a burst of autophagic degradation was evoked when animals were abruptly changed from a low-light habitat to a high-light habitat. The contents of many of the autophagic vacuoles showed distinct staining for opsin, suggesting that "surplus" opsin might be degraded before its insertion into new disc membranes. As hypothesized, the number of autophagic vacuoles was significantly increased during the first three days of the new light environment (Remé, Wolfrum, Huber, Imsand, and Williams, in preparation).

The disc-shedding and autophagy bursts following an unscheduled exposure to bright light may thus be seen as an emergency adaptation to a sudden change into higher illuminance levels. Using a dawn and dusk simulator (Terman, in press), we have analyzed the onset and increase of disc shedding in the morning (Bush *et al.,* 1990) (Fig. 3). Under twilight conditions, disc shedding is triggered at intensities about 5 log units lower than that used in the laboratory with an abrupt light onset. Shedding begins long before the final illuminance for either the twilight or the abrupt-onset regimen is reached: in the lowest scotopic range at 0.002 lux. The peak number of phagosomes was equal under the two paradigms, perhaps because photostatic adjustment of the total amount of shedding operates under both conditions. Interestingly, phagosome counts were distinctly higher in the upper than that in the lower half of the retina, confirming the sensitive area defined by Williams and coworkers (Rapp and Williams, 1980). These observations may indicate that laboratory conditions with abrupt changes of light levels imply a potential danger for photoreceptors and furthermore that the rod signal may represent an important natural Zeitgeber that prepares the system for mesopic and photopic vision. Remnants of this shedding initiation by low light levels may be reflected in the "anticipatory" rise of shedding during the end of the dark period, which is frequently noticed in our experiments using rectangular light regimens. Additional evidence comes from our experiments showing that the sensitivity of shedding to an acute light pulse is greatest around the time of late dark and potential early dawn (Remé *et al.*, 1986).

3. BEYOND ADAPTATION: CELL DEATH

We have seen that acute exposure to light levels above those normally encountered in a given laboratory environment evoke "adaptive" disc shedding and autophagy. The previous light history of an animal defines their thresholds (Penn and Williams, 1986). Is there an upper limit of protection provided by those processes? When a situation presents "too much light within too short a time period" for an organism's photoreceptors, adaptation fails and damage occurs. What is the structural correlate of such acutely inflicted lesions?

We have developed a model for acute light-induced damage to photoreceptors, which is reflected in vesiculations and disruptions of rod outer segment discs (Remé *et al.*, 1994, in press). The amount of those changes is a function of illuminance level and can be monitored exactly with a computerized morphometric method (Wei *et al.*, 1993; Remé *et al.*, 1994). Above certain illuminance levels (which are defined by habitat illuminance), there is a point of no return, when the lesions comprise the entire photoreceptor cell. At this stage, apoptosis is seen as a synchronized burst of cell death (Remé *et al.*, 1995; Szczesny *et al.*, 1995; Hafezi *et al.*, 1997, in press) (Fig. 4).

In many cell systems, apoptosis is a physiological regulator; for example, in development it serves to remove surplus or transitory cells. Blood cells are continually eliminated by apoptosis, and in many tissues, an equilibrium is maintained by apoptosis between proliferation and degradation. One component of tumor growth is the lack of apoptotic cell removal. An evolutionary advantage of apoptosis is no harm is done to neighboring cells and tissues, in contrast to necrotic cell death where the environment is damaged by proteolytic enzymes and toxic components of lytic and dying cells (Wyllie *et al.*, 1980; Kroemer *et al.*, 1995; Steller, 1995).

Apoptosis also occurs in ocular tissues under various conditions. Apoptotic photoreceptor cell death is seen in retinal development (Young, 1984), in animal models of retinal dystrophies (Chang *et al.*, 1993), in other retinal pathologies, and after light exposure (Remé *et al.*, 1995; Szczesny *et al.*, 1995). In our model, apoptosis, autophagy, and disc shedding are frequently seen simultaneously during the acute stages after light exposure and may, in this context, represent the aforementioned "adaptational struggle" for survival.

Apoptosis is a gene-regulated process, and much research is concerned with its regulative mechanisms. We have exposed mice lacking the protooncogene *c-fos* to light levels that were distinctly damaging to control wild-type animals. Almost no apoptotic cell death was found in *c-fos* knockout mice, even after extremely high light levels (15,000 lux for 2 hours) and long postexposure periods (10 days) (Hafezi *et al.*, 1997; Marti, Hafezi, Wenzel, Munz, and Remé, in preparation). In those experiments, almost no necrotic cell death was found in knockout animals; the *c-fos*$^{-/-}$ mice showed normal retinae at all time points investigated. These experiments therefore indicate that, under our conditions, apoptosis appears to be the major correlate of acute light damage.

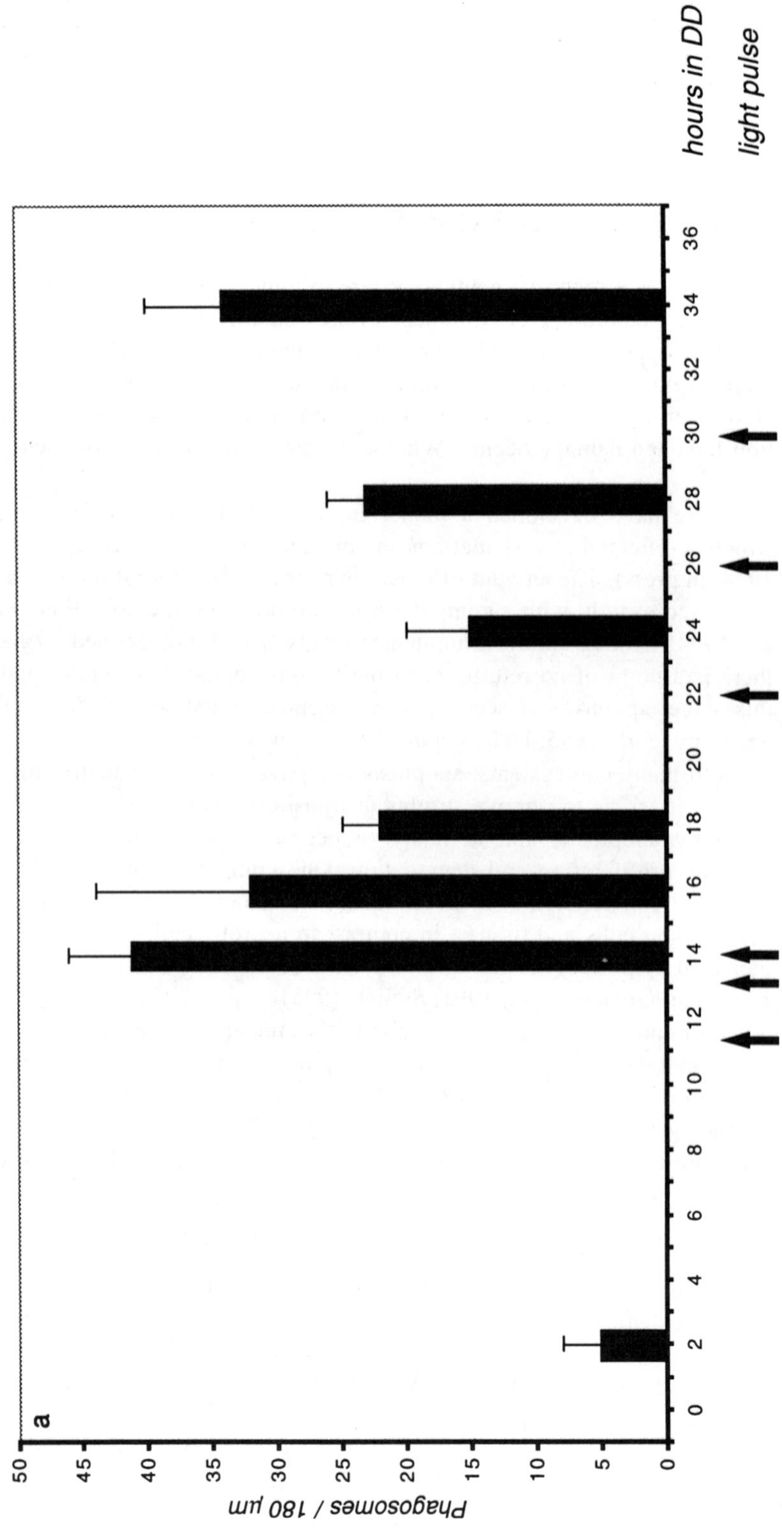
a
Phagosomes / 180 µm
0 5 10 15 20 25 30 35 40 45 50
0 2 4 6 8 10 12 14 16 18 20 22 24 26 28 30 32 34 36
hours in DD
light pulse

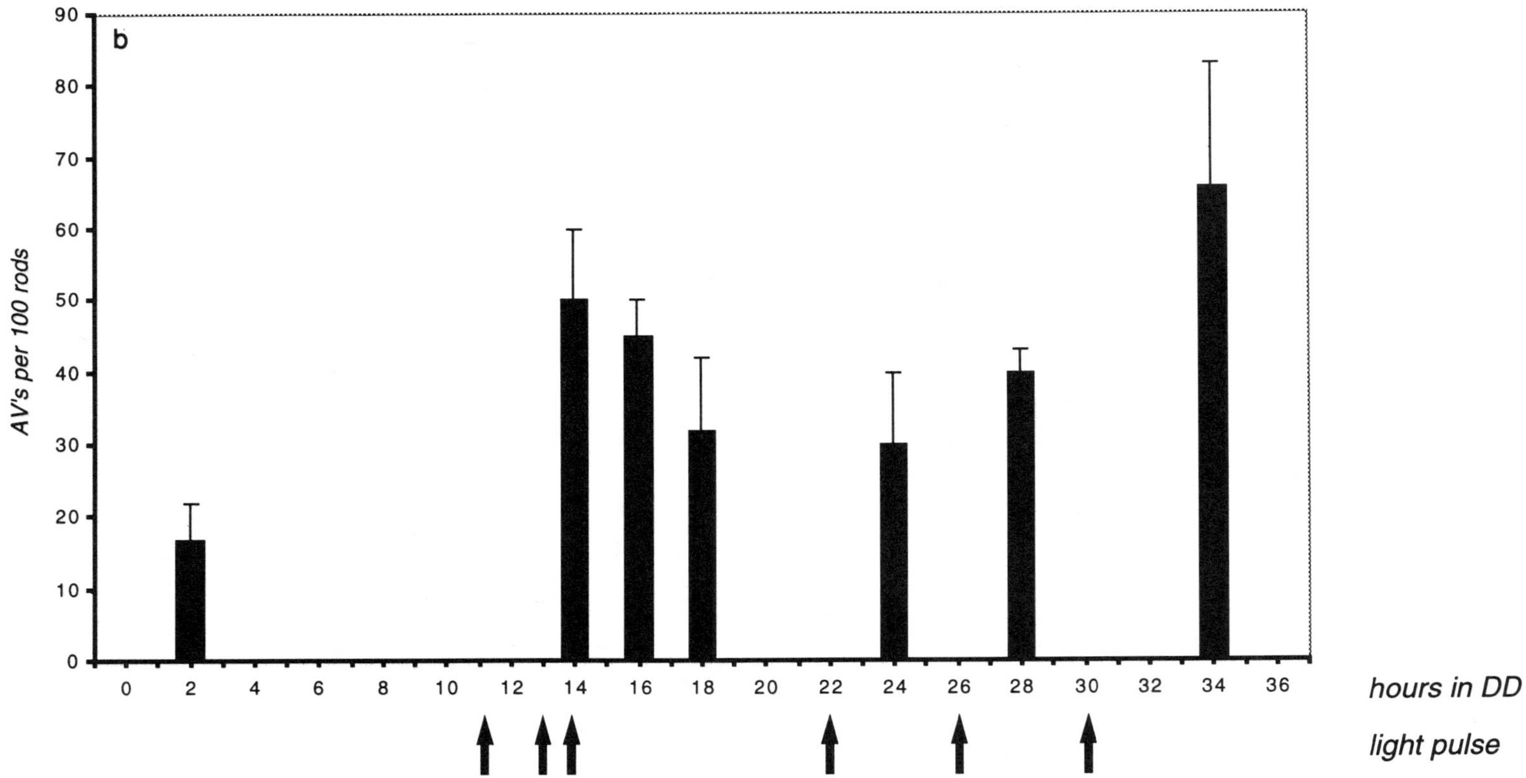

Figure 2. Stimulation of shedding and autophagy responses by 30-minute light exposures to 400 lux during 36 hours of constant darkness. Animals had been kept on a 30- to 50-lux light cycle since birth. The bars represent phagosome and autophagy counts taken from animals sacrificed 1.5–2 hours after the end of the light pulse. Counts at 2 hours of DD represent unstimulated baseline values. Error bars: 1 SD. (a) The highest disc-shedding response is seen after light stimulation in the late dark phase; the lowest response occurs after stimulation late in the subjective day. (b) The highest autophagy response is monitored after stimulation at midnight, the lowest one after stimulation early in the subjective light phase. (Redrawn and modified from Remé *et al.*, 1986.)

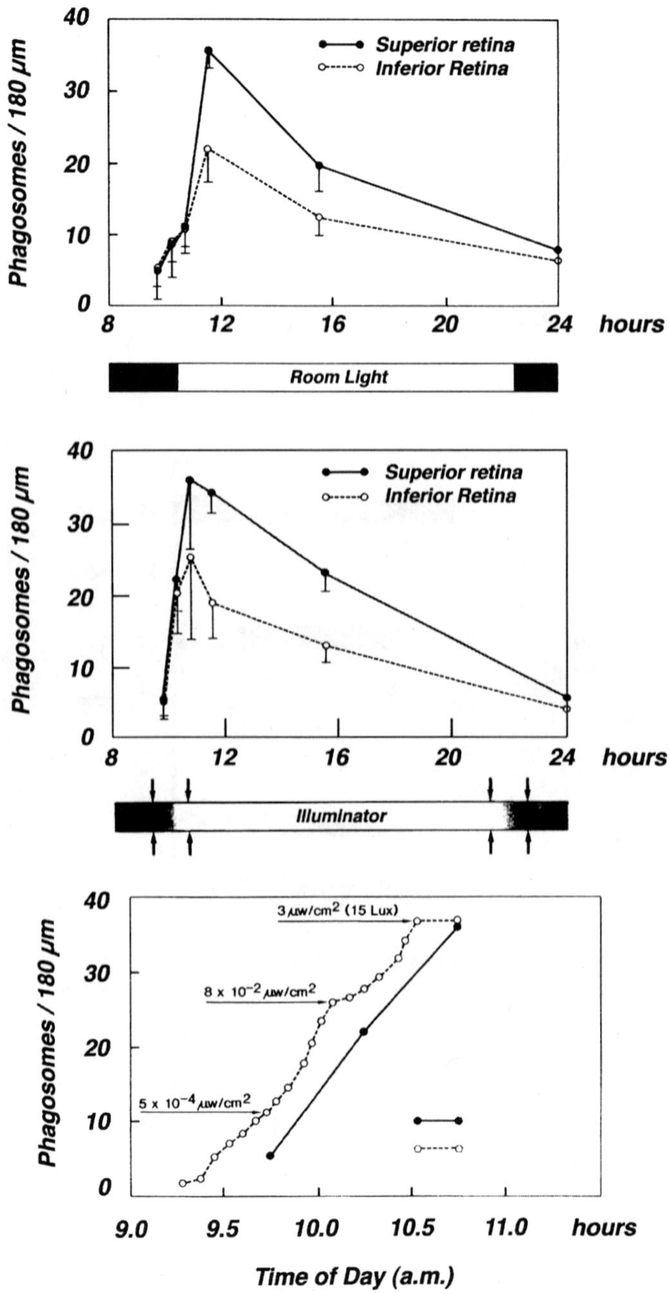

Figure 3. Analysis of disc shedding under laboratory conditions with abrupt light onset and with simulated dawn. (Top) In room light conditions, the well-known burst of shedding is seen about 1 hour after light onset. (Middle) Under simulated dawn, shedding is triggered at light levels far below those encountered under room light conditions at about 0.002 lux and reaches its peak significantly earlier than under room light. The shedding burst reached the same peak number of phagosomes under the two conditions and showed the same rate of phagosome accumulation. (Bottom) Detailed analysis of the intensity transition and phagosome accumulation in simulated dawn. Filled circles and black line: phagosome numbers; open circles and dotted line: illuminance levels.

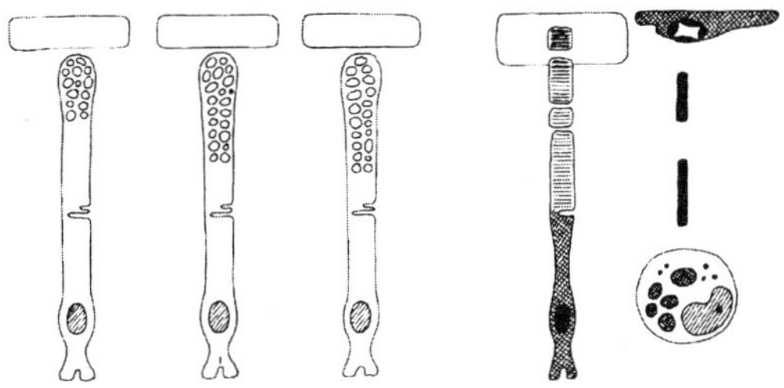

Figure 4. Schematic drawing indicating the light-induced alterations of photoreceptors as a function of illuminance levels. Up to exposures of 1000 lux for 30–45 minutes, changes are confined to rod outer segments and consist of vesiculations and disruption of discs. Those changes are reversible within the physiological turnover of rod outer segments. For illuminances of 1000–3000 lux for 2 hours, apoptotic death of entire rod cells is encountered. Apoptotic bodies, the fragments of apoptotic cells, are removed by macrophages and Müller cells. (Modified from Remé *et al.*, in press).

4. CONCLUSION

We have suggested that light can modulate photoreceptor structure and functional properties. The bursts of disc shedding and autophagy evoked by acute exposures to light that is brighter than that normally encountered in a given animal's environment represent an immediate response (Remé *et al.,* 1991). The constituents necessary for the execution may be continually present within the cells. In contrast, the long-term photostatic adaptation to changing environmental illuminances is likely to be mediated by gene expression.

Thus both the acute and the chronic responses may be seen as adaptive, as having the "aim" of avoiding cell damage and capturing an adequate number of photons within a light-dark cycle. It will be interesting to apply the dawn simulation to the light regimens developed in the laboratory of Williams: will the adaptive bursts of disc shedding and autophagy occur when rats reared in a 3-lux environment are transferred to a 200-lux environment but with the protracted, gradual increase of the dawn simulator?

Apoptotic cell death may also be envisioned as adaptive, but in only one direction, because it is an irreversible process. It is adaptive in the sense that the tissue as a whole is not injured or destroyed, because only single cells are dying. Necrosis in contrast would destroy large areas or the entire tissue in an unregulated fashion. Thus in general, adaptation may fail under conditions when there is "no time" for any process that reduces light sensitivity, including depletion of protective molecules and enzyme systems, in addition to the generation of overwhelming amounts of toxic photochemical products.

ACKNOWLEDGMENTS

Our work was supported by the Swiss National Science Foundation, Nr. 31-40791.94 and Bruppacher Foundation, Zürich, Switzerland.

REFERENCES

Boulos, Z., Macchi, M., and Terman, M. (1996). Twilight transitions promote circadian entrainment to lengthening light-dark cycles. *American Journal of Physiology* 271:R813–R818.

Bush, R. A., Remé, C. E., Terman, M., and Malnoë, A. (1990). Disk shedding and dopamine rhythms under simulated dawn and dusk. *Investigative Ophthalmology and Visual Science Supplement* 31:6.

Chang, G. Q., Hao, Y., and Wong, F. (1993). Apoptosis: final common pathway of photoreceptor death in rd, rds, and rhodopsin mutant mice. *Neuron* 11:595–605.

Hafezi, F., Steinbach, J., Marti, A., Munz, K., Wang, Z.-Q., Wagner, E. F., Aguzzi, A., and Remé, C. E. (1997). The absence of *c-fos* prevents light-induced apoptotic cell death of photoreceptors in retinal degeneration in vivo. *Nature Medicine* 3:346–349.

Hafezi, F., Marti, A., Munz, K., and Remé, C. E. (1997). Light-induced apoptosis: differential timing in the retina and pigment epithelium. *Experimental Eye Research* 64:963–970.

Kroemer, G., Petit, P., Zamzami, N., Vayssière, J. L., Mignotte, B. (1995). The biochemistry of programmed cell death. *FASEB Journal* 9:1277–1287.

LaVail, M. M. (1976). Rod outer segment disc shedding in the rat retina: relationship to cyclic lighting. *Science* 194:1071–1073.

Organisciak, D. T., Xie, A., Wang, H.-M., Jiang, Y.-L., Darrow, R. M., and Donoso, L. A. (1991). Adaptive changes in visual cell transduction protein levels: effect of light. *Experimental Eye Research* 53:773–779.

Penn, J. S., and Anderson, R. E. (1987). Effect of light history on rod outer segment membrane composition in the rat. *Experimental Eye Research* 44:767–778.

Penn, J. S., and Williams, T. P. (1986). Photostasis: regulation of daily photon-catch by rat retinas in response to various cyclic illuminances. *Experimental Eye Research* 44:915–928.

Rapp, L. M., and Williams, T. P. (1980). A parametric study of retinal light damage in albino and pigmented rats. In: Williams, T. P., and Baker, B. N., eds. *The Effects of Constant Light on Visual Processes.* Plenum Press, New York. pp. 135–159.

Remé, C. E., Aeberhard, B., and Schoch, M. (1985). Circadian rhythms of autophagy and light responses of autophagy and disc shedding in the rat retina. *Journal of Comparative Physiology A* 155:669–678.

Remé, C. E., Wirz-Justice, A., Rhyner, A., and Hofmann, S. (1986). Circadian rhythm in the light response of rat retinal disk-shedding and autophagy. *Brain Research* 369:356–360.

Remé, C. E., Wirz-Justice, A., and Terman, M. (1991). The visual input stage of the mammalian circadian pacemaking system. I. Is there a clock in the mammalian eye? *Journal of Biological Rhythms* 6:5–29.

Remé, C. E., Malnoë, A., Jung, H. H., Wei, Q., Munz, K. (1994). Effect of dietary fish oil on acute light-induced photoreceptor damage in the rat retina. *Investigative Ophthalmology and Visual Science* 35:78–90.

Remé, C. E., Weller, M., Szczesny, P., Munz, K., Hafezi, F., Reinboth, J.-J., and Clausen, M. (1995). Light induced apoptosis in the rat retina *in vivo:* morphological features, threshold on a time course. In: Anderson, R. E., LaVail, M. M., and Hollyfield, J. G., eds. *Degenerative Diseases of the Retina.* Plenum Press, New York. pp. 19–25.

Remé, C. E., Hafezi, F., Marti, A., Munz, K., and Reinboth, J. J. (1998). Light damage to retina and pigment epithelium. In: Wolfensberger, M. F. M. T. J., ed. *The Retinal Pigment Epithelium, Current Aspects of Function and Disease.* Oxford University Press, Oxford.

Schremser, J. L., and Williams, T. P. (1995). Rod outer segment (ROS) renewal as a mechanism for adaptation to a new intensity environment. II. Rhodopsin synthesis and packing density. *Experimental Eye Research* 61:25–32.

Steller, H. (1995). Mechanisms and genes of cellular suicide. *Science* 267:1445–1449.

Szczesny, P. J., Munz, K., and Remé, C. E. (1995). Light damage in the rat retina: patterns of acute lesions and recovery. In: Pleyer, U., Schmidt, K., and Thiel, H. J., eds. *Cell and Tissue Protection in Ophthalmology.* Hippokrates Verlag, Stuttgart. pp. 163–175.

Terman, M. (in press). Light on sleep. In: Schwartz, W. J., ed. *Scientific Foundations of Sleep Therapies.* Karger, Basel.

Wei, Q., Remé, C., and Stucki, P. (1993). An advanced image processing and modelling system for the analysis of cell micrographs in morphology. In: Knox, K. T., and Granger, E., Eds. *IS&T/SPIE's Symposium on Electronic Imaging: Science and Technology.* The Society for Imaging Science and Technology, San Jose, California. p. 71.

Wirz-Justice, A., Terman, M., Danilenko, K. V., Terman, J. S., Boulos, Z., and Remé, C. E. (in press). Dawn simulation in animals and humans: effects on melatonin secretion, depression and sleep. *Proceedings of the International Congress on Chronobiology,* Paris, September 1997.

Wyllie, A. H., Kerr, J. F. R., and Currie, A. R. (1980). Cell death: the significance of apoptosis. *International Review of Cytology* 68:251–300.

Young, R. W. (1984). Cell death during differentiation of the retina in the mouse. *Journal of Comparative Neurology* 229:362–373.

13

PHOTOSTASIS IN PLANTS

Mechanisms and Regulation

Anastasios Melis

Department of Plant Biology
411 Koshland Hall
University of California
Berkeley, California 94720-3102

1. INTRODUCTION

Life on earth is sustained by oxygenic photosynthesis, a process that involves the absorption and utilization of light energy from the sun. The chemical energy released by this endergonic process results in the oxidation of water molecules, the release of oxygen, and the generation of reductant (reduced ferredoxin and NADPH) and high-energy phosphate bond (ATP). This process takes place in the thylakoid membranes of chloroplasts. It requires coordinated interaction between a large number of electron carrier compounds and enzymatic proteins that facilitate the lateral transport of electrons in the thylakoid membrane from H_2O molecules to ferredoxin and $NADP^+$ and the transverse transport of H^+ from the surrounding stroma to the enclosed space of the thylakoid lumen. The electron-transport components are localized in the thylakoid membrane and catalyze the transfer of electrons laterally in the plane of the membrane from the grana regions, where oxidation of H_2O takes place, to the stroma-exposed thylakoid regions, where the reduction of ferredoxin and the generation of ATP take place (Andersson and Anderson, 1980; Anderson and Melis, 1983). Functionally, electron transport occurs from intermediate to intermediate in a sequential manner formulated as the Z-scheme of photosynthesis (Hill and Bendall, 1960). The overall process of electron transport from H_2O to $NADP^+$ is strongly endergonic and is realized through the input and utilization of light energy in two distinct steps. The absorption of light and the conversion of excitation energy to chemical energy takes place in photosystem II (PSII) and photosystem I (PSI) in the thylakoid membrane (Duysens *et al.,* 1961). Light energy in PSII facilitates the generation of a strong oxidant capable of oxidizing H_2O molecules. Light energy in PSI facilitates the generation of a strong reductant capable of reducing ferredoxin and $NADP^+$.

This chapter presents a review of the functions of three distinct acclimation mechanisms designed to maintain the high quantum efficiency of photosynthesis and sustain

Photostasis and Related Phenomena, edited by Williams and Thistle
Plenum Press, New York, 1998

growth and productivity in plants under contrasting light environments. A discussion of the significance of such photostasis in plants is also offered.

2. PHOTOSTASIS IN THE ABSORPTION OF LIGHT BY CHLOROPLASTS: ADJUSTMENTS IN THE LIGHT-HARVESTING CHLOROPHYLL ANTENNA SIZE OF PHOTOSYNTHESIS

Photochemistry in PSII and PSI depends on the absorption of light by chlorophyll molecules. In each photosystem, chlorophylls and other accessory pigments act cooperatively in the absorption of incoming electromagnetic radiation (Emerson and Arnold, 1932a, b). The concept of the photosynthetic unit, proposed by Gaffron and Wohl (1936), stipulates that distinct assemblies of photosynthetic pigment molecules serve as antennae for the collection of light and as a conducting medium for excitation migration toward a photochemical reaction center. It is now recognized that distinct chlorophyll-protein complexes are associated with the reaction-center proteins of PSII and PSI and perform the functions of light absorption and excitation-energy transfer to the respective photochemical reaction center (Pichersky and Jansson, 1996; Simpson and Knoetzel, 1996). Up to 500 chlorophyll (Chl) *a* and Chl *b* molecules are found in association with PSII, whereas the Chl antenna of PSI may contain up to 250 mostly Chl *a* molecules. These numbers are not fixed, however. Rather, the Chl antenna size of the photosystems is variable, and it changes in a manner that is inversely related to the level of irradiance under which a plant is grown, suggesting a photostasis in plants, i.e., the existence of a photoacclimation mechanism in chloroplasts designed to maintain a fairly constant absorption of light on a per-chloroplast basis.

Photosynthetic light-harvesting complexes have been the subject of extensive research because of interest in their primary function, namely the absorption of light and the transfer of excitation energy to a photochemical reaction center. As a result, a great deal of information has accumulated on the pigment cofactors, structure, and folding of the corresponding proteins, molecular and supramolecular organization in the thylakoid membrane, and their association with photosystem II and photosystem I. In brief, associated with the reaction-center core of PSII is the sizable auxiliary chlorophyll *a-b* light-harvesting complex (LHC-II). The LHC-II is composed of up to ~30 individual chlorophyll-proteins, each binding seven Chl *a* and six Chl *b* molecules (Thornber *et al.,* 1988; Morrissey *et al.,* 1989; Kühlbrandt and Wang, 1991; Kühlbrandt *et al.,* 1994). Polypeptides of the LHC-II are coded for by the nuclear *Lhcb* gene family (Jansson *et al.,* 1992; Pichersky and Jansson, 1996; Simpson and Knoetzel, 1996).

When denatured, proteins of the LHC-II migrate in the 30–21 kDa region in SDS-PAGE and contain six distinct subunit polypeptides, the product of nuclear genes *Lhcb1* through *Lhcb6* (Darr *et al.,* 1986; Peter and Thornber, 1991; Harrison and Melis, 1992; Jansson *et al.,* 1992). Collectively, the LHC-II macrocomplex may bind over 400 Chl (*a* and *b*) molecules in the fully assembled light-harvesting antenna of PSII (Melis and Anderson, 1983; Glazer and Melis, 1987; Melis, 1996). Figure 1 shows a model of the organization and protein subunit stoichiometry of the Chl *a-b* light-harvesting antenna (LHC-II) of PSII. Noteworthy is the association of monomeric Lhcb3, Lhcb4, Lhcb5, and Lhcb6 subunits with the core complex of PSII (CP47 and CP43) and the trimer $((b1)_2(b2)_1)$ configuration of the more abundant but peripheral Lhcb1 and Lhcb2 protein subunits. Note also the presence of a unique configuration trimer $((b1)_3)$ localized in close proximity to the Lhcb3 protein subunit(s).

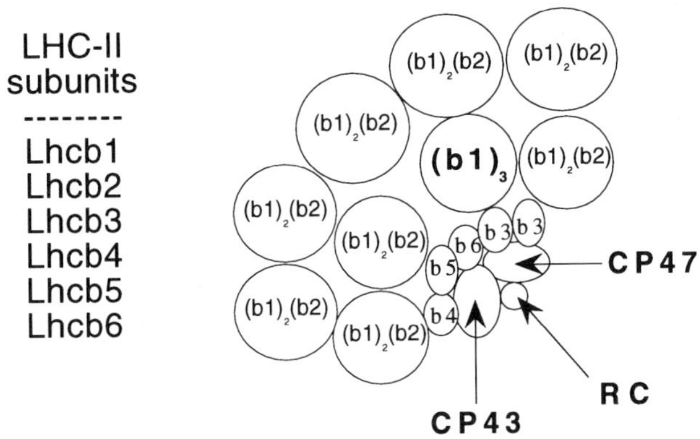

Figure 1. Schematic showing the organization and protein subunit stoichiometry of the Chl *a-b* light-harvesting antenna (LHC-II) in the green alga *Dunaliella salina.* Each polypeptide subunit, from Lhcb1 (b1) through Lhcb6 (b6), binds ~13 Chl (*a* and *b*) molecules. The PSII core complex, including chlorophyll proteins CP47 and CP43, and reaction-center proteins (RC) collectively bind ~40 Chl *a* molecules (Glick and Melis, 1988). The above Chl antenna configuration of PSII is encountered in cells grown in low light and contains about 460 Chl (*a*+*b*) molecules. Excitation energy absorbed by any Chl molecule in the LHC-II supramolecular structure is unidirectionally transferred to the photochemical reaction center (RC).

Associated with the reaction center core of PSI is a smaller auxiliary chlorophyll *a-b* light-harvesting complex (LHC-I). The LHC-I contains only 100–150 Chl (*a* and *b*) molecules. Chlorophyll-protein subunits of the LHC-I also bind ~13 chlorophylls, but in contrast to the Chl composition of the LHC-II, there are 10–11 Chl *a* and only 2–3 Chl *b* molecules in the LHC-I (Lam *et al.,* 1984). Polypeptides of the LHC-I are coded for by the nuclear *Lhca* gene family (Jansson *et al.,* 1992; Pichersky and Jansson, 1996; Simpson and Knoetzel, 1996). When denatured, they migrate in the 25–19 kDa region in SDS-PAGE, are immunologically distinct from those of the LHC-II, and may contain up to four distinct members (the products of nuclear genes *Lhca1* through *Lhca4*) (Anderson *et al.,* 1983; Haworth *et al.,* 1983; Lam *et al.,* 1984; Ikeuchi *et al.,* 1991; Jansson *et al.,* 1992; Knoetzel *et al.,* 1992).

The level of irradiance during plant growth modulates the size and composition of the light-harvesting antennae of the photosystems (reviewed by Melis, 1996). In general, low light intensity promotes larger Chl antenna size for both PSII and PSI (larger photosynthetic unit size). For example, Figure 1 depicts the Chl *a-b* light-harvesting antenna size encountered in *Dunaliella salina* (a green alga) grown in low light. Conditions of higher light intensity elicit a smaller Chl antenna. This adjustment in the Chl antenna size of the photosystems comes about because of changes in the size of LHC-II and LHC-I (Leong and Anderson, 1984; Larsson *et al.,* 1987; Sukenik *et al.,* 1988; Smith *et al.,* 1990). In PSII, these changes are implemented through the assembly/association of a variable number of $((b1)_2(b2)_1)$ trimers with the LHC-II (Morrissey *et al.,* 1989; Mawson *et al.,* 1994). Figure 2 shows the Chl antenna organization in *D. salina* grown under moderate irradiance. It is seen that, under such conditions, the peripheral $((b1)_2(b2)_1)$ trimers have not assembled. Figure 3 shows the Chl antenna organization in *D. salina* grown under high irradiance. In this case, none of the $((b1)_2(b2)_1)$ trimers assemble in the thylakoid membrane. Under high irradiance, light harvesting for PSII takes place through Chl (*a* and *b*) molecules bound within the $(b1)_3$ trimer and within the less abundant Lhcb3-Lhcb6 pro-

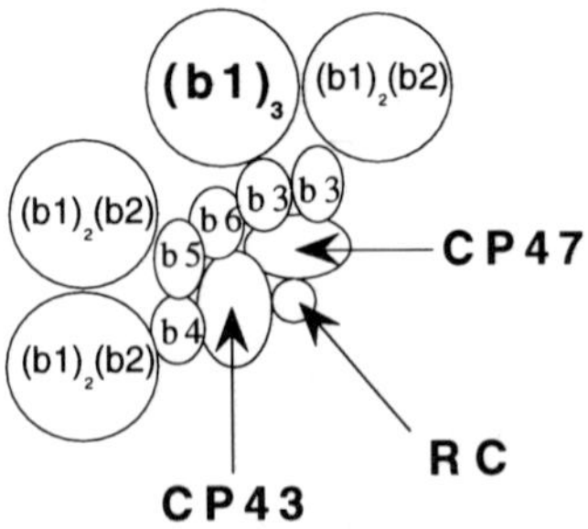

Figure 2. Schematic showing the organization and protein subunit stoichiometry of the Chl *a-b* light-harvesting antenna (LHC-II) of PSII in cells grown in medium light, containing about 250 Chl (*a*+*b*) molecules. For other details see legend of Figure 1.

teins (Ley and Mauzerall, 1982; Sukenik *et al.,* 1988; Smith et al., 1990; Harrison and Melis, 1992; Mawson *et al.,* 1994; Webb and Melis, 1995). Under chronically high irradiance, a very small PSII Chl antenna contains only the Lhcb3 subunits of the LHC-II and is limited to a total of about 60 Chl molecules (not shown) (Smith *et al.,* 1990; Harrison and Melis, 1992).

It has been demonstrated that, in green algae, the Chl antenna size of PSII can vary from a maximum of ~500 to as low as ~60 chlorophyll molecules. Smaller-amplitude variations in the Chl antenna size of PSI (from ~250 to ~100) have also been reported (Ley and Mauzerall, 1982; Smith *et al.,* 1990). Figure 4 shows the kinetics of such change for PSII in light-shift experiments with *D. salina.* It is interesting to observe that cells promptly increase the Chl antenna size when light intensity is lowered. The converse is not true, however, suggesting that, once assembled, the LHC-II is stable and that a lowering of the Chl antenna size would occur only over time during further growth and development of the organism. These observations indicate an organismal readiness to increase the biosynthesis/assembly of the LHC-II and an apparent inability to degrade assembled LHC-II proteins actively.

The response described in Figure 4 appears to be highly conserved in all photosynthetic organisms examined. The phenomenon is a compensation response in chloroplasts—adjustments in the chlorophyll antenna size of the photosystems are inversely related to the incident irradiance—but a rigorous analysis of this phenomenon has not yet been undertaken that would test for the reciprocity between light intensity and the antenna-size response. Moreover, little is known about the regulation of this phenomenon at the molecular and membrane levels: work by Falkowski and coworkers (LaRoche *et al.,* 1991; Escoubas *et al.,* 1995) suggested that changes in the abundance of LHC-II are preceded by changes in the abundance of the mRNA coding for these proteins. Independently, there are indications that light intensity down-regulates the rate of chlorophyll

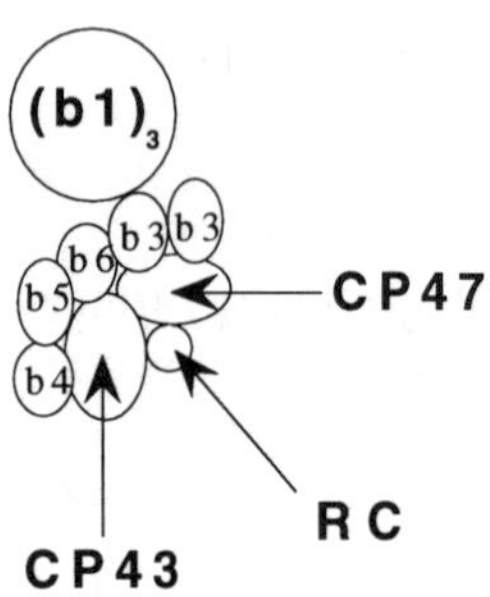

Figure 3. Schematic showing the organization and protein subunit stoichiometry of the Chl *a-b* light-harvesting antenna (LHC-II) of PSII in cells grown in high light, containing about 140 Chl (*a*+*b*) molecules. For other details see legend of Figure 1.

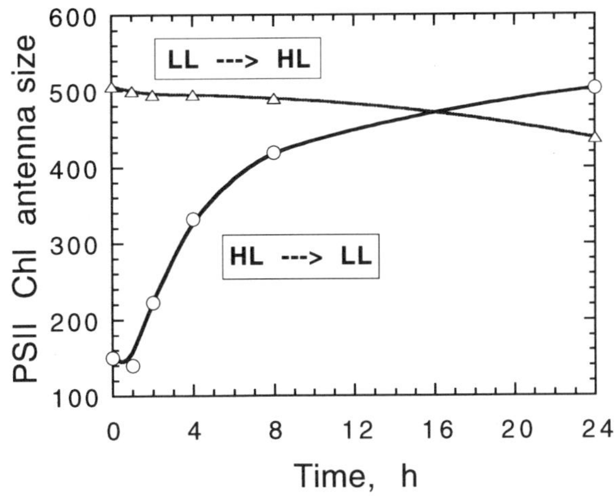

Figure 4. Adjustments in the PSII chlorophyll antenna size of the green alga *Dunaliella salina* in response to changes in the level of irradiance. LL: 50 µmol photons m^{-2} s^{-1} incident to the culture. HL: 2000 µmol photons m^{-2} s^{-1} incident to the culture. Note the prompt increase in the Chl antenna size following a shift from HL to LL and the extended lag following a shift from LL to HL.

biosynthesis. In turn, Chl availability may define the Chl-protein species that would assemble in the chloroplast (Falbel *et al.*, 1996; Ohtsuka *et al.*, 1997). According to this concept, Chl-protein assembly in the chloroplast occurs with the following distinct hierarchy: PSII core > PSI core > LHC monomers > LHC inner trimers > LHC peripheral trimers. A limited availability of Chl under moderate or high irradiance will permit the assembly of the PSII and PSI core complexes, but lack of sufficient Chl will not be conducive to the assembly of LHC peripheral trimers, resulting in a smaller Chl antenna.

3. PHOTOSTASIS IN THE LIGHT UTILIZATION EFFICIENCY OF PHOTOSYNTHESIS: ADJUSTMENT AND OPTIMIZATION OF THE PSII AND PSI STOICHIOMETRY IN CHLOROPLASTS

The light-utilization efficiency of oxygenic photosynthesis depends on the coordinated interaction of the two photosystems in the linear electron-transport process. In green plants, PSII absorbs light by means of Chl *b* and short-wavelength absorbing Chl *a* molecules, whereas in cyanobacteria and red algae phycocyanin and phycoerythrin are the main pigments absorbing light for PSII. Photosystem II uses the excitation energy to drive the oxidation of water and to reduce plastoquinone (PQ). Photosystem I employs mostly Chl *a* pigments, and especially long-wavelength absorbing Chl *a*, to acquire the energy needed for the transport of electrons from plastohydroquinone, and from the cytochrome *b-f* complex, to ferredoxin. The quantum efficiency of photosynthesis in many plant species is very close to a theoretical upper limit of 0.125 mol O_2 evolved per mol photon absorbed (Ley and Mauzerall, 1982; Björkman and Demmig, 1987; Evans, 1987), independent of the light climate in which plants grow. This is a remarkable feature of the photosynthetic apparatus, given the requirement for a coordinated transport of electrons by the two photosystems under the variety of contrasting light intensities that prevail in different plant environments (Björkman and Ludlow, 1972; Kirk, 1983; Terashima and Saeki, 1983) and the fact that spectroscopically different pigments absorb light for PSII and PSI in the thylakoid membrane of oxygenic photosynthesis. This feature suggested the existence of a mechanism in chloroplasts designed to correct for the uneven absorption of light by the two photosystems and, thus, to maintain the coordination of electron transport between the two photoreactions.

The spectral distribution of solar quantum irradiance at the surface of the earth is characterized by approximately constant emission in the visible wavelength region between 450 and 700 nm and by progressively declining intensities in the violet and ultraviolet regions (400–280 nm). This emission profile, however, is attenuated extensively as filtering creates strong gradients in light quality within a single leaf (Terashima and Saeki, 1983; Vogelmann, 1989, 1993), within the canopy of a single tree or the canopy of a forest (Björkman and Ludlow, 1972), and in the aquatic environment (Kirk, 1983).

Direct sunlight is more or less evenly distributed among the photosynthetic pigments, but in the aquatic environment, where blue-green and red algae live, red wavelengths are limited, so light would sensitize the phycocyanin and phycoerythrin antenna of PSII (480–650 nm region) preferentially over the Chl *a* antenna of PSI (670–720 nm). Clearly, in this light environment, PSII will have a relative advantage over PSI in the processes of light absorption and electron transport. In the terrestrial environment, sunlight filtered through the leaf, or the canopy, is selectively depleted in wavelengths where the PSII accessory pigments absorb and contains greater relative emissions in the long-wavelength region (e.g., 690–720 nm) where PSI pigments absorb exclusively. Thus, in the inner leaf, and within the canopy, absorption of light and electron transport by PSI will be favored over those by PSII.

Imbalance in the absorption of light between PSII and PSI in the photosynthetic apparatus would cause inefficiency in the linear electron-transport process and would lower the productivity of photosynthesis. Photosynthetic organisms alleviated this potential inefficiency by acquiring variable photosystem stoichiometries in a way that compensates for the uneven absorption of light by PSII and PSI in a given environment. The major classes of photosynthetic organisms, depending on the light-harvesting antennae they possess and the light environments they occupy, therefore show photosystem stoichiometry ratios that deviate significantly from unity. Table 1 presents a summary of the PSII/PSI ratio in diverse photosynthetic organisms. It is evident that cyanophytes and rhodophytes, which possess phycocyanin and phycoerythrin as the major PSII pigments, need less PSII than PSI in their thylakoid membranes. Conversely, green algae and higher plants require more PSII than PSI in their thylakoid membrane because PSI has a relative advantage over PSII in the processes of light absorption and electron transport.

Phycocyanin- and phycoerythrin-containing organisms have a PSII/PSI < 1, whereas Chl *b*–containing organisms have, in general, a PSII/PSI > 1 (from Fujita *et al.*, 1985; Cunningham *et al.*, 1990; Kim *et al.*, 1993a; Melis *et al.*, 1996).Beyond the apparent lack of uniformity in the PSII/PSI ratio among diverse photosynthetic organisms, plants may encounter changing light qualities in the course of growth. This again is a potentially suboptimal situation that would lower the quantum yield and productivity of photosynthesis. Chloroplasts compensate for shifting light qualities by actively adjusting and optimiz-

Table 1. The PSII/PSI ratios of diverse photosynthetic organisms

Organism	PSII/PSI ratio
Cyanophytes (blue-green algae)	0.45/1
Rhodophytes (red algae)	0.5/1
Chlorophytes (green algae)	1.2/1
Angiosperms (green plants)	1.8/1

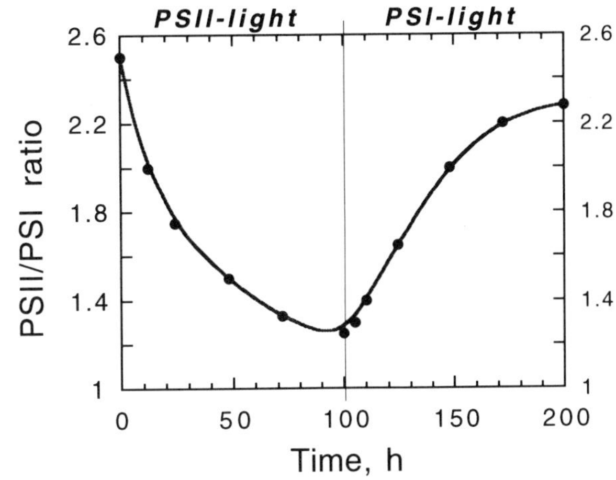

Figure 5. The response of photosystem stoichiometry (PSII/PSI) to changes in the light regime during plant growth. Pea plants already acclimated to PSI light were switched (at zero time) to a predominantly PSII-light environment. After 100 hours, the same plants were switched back to a predominantly PSI-light environment. Under direct sunlight, the PSII/PSI ratio was 1.8/1.

ing the photosystem stoichiometry in the thylakoid membrane. This adjustment is a long-term but fully reversible response. Figure 5 shows results of an experiment in which pea plants were grown under predominantly PSI light (PSII/PSI = 2.5:1). At zero time, they were switched to a predominantly PSII light environment (Fig. 5, left panel). The PSII/PSI stoichiometry ratio of nearly expanded leaves was decreased as a function of time until it reached a new steady state of about 1.3:1 with a half-time of about 20 hours. Upon transition to PSI light conditions (Fig. 5, right panel), the PSII/PSI ratio gradually increased to the value typical for these plants grown in PSI light (Kim *et al.*, 1993a). A similar response by photosystem stoichiometry to light quality was obtained with cyanophytes (Melis *et al.*, 1989; Murakami *et al.*, 1997).

Changes in the PSII/PSI ratio are accompanied by structural alterations in thylakoid membranes. Plants grown under PSI light have extensive grana stacks, where PSII is localized, and short stroma-exposed thylakoids, where PSI is localized (Anderson and Melis, 1983). Plants grown under PSII light have smaller grana stacks and extensive stroma-exposed lamellae (Melis and Harvey, 1981; Melis, 1984; Deng *et al.*, 1989; Anderson *et al.*, 1992). Similarly, cyanophytes and red algae under PSI light have higher ratios of phycobilin to Chl, whereas the converse is true under PSII light conditions (Myers *et al.*, 1980; Fujita *et al.*, 1985; Manodori and Melis, 1986; Melis *et al.*, 1989; Cunningham *et al.*, 1990).

These results suggested the existence of a widely conserved mechanism in all photosynthetic organisms that is designed to sense and correct long-term imbalance in light absorption by the two photosystems through adjustment and optimization of the PSII/PSI stoichiometry ratio. From the evolutionary standpoint, photosynthetic organisms possessing such acclimation mechanism might enjoy a significant selective advantage over others with a fixed photosystem stoichiometry in their thylakoid membranes (Chow *et al.*, 1990b).

Table 2 provides examples of the PSII/PSI ratio in predominantly in plants grown in PSI or PSII light and illustrates the effect of PSII/PSI ratio adjustments on the quantum yield (Φ) of photosynthesis.

Plants were allowed to acclimate for several days under either predominantly PSI-light or PSII-light conditions. These light qualities are defined (Glick *et al.*, 1986) as light absorbed preferentially by PSI (i.e., light absorbed by Chl *a* more than by Chl *b*) or by PSII (i.e., by Chl *b* or phycocyanin more than by Chl *a*). The numbers in parentheses refer to the PSII/PSI ratios of the samples. The schematic in Figure 6 summarizes aspects of the

Table 2. Quantum yield (Φ) of photosynthesis in PSI-light- and PSII-light-acclimated *Pisum sativum* and *Synechococcus* 6301[a]

	Φ Measured under PSI irradiance	Φ Measured under PSII irradiance
Pisum sativum (pea)[b]		
PSI-light leaves	0.108 (2.5/1)	0.080 (2.5/1)
PSII-light leaves	0.091 (1.2/1)	0.097 (1.2/1)
Difference in Φ	19%	21%
Synechococcus 6301[c]		
PSI-light cells	1.00 (0.7/1)	0.60 (0.7/1)
PSII-light cells	0.75 (0.3/1)	1.00 (0.3/1)
Difference in Φ	25%	40%

[a]From Murakami and Fujita, 1988; Melis *et al.*, 1989; Chow *et al.*, 1990b.
[b]Absolute Φ units in mol oxygen evolved per mol photon absorbed.
[c]Relative Φ units.

mechanism that regulates the photosystem stoichiometry in chloroplasts. Light-quality (signal) perception for photosystem stoichiometry adjustment occurs at the thylakoid membrane level as differential sensitization of the auxiliary pigments associated with PSII and PSI. This conclusion is supported by the observation that, in the absence of accessory pigments (in Chl *b*-less and phycobilin-less mutants), the photosystem ratio cannot be influenced by the quality of illumination during plant growth (Kim *et al.,* 1993a).

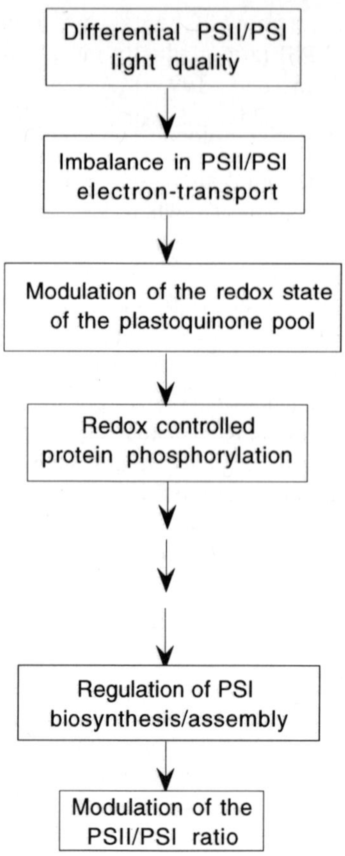

Figure 6. A feedback control mechanism for the adjustment and optimization of the PSII/PSI stoichiometry in thylakoid membranes. Light-quality (signal) is perceived as a differential sensitization of PSII and PSI pigments. The first step in the signal transduction pathway involves imbalance in the rates of electron transport between PSII and PSI, which effects a modulation in the oxidation-reduction state of the plastoquinone pool. The latter regulates the activity of a redox-controlled protein kinase. The signal-transduction pathway involves regulation of the biosynthesis/assembly of PSI in chloroplasts, whereas PSII is constitutively expressed.

Uneven absorption of light by auxiliary pigments creates imbalance in the rates of electron transport by PSII and PSI. Light absorbed preferentially by PSII leads to an overreduction of the intermediate plastoquinone (PQ) pool, whereas preferential sensitization of PSI causes oxidation of the intermediate PQ pool. The redox state of the PQ pool in steady-state electron transport is believed to regulate the function of protein kinase(s) (Coughlan, 1988; Allen *et al.,* 1981; Allen, 1992) that reversibly phosphorylate a number of chloroplast proteins, a condition required for adjustments in the PSII/PSI ratio (Chow *et al.,* 1987, 1990a; Fujita *et al.,* 1989; Murakami and Fujita, 1988, 1991; Bennett, 1991; Anderson and Chow, 1992). Reduction of the PQ pool promotes the protein kinase activity, whereas oxidation of the PQ pool inhibits it. Eventually, photosystem stoichiometry adjustments are implemented as adjustments in the relative amount of PSI in thylakoids, whereas PSII is constitutively expressed, and its concentration remains fairly constant in the chloroplast (Fujita and Murakami, 1987; Aizawa *et al.,* 1992; Melis *et al.,* 1996).

4. PHOTOSTASIS IN THE FUNCTION OF PHOTOSYSTEM II: THE PSII DAMAGE AND REPAIR CYCLE

Development of the ability of PSII to utilize water molecules from which to extract electrons and protons was undoubtedly a significant event in the evolution of life on earth. It contributed to the gradual accumulation of oxygen in the atmosphere, thereby permitting the evolution of oxidative phosphorylation. Many scientists rightfully refer to PSII as "the engine of life on earth."

From the biochemical point of view, PSII is a specialized water-to-plastoquinone oxidoreductase. This specialized enzyme features a rather sizable holocomplex consisting of up to 40 transmembrane and peripheral proteins (Fig. 7). Most of the transmembrane proteins function as chlorophyll-protein light-harvesting complexes (Fig. 1). The functional center of the holocomplex is the so-called D1/D2 heterodimer reaction center proteins (Nanba and Satoh, 1987) that perform the light utilization, water oxidation and primary electron transport reactions in PSII. (D1 is the 32-kDa reaction-center protein of photosystem II coded by the chloroplast *psbA* gene; D2 is the 34-kDa reaction-center protein of photosystem II coded by the chloroplast *psbD* gene.) These highly specialized functions of PSII take place in a microenvironment where oxygen abounds and where photons, in the form of excitation energy, may arrive at a rate of up to 10,000 per second. The transient formation of strong oxidants, the abundance of oxygen, and the presence of excitation energy are conditions that may lead to photo-oxidative damage. Indeed, such damage occurs frequently within the reaction center of PSII. It causes an irreversible inhibition in the function of the D1 reaction-center protein and stops photosynthesis (Powles, 1984; Demeter *et al.,* 1987).

Despite 2–3 billion years of evolution, organisms with oxygenic photosynthesis have not been able to prevent this photodamage from occurring. To date, every oxygen-evolving photosynthetic organism known, from cyanophytes to C4 plants, is subject to this irreversible photodamage. Nature, however, has devised a repair mechanism that restores the functional status of PSII. This PSII repair cycle is of great importance for the maintenance and productivity of photosynthesis. It has been estimated that, without it, photodamage would lower the yield of photosynthesis to less than about 5% of that measured presently. Clearly, life on earth would have been quite different in the absence of the PSII repair process. The mechanism of the process is the selective removal (degradation) and

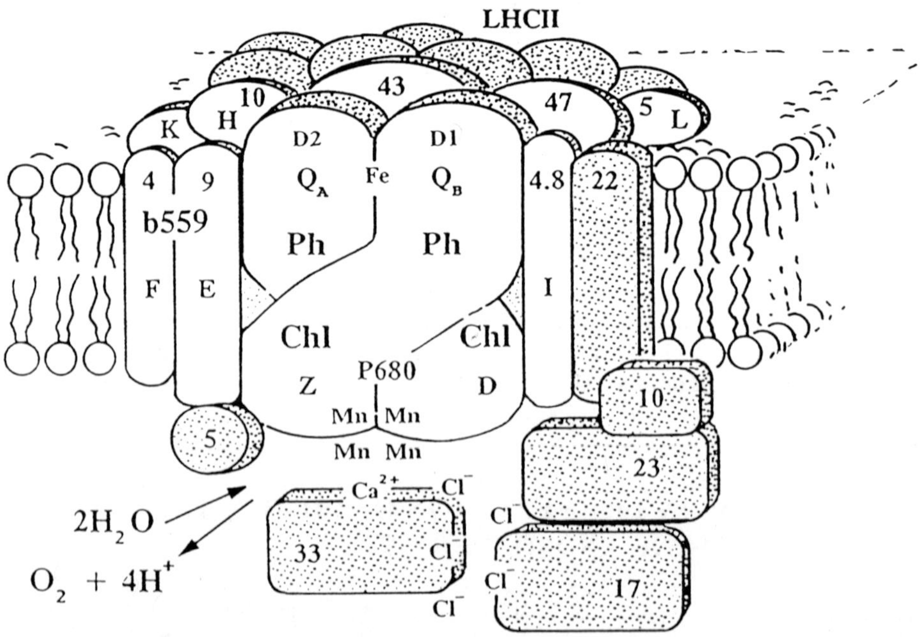

Figure 7. A three-dimensional schematic of the PSII holocomplex showing polypeptide constituents and electron-transport intermediates. The reaction-center heterodimer proteins (D1/D2) bind the tetranuclear Mn cluster, the photochemical Chl *a* molecule (P680), the primary electron-acceptor pheophytin (Ph), and the bound quinones Q_A and Q_B. Also shown are the Chl *a*–containing 43= and 47-kDa proteins (known as CP43 and CP47) and the many Chl proteins of the LHC-II (reprinted by permission from Barber, 1989).

replacement (via *de novo* biosynthesis) of an individual protein (the photodamaged D1 reaction-center polypeptide) in the PSII multi-protein holocomplex, which is localized in the inaccessible region of the grana membranes in chloroplasts.

The PSII repair cycle is a multi-step enzymatic process that is initiated upon D1 photodamage and is terminated with the activation of newly reconstituted PSII reaction centers. Photodamage to a reaction-center D1 protein is followed by partial disassembly of the PSII holocomplex, degradation of photodamaged D1, *de novo* D1 biosynthesis and insertion into the thylakoid membrane, reassembly of the PSII reaction-center complex, and activation of the electron-transport process through the reconstituted D1/D2 heterodimer. The sequence of events in the PSII damage and repair cycle is presented in Figure 8. A number of known biochemical reactions are associated with the mechanism of the cycle:

(a) A reversible phosphorylation of the D1 protein may regulate the rate of inactive D1 degradation in higher-plant chloroplasts (Callahan *et al.,* 1990; Aro *et al.,* 1992; Elich *et al.,* 1992).
(b) The PSII holocomplex is disassembled and, and a smaller PSII core complex is excluded from the membrane of the grana partition regions. This step serves both to unfold the PSII holocomplex and to expose the D1/D2 heterodimer to the aqueous stroma-phase where removal of the photodamaged D1 and replacement by a *de novo* synthesized protein occurs.
(c) Photodamaged reaction centers have been identified as distinct 160-kDa complexes (Kim *et al.,* 1993b; Baroli and Melis, 1996). The 160-kDa complex is

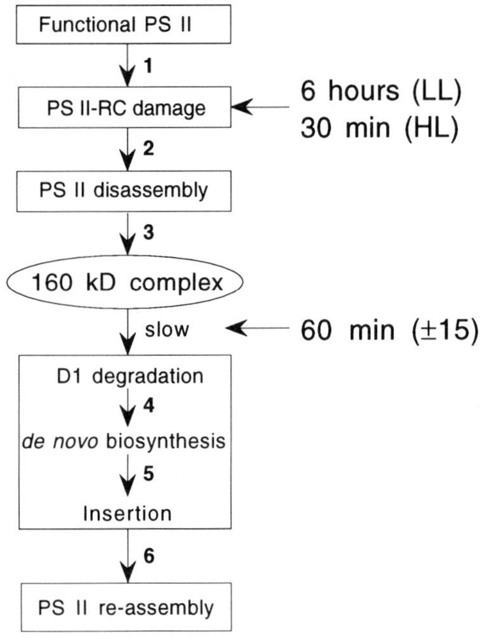

Figure 8. Temporal sequence of events in the PS II damage, disassembly, and D1 degradation and replacement. The rate of photodamage to PS II (step 1) is directly proportional to the incident light intensity (Baroli and Melis, 1996). The rate of PS II disassembly (step 2) is not limiting under a broad range of incident intensities. Direct D1 degradation (step 3) and *de novo* D1 biosynthesis (step 4) become rate limiting under high light intensities. Under these conditions, 160-kDa complexes accumulate in thylakoids (step 3). D1 degradation (slow) is the rate-limiting step of the PSII repair process under high light intensities. It is estimated to occur with a half-time of 60 ± 15 min.

clearly an intermediate in the PSII damage and repair cycle; it presumably helps to hold disassembled PSII in a stable configuration until the D1 degradation and replacement can safely take place.

(d) D1-specific protease(s) is (are) probably involved in the degradation and removal of photodamaged D1 (Salter *et al.,* 1992).

(e) A reversible D1 palmitylation step facilitates the insertion of the nascent D1 polypeptide into the PSII reaction-center complex (Mattoo and Edelman, 1987).

(f) The plastoquinone reduction activity is activated (Guenther *et al.,* 1990; Neale and Melis, 1990), concomitant with the reassembly of the PSII holocomplex and incorporation into the grana.

This temporal and spatial sequence of events is consistent with the known properties of the D1 protein turnover (Mattoo *et al.,* 1984; Mattoo and Edelman, 1987) and also with the heterogeneity in PSII configuration and function (Melis, 1991; Aro *et al.,* 1993). Our limited knowledge does not, however, include information about the identity and regulation of enzymes that perform the PSII repair or about the genes that code for the repair proteins. Moreover, it is very likely that additional (currently unknown) biochemical steps are involved. Identification of the PSII repair enzymes and elucidation of the mechanistic and regulatory aspects of the PSII repair process is a long-term goal of investigators in this field.

ACKNOWLEDGMENTS

Related work in the laboratory of the author is supported by grant 96-35100-3201 from the USDA National Research Initiative, Competitive Grants Program.

LITERATURE CITED

Aizawa, K., Shimizu, T., Hiyama, T., Satoh, K., Nakamura, Y., and Fujita, Y. (1992). Changes in composition of membrane proteins accompanying the regulation of PSI/PSII stoichiometry observed with *Synechocystis* PCC 6803. *Photosynthesis Research* 32:131–138.

Allen, J. F. (1992). Protein phosphorylation in regulation of photosynthesis. *Biochimica et Biophysica Acta* 1098:275–335.

Allen, J. F., Bennett, J., Steinback, K. E., and Arntzen, C. J. (1981). Chloroplast protein phosphorylation couples plastoquinone redox state to distribution of excitation energy between photosystems. *Nature* 291:21–25.

Anderson, J. M., and Chow, W. S. (1992). A regulatory feedback mechanism for light acclimation of the photosynthetic apparatus: Are photosystems II and I self-regulatory sensors? In: Argyroudi-Akoyunoglou, J. H., ed. *Regulation of Chloroplast Biogenesis.* Plenum Press, New York. pp. 475–482.

Anderson, J. M., and Melis, A. (1983). Localization of different photosystems in separate regions of chloroplast membranes. *Proceedings of the National Academy of Sciences of the USA* 80:745–749.

Anderson, J. M., Brown, J. S., Lam, E., and Malkin, R. (1983). Chlorophyll *b:* an integral component of photosystem I of higher plant chloroplasts. *Photochemistry and Photobiology* 38:205–210.

Anderson, J. M., Chow, W. S., Adamson, H., and Melis, A. (1992). Dynamic light acclimation of the photosynthetic apparatus of higher plants. In: Argyroudi-Akoyunoglou, J. H., ed. *Regulation of Chloroplast Biogenesis.* Plenum Press, New York. pp. 483–490.

Andersson, B., and Anderson, J. M. (1980). Lateral heterogeneity in the distribution of chlorophyll-protein complexes of the thylakoid membranes of spinach chloroplasts. *Biochimica et Biophysica Acta* 593:427–440.

Aro, E.-M., Kettunen, R., Tyystjärvi, E. (1992). ATP and light regulate D1 protein modification and degradation: role of D1 in photoinhibition. *FEBS Letters* 297:29–33.

Aro, E.-M., Virgin, I., and Andersson, B. (1993). Photoinhibition of photosystem II. Inactivation, protein damage and turnover. *Biochimica et Biophysica Acta* 1143:113–134.

Barber, J. (1989). Function and molecular biology of photosystem II. *Oxford Surveys of Plant Molecular and Cell Biology* 6:115–162.

Baroli, I., and Melis, A. (1996). Photoinhibition and repair in *Dunaliella salina* acclimated to different growth irradiances. *Planta* 198:640–646.

Bennett, J. (1991). Protein phosphorylation in green plant chloroplasts. *Annual Review of Plant Physiology and Plant Molecular Biology* 42:281–311.

Björkman, O., and Demmig, B. (1987). Photon yield of O_2 evolution and chlorophyll fluorescence characteristics at 77 K among vascular plants of diverse origins. *Planta* 170:489–504.

Björkman, O., and Ludlow, M. M. (1972). Characterization of the light climate on the floor of a Queensland rainforest. *Carnegie Institution of Washington Yearbook* 71:85–94.

Callahan, F. E., Ghirardi, M. L., Sopory, S. K., Mehta, A. M., Edelman, M., Mattoo, A. K. (1990). A novel metabolic form of the 32 kDa-D1 protein in the grana-localized reaction center of photosystem II. *Journal of Biological Chemistry* 265:15357–15360.

Chow, W. S., Haehnel, W., and Anderson, J. M. (1987). The composition and function of thylakoid membranes from pea plants grown under white or green light with or without far-red light. *Physiologia Plantarum* 70:196–202.

Chow, W. S., Goodchild, D. J., Miller, C., and Anderson, J. M. (1990a). The influence of high levels of brief or prolonged supplementary far-red illumination during growth on the photosynthetic characteristics, composition and morphology of *Pisum sativum* chloroplasts. *Plant, Cell and Environment* 13:135–145.

Chow, W. S., Melis, A., and Anderson, J. M. (1990b). Adjustments of photosystem stoichiometry in chloroplasts improve the quantum efficiency of photosynthesis. *Proceedings of the National Academy of Sciences of the USA* 87:7502–7506.

Coughlan, S. J. (1988). Chloroplast thylakoid protein phosphorylation is influenced by mutations in the cytochrome *b-f* complex. *Biochimica et Biophysica Acta* 933:413–422.

Cunningham, F. X., Dennenberg, R. J., Jursinic, P. A., and Gantt, E. (1990). Growth under red light enhances photosystem II relative to photosystem I and phycobilisomes in the red algae *Porphyridium cruentum. Plant Physiology* 93:888–895.

Darr, S. C., Somerville, S. C., and Arntzen, C. J. (1986). Monoclonal antibodies to the light harvesting chlorophyll *a/b* protein complex of photosystem II. *Journal of Cell Biology* 103:733–740.

Demeter, S., Neale, P. J., and Melis, A. (1987). Photoinhibition: impairment of the primary charge separation between P680 and pheophytin in photosystem II of chloroplasts. *FEBS Letters* 214:370–374.

Deng, X.-W., Tonkyn, J. C., Peter, G. F., Thornber, J. P., Gruissem, W. (1989). Post-transcriptional control of plastid mRNA accumulation during adaptation of chloroplasts to different light quality environments. *Plant Cell* 1:645–654.

Duysens, L. N. M., Amesz, J., and Kamp, B. M. (1961). Two photochemical systems in photosynthesis. *Nature* 190:510–511.

Elich, T. D., Edelman, M., and Mattoo, A. (1992). Identification, characterization, and resolution of the *in vivo* phosphorylated form of the D1 photosystem II reaction center protein. *Journal of Biological Chemistry* 267:3523–3529.

Emerson, R., and Arnold, W. (1932a). A separation of the reactions in photosynthesis by means of intermittent light. *Journal of General Physiology* 15:391–420.

Emerson, R., and Arnold, W. (1932b). The photochemical reactions in photosynthesis. *Journal of General Physiology* 16:191–205.

Escoubas, J. M., Lomas, M., LaRoche, J., and Falkowski, P. G. (1995). Light intensity regulation of cab gene transcription is signalled by the redox state of the plastoquinone pool. *Proceedings of the National Academy of Sciences of the USA* 92:10237–10241.

Evans, J. R. (1987). The dependence of quantum yield on wavelength and growth irradiance. *Australian Journal of Plant Physiology* 14:69–79.

Falbel, T., Meehl, J. B., and Staehelin, L. A. (1996). Severity of mutant phenotype in a series of chlorophyll-deficient wheat mutants depends on light intensity and the severity of the block in chlorophyll synthesis. *Plant Physiology* 112:821–832.

Fujita, Y., and Murakami, A. (1987). Regulation of electron transport composition in cyanobacterial photosynthetic system: stoichiometry among photosystem I and II complexes and their light-harvesting antennae and cytochrome b_6/f complex. *Plant and Cell Physiology* 28:1547–1553.

Fujita, Y., Ohki, K., and Murakami, A. (1985). Chromatic regulation of photosystem composition in the photosynthetic system of red and blue-green algae. *Plant and Cell Physiology* 26:1541–1548.

Fujita, Y., Iwama, Y., Ohki, K., Murakami, A., Hagiwara, N. (1989). Regulation of the size of light-harvesting antennae in response to light intensity in the green alga *Chlorella pyrenoidosa. Plant and Cell Physiology* 30:1029–1037.

Gaffron, H., and Wohl, K. (1936). Zur Theorie der Assimilation. *Naturwissenschaften* 24:81–90.

Glazer, A. N., and Melis, A. (1987). Photochemical reaction centers: structure, organization, and function. *Annual Review of Plant Physiology* 38:11–45.

Glick, R. E., and Melis, A. (1988). Minimum photosynthetic unit size in system-I and system-II of barley chloroplasts. *Biochimica et Biophysica Acta* 934:151–155.

Glick, R. E., McCauley, S. W., Gruissem, W., and Melis, A. (1986). Light quality regulates expression of chloroplast genes and assembly of photosynthetic membrane complexes. *Proceedings of the National Academy of Sciences of the USA* 83:4287–4291.

Guenther, J. E., Nemson, J. A., and Melis, A. (1990). Development of PSII in dark grown *Chlamydomonas reinhardtii.* A light-dependent conversion of $PSII_\beta$, Q_B-nonreducing centers to the $PSII_\alpha$, Q_B-reducing form. *Photosynthesis Research* 24:35–46.

Harrison, M. A., and Melis, A. (1992). Organization and stability of polypeptides associated with the chlorophyll *a-b* light-harvesting complex of photosystem II. *Plant and Cell Physiology* 33:627–637.

Haworth, P., Watson, J. L., and Arntzen, C. J. (1983). The detection, isolation and characterization of a light-harvesting complex which is specifically associated with photosystem I. *Biochimica et Biophysica Acta* 724:151–158.

Hill, R., and Bendall, F. (1960). Function of the two cytochrome components in chloroplasts: a working hypothesis. *Nature* 186:136–137.

Ikeuchi, M., Hirano, A., and Inoue, Y. (1991). Correspondence of apoproteins of light-harvesting chlorophyll *a/b* complexes associated with photosystem-I to *cab* genes: evidence for a novel Type IV apoprotein. *Plant and Cell Physiology* 32:103–112.

Jansson, S., Pichersky, E., Bassi, R., Green, B. R., Ikeuchi, M., Melis, A., Simpson, D. J., Spangfort, M., Staehelin, L. A., Thornber, J. P. (1992). A nomenclature for the genes encoding the chlorophyll *a/b*-binding proteins of higher plants. *Plant Molecular Biology Reports* 10:242–253.

Kim, J. H., Glick, R. E., and Melis, A. (1993a). Dynamics of photosystem stoichiometry adjustment by light-quality in chloroplasts. *Plant Physiology* 102:181–190.

Kim, J. H., Nemson, J. A., and Melis, A. (1993b). Photosystem-II reaction center damage and repair in the green alga *Dunaliella salina:* analysis under physiological and adverse irradiance conditions. *Plant Physiology* 103:181–189.

Kirk, J. T. O. (1983). *Light and Photosynthesis in Aquatic Ecosystems.* Cambridge University Press, New York.

Knoetzel, J., Svendsen, I., and Simpson, D. J. (1992). Identification of the photosystem-I antenna polypeptides in barley Isolation of 3 pigment-binding antenna complexes. *European Journal of Biochemistry* 206:209–215.

Kühlbrandt, W., and Wang, D. N. (1991). Three-dimensional structure of plant light-harvesting complex determined by electron crystallography. *Nature* 350:130–134.

Kühlbrandt, W., Wang, D. N., and Fujiyoshi, Y. (1994). Atomic model of plant light-harvesting complex determined by electron crystallography. *Nature* 367:614–621.

Lam, E., Ortiz, W., Mayfield, S., and Malkin, R. (1984). Isolation and characterization of a light-harvesting chlorophyll *a/b* complex associated with photosystem I. *Plant Physiology* 74:650–655.

LaRoche, J., Mortain-Bertrand, A., and Falkowski, P. G. (1991). Light-intensity-induced changes in *cab* mRNA and light-harvesting complex II apoprotein levels in the unicellular chlorophyte *Dunaliella tertiolecta. Plant Physiology* 97:147–153.

Larsson, U. K., Anderson, J. M., and Andersson, B. (1987). Variations in the relative content of the peripheral and inner light-harvesting chlorophyll a/b-protein complex (LHC-II) subpopulations during thylakoid light adaptation and development. *Biochimica et Biophysica Acta* 894:69–75.

Leong, T. A., and Anderson, J. M. (1984). Adaptation of the thylakoid membranes of pea chloroplasts to light intensities. I. Study on the distribution of chlorophyll-protein complexes. *Photosynthesis Research* 5:105–115.

Ley, A. C., and Mauzerall, D. C. (1982). Absolute absorption cross sections for photosystem II and the minimum quantum requirement for photosynthesis in *Chlorella vulgaris. Biochimica et Biophysica Acta* 680:95–106.

Manodori, A., and Melis, A. (1986). Cyanobacterial acclimation to photosystem I or photosystem II light. *Plant Physiology* 82:185–189.

Mattoo, A. K., and Edelman, M. (1987). Intramembrane translocation and posttranslational palmitoylation of the chloroplast 32-kDa herbicide-binding protein. *Proceedings of the National Academy of Sciences of the USA* 84:1497–1501.

Mattoo, A., Hoffman-Falk, H., Marder, J., and Edelman, M. (1984). Regulation of protein metabolism: coupling of photosynthetic electron-transport to in vivo degradation of the rapidly metabolized 32-kDa protein of the chloroplast membranes. *Proceedings of the National Academy of Sciences of the USA* 81:1380–1384.

Mawson, B. T., Morrissey, P. J., Gomez, A., and Melis, A. (1994). Thylakoid membrane development and differentiation: assembly of the chlorophyll *a-b* light-harvesting complex and evidence for the origin of M_r = 19, 17.5 and 13.4 kDa proteins. *Plant and Cell Physiology* 35:341–151.

Melis, A. (1984). Light regulation of photosynthetic membrane structure, organization and function. *Journal of Cellular Biochemistry* 24:271–285.

Melis, A. (1991). Dynamics of photosynthetic membrane composition and function. *Biochimica et Biophysica Acta* 1058:87–106.

Melis, A. (1996). Excitation energy transfer: functional and dynamic aspects of *Lhc (cab)* proteins. In: Ort, D. R., and Yocum, C. F., eds. *Oxygenic Photosynthesis: The Light Reactions.* Kluwer Academic Publishers, Dordrecht, The Netherlands. pp. 523–538.

Melis, A., and Anderson, J. M. (1983). Structural and functional organization of the photosystems in spinach chloroplasts: antenna size, relative electron transport capacity, and chlorophyll composition. *Biochimica et Biophysica Acta* 724:473–484.

Melis, A., and Harvey, G. W. (1981). Regulation of photosystem stoichiometry, chlorophyll *a* and chlorophyll *b* content and relation to chloroplast ultrastructure. *Biochimica et Biophysica Acta* 637:138–145.

Melis, A., Mullineaux, C. W., and Allen, J. F. (1989). Acclimation of the photosynthetic apparatus to photosystem I or photosystem II light: evidence from quantum yield measurement and fluorescence spectroscopy of cyanobacterial cells. *Zeitschrift für Naturforschung Teil C* 44:109–118.

Melis, A., Murakami, A., Nemson, J. A., Aizawa, K., Ohki, K., and Fujita, Y. (1996). Chromatic regulation in *Chlamydomonas reinhardtii* alters photosystem stoichiometry and improves the quantum efficiency of photosynthesis. *Photosynthesis Research* 47:253–265.

Morrissey, P. J., Glick, R. E., and Melis, A. (1989). Supramolecular assembly and function of subunits associated with the chlorophyll *a/b* light-harvesting complex II (LHC-II) in soybean chloroplasts. *Plant and Cell Physiology* 30:335–344.

Murakami, A., and Fujita, Y. (1988). Steady state of photosynthesis in cyanobacterial photosynthetic systems before and after regulation of electron transport composition: overall rate of photosynthesis and PS I/PS II composition. *Plant and Cell Physiology* 29:305–311.

Murakami, A., and Fujita, Y. (1991). Steady state of photosynthetic electron transport in cells of the cyanophyte *Synechocystis* PCC 6714 having different stoichiometry between PSII and PSI: analysis of flash-induced oxidation-reduction of cytochrome *f* and P_{700} under steady state of photosynthesis. *Plant and Cell Physiology* 32:213–222.

Murakami, A., Nemson, J. A., Fujita, Y., and Melis, A. (1997). Chromatic regulation in *Chlamydomonas reinhardtii:* time course of photosystem stoichiometry adjustment following a shift in growth light quality. *Plant and Cell Physiology* 38:188–193.

Myers, J., Graham, J. R., and Wang, R. T. (1980). Light-harvesting in *Anacystis nidulans* studied in pigment mutants. *Plant Physiology* 66:1144–1149.

Nanba, O., and Satoh, K. (1987). Isolation of a photosystem-II reaction center consisting of D-1 and D-2 polypeptides and cytochrome *b*-559. *Proceedings of the National Academy of Sciences of the USA* 84:109–112.

Neale, P. J., and Melis, A. (1990). Activation of a reserve pool of photosystem II in *Chlamydomonas reinhardtii* counteracts photoinhibition. *Plant Physiology* 92:1196–1204.

Ohtsuka, T., Ito, H., and Tanaka, A. (1997). Conversion of chlorophyll *b* to chlorophyll *a* and the assembly of chlorophyll with apoproteins by isolated chloroplasts. *Plant Physiology* 113:137–147.

Peter, G. F., and Thornber, J. P. (1991). Biochemical composition and organization of higher plant photosystem-II light-harvesting pigment-proteins. *Journal of Biological Chemistry* 266:16745–16754.

Pichersky, E., and Jansson, S. (1996). The light-harvesting chlorophyll *a/b*-binding polypeptides and their genes in angiosperm and gymnosperm species. In: Ort, D. R., and Yocum, C. F., eds. *Oxygenic Photosynthesis: The Light Reactions.* Kluwer Academic Publishers, Dordrecht, The Netherlands. pp. 507–521.

Powles, S. B. (1984). Photoinhibition of photosynthesis induced by visible light. *Annual Review of Plant Physiology* 35:15–44.

Salter, A. H., Virgin, I., Hagman, Å., Andersson, B. (1992). On the molecular mechanism of light-induced D1-protein degradation in photosystem-II core particles. *Biochemistry* 31:3990–3998.

Simpson, D. J., and Knoetzel, J. (1996). Light-harvesting complexes of plants and algae: introduction, survey and nomenclature. In: Ort, D. R., and Yocum, C. F., eds. *Oxygenic Photosynthesis: The Light Reactions.* Kluwer Academic Publishers, Dordrecht, The Netherlands. pp. 493–506.

Smith, B. M., Morrissey, P. J., Guenther, J. E., Nemson, J. A., Harrison, M. A., Allen, J. F., and Melis, A. (1990). Response of the photosynthetic apparatus in *Dunaliella salina* (green algae) to irradiance stress. *Plant Physiology* 93:1433–1440.

Sukenik, A., Bennett, J., and Falkowski, P. G. (1988). Changes in the abundance of individual apoproteins of light-harvesting chlorophyll *a/b*-protein complexes of photosystem I and II with growth irradiance in the marine chlorophyte *Dunaliella tertiolecta. Biochimica et Biophysica Acta* 932:206–215.

Terashima, I., and Saeki, T. (1983). Light environment within a leaf. I. Optical properties of paradermal sections of *Camelia* leaves with special reference to differences in the optical properties of palisade and spongy tissues. *Plant and Cell Physiology* 24:1493–1501.

Thornber, J. P., Peter, G. F., Chitnis, P. R., Nechushtai, R., and Vainstein, A. (1988). The light-harvesting complex of photosystem II of higher plants. In: Stevens, S. E., Jr., and Bryant, D. A., eds. *Light-energy Transduction in Photosynthesis: Higher Plant and Bacterial Models.* The American Society of Plant Physiologists, Rockville, Maryland. pp. 137–154.

Vogelmann, T. (1989). Penetration of light into plants. *Photochemistry and Photobiology* 50:895–902.

Vogelmann, T. C. (1993). Plant tissue optics. *Annual Review of Plant Physiology and Plant Molecular Biology* 44:231–251.

Webb, M. R., and Melis, A. (1995). Chloroplast response in *Dunaliella salina* to irradiance stress. Effect on thylakoid membrane assembly and function. *Plant Physiology* 107:885–893.

INDEX